Advancing Methods for Biomolecular Crystallography

NATO Science for Peace and Security Series

This Series presents the results of scientific meetings supported under the NATO Programme: Science for Peace and Security (SPS).

The NATO SPS Programme supports meetings in the following Key Priority areas: (1) Defence Against Terrorism; (2) Countering other Threats to Security and (3) NATO, Partner and Mediterranean Dialogue Country Priorities. The types of meeting supported are generally "Advanced Study Institutes" and "Advanced Research Workshops". The NATO SPS Series collects together the results of these meetings. The meetings are co-organized by scientists from NATO countries and scientists from NATO's "Partner" or "Mediterranean Dialogue" countries. The observations and recommendations made at the meetings, as well as the contents of the volumes in the Series, reflect those of participants and contributors only; they should not necessarily be regarded as reflecting NATO views or policy.

Advanced Study Institutes (ASI) are high-level tutorial courses intended to convey the latest developments in a subject to an advanced-level audience

Advanced Research Workshops (ARW) are expert meetings where an intense but informal exchange of views at the frontiers of a subject aims at identifying directions for future action

Following a transformation of the programme in 2006 the Series has been re-named and re-organised. Recent volumes on topics not related to security, which result from meetings supported under the programme earlier, may be found in the NATO Science Series.

The Series is published by IOS Press, Amsterdam, and Springer, Dordrecht, in conjunction with the NATO Emerging Security Challenges Division.

Sub-Series

A.	Chemistry and Biology	Springer
B.	Physics and Biophysics	Springer
C.	Environmental Security	Springer
D.	Information and Communication Security	IOS Press
E.	Human and Societal Dynamics	IOS Press

http://www.nato.int/science
http://www.springer.com
http://www.iospress.nl

Series A: Chemistry and Biology

Advancing Methods for Biomolecular Crystallography

edited by

Randy Read
University of Cambridge, UK

Alexandre G. Urzhumtsev
Université de Lorraine, Vandoeuvre-lès-Nancy, France
IGBMC, Illkirch, France

and

Vladimir Y. Lunin
Russian Academy of Sciences, Pushchino, Russia

Springer

Published in Cooperation with NATO Emerging Security Challenges Division

Proceedings of the NATO Advanced Study Institute on Present and Future Methods for Biomolecular Crystallography: the Structural Path to Defence against CBRN Agents
Erice, Italy
1–10 June 2012

Library of Congress Control Number: 2013935871

ISBN 978-94-007-6319-7 (PB)
ISBN 978-94-007-6231-2 (HB)
ISBN 978-94-007-6232-9 (e-book)
DOI 10.1007/978-94-007-6232-9

Published by Springer,
P.O. Box 17, 3300 AA Dordrecht, The Netherlands.

www.springer.com

Printed on acid-free paper

We wish to dedicate these proceedings to the memory of Lodovico Riva di Sanseverino, who was the linchpin of the Erice schools for many years. His presence was constantly felt in the traditions he established and the special atmosphere he did so much to create.

Preface

This volume comprises papers presented at the 2012 edition of the "Crystallography of Molecular Biology" series. These are part of a wide-ranging set of crystallography courses held since 1974 in the hilltop town of Erice, Italy, in the Ettore Majorana Centre for Scientific Culture, based in several old monasteries. The series of courses is renowned for bringing leaders in the field of macromolecular crystallography together with highly motivated students, in a beautiful and intimate location. The warm and informal atmosphere of Erice encourages a level of interaction that was rewarded, this year, by the determination of at least three new structures during the school.

Lecturers were chosen from world leaders in the field of structural biology, and all made great efforts to present cutting-edge science at a level accessible to participants with limited experience. Jane Richardson opened the meeting by reminding everyone that structure validation is a continuous process of quality control that should play a role in every step of structure determination, not just at the end.

Structure determination starts with protein production, and Stephen Kent showed how it is possible to synthesize proteins chemically, which opens the door to exciting new approaches such as crystallization from racemic mixtures of the protein and its mirror image. Todd Yeates discussed how crystallization can go wrong with pathologies such as twinning and lattice translocation disorders. Growing crystals is especially difficult for membrane proteins, and Martin Caffrey showed how the special properties of lipidic mesophases make them particularly useful, as well as how the resulting crystal structures give important insights into biology.

Collecting good data from a hard-won crystal is not necessarily straightforward. Sean McSweeney showed how to find the best crystals, or even the best parts of the best crystals, and how to merge compatible data from multiple crystals. Kay Diederichs suggested an approach to answering the perpetual thorny question about how to choose a resolution limit for a data set. Elspeth Garman explained why radiation damage limits how much can be collected from one crystal, as well as ways in which the severity of radiation damage might be reduced. Radiation damage depends in large part on the elemental composition of the crystal, and

Elspeth also showed how the technique of microPIXE can be used to determine elemental composition for this and other purposes. Tatiana Petrova presented some case studies showing the physical consequences of radiation damage on the proteins inside crystals. Dominika Borek showed that, to a great extent, careful data processing can ameliorate the effect of radiation damage on the data.

As the Protein Data Bank expands, fewer truly novel structures are determined so the majority can be solved using the molecular replacement method. Randy Read showed ways of extending the reach of this method in Phaser, including approaches that combine molecular replacement with other phasing methods. Isabel Usón presented the *ab initio* Arcimboldo procedure, which uses molecular replacement to place small fragments such as helices that can then be expanded to a complete structure.

More typically, novel structures are solved by experimental phasing methods such as SAD or MAD (reviewed by Zbyszek Dauter), starting from the substructure of anomalous scatterers (SHELXD: Tim Gruene). Felix Frolow presented several case studies of structures solved with a weak anomalous signal. Vladimir Lunin showed how reliable phase information can be obtained just from the native structure factor amplitudes, albeit only to low resolution.

Tim Gruene and Tom Terwilliger presented different approaches to using a combination of density modification and automated building to improve the electron density and interpret it as an atomic model. Often human abilities are needed to supplement automated building algorithms, and Paul Emsley showed how coot provides a large number of intuitive tools to interpret electron density manually.

Pavel Afonine described the tools in phenix.refine to refine macromolecular structures to give better agreement with the diffraction data, at a variety of resolutions, and Sacha G. Urzhumtsev concentrated on the special considerations of interpreting data at both extremes of very high and very low resolution. Model-building and refinement are particularly difficult at low resolution, and Garib Murshudov discussed approaches in Refmac to dealing with these problems. Mariusz Jaskolski, in turn, dealt with the particular challenges and opportunities presented by very high resolution.

Once the structure has been obtained and refined, it is ready to be presented to the world. Jaime Prilusky showed how this can be done interactively on the Proteopedia website, and he challenged the participants to prepare their own Proteopedia pages during the course.

A number of speakers presented the structural fruits of their research. Included here are contributions on complement proteins (Piet Gros), monoamine oxidase inhibitors (Andrea Mattevi), and the eukaryotic ribosome (Sergey Melnikov).

Although the course concentrated on single-crystal X-ray diffraction, the horizons were broadened with a number of complementary approaches. Nobuo Niimura described how the emergence of new neutron radiation sources is leading to a renaissance of neutron diffraction; by showing the positions of protons and deuterons it nicely complements the information on heavier atoms from X-ray diffraction. Tatiana Latychevskaya showed that diffraction with low-energy electrons can be used for image reconstruction. Dmitri Svergun explained how molecular shapes can be determined at a surprising level of detail with the one-dimensional information obtained from small-angle X-ray scattering. Finally, Frank DiMaio talked about how the modelling program Rosetta can be used in computational enzyme design, in the absence of experimental data, and also how it can leverage small amounts of data by improving models for molecular replacement or in extending the convergence radius of refinement.

Most of the real organizational work for the course was done by Paola Spadon and Annalisa Guerri who, between them, found most of the funding, corresponded with applicants, and coordinated the selection of participants. John Irwin played an essential role, organizing all the computing facilities necessary to conduct tutorials and demonstrations, and providing a web-based video feed to people unable to attend the meeting. In the local organization and logistics, Paola, Annalisa and John were ably assisted by the traditional team of "orange scarves": Sara Giannetti, Gianni Grassi, Agata Impellizzieri, Matteo Lusi, Claudia Minici, Fabio Nicoli, Elisa Pasqualetto and Giovanna Scapin. We also thank Claire Chapman for her patient and expert assistance in corresponding with authors, collating their contributions, and coordinating the production of this book.

The course was financed by NATO as an Advanced Study Institute. In addition to the essential support from NATO, generous financial support was received from the European Crystallographic Association, the International Union of Biochemistry and Molecular Biology, the International Union of Crystallography, CCP4, the Organisation for the Prohibition of Chemical Weapons, Bruker AXS and Advanced Design Consulting.

Randy J. Read
Sacha G. Urzhumtsev
Vladimir Y. Lunin

Contents

Chapter 1
The Zen of Model Anomalies – Correct Most of Them. Treasure the Meaningful Valid Few. Live Serenely with the Rest!

Jane S. Richardson and David C. Richardson

Abstract Historically, validation has been considered primarily as a gatekeeping function done at the end of a structure solution. Currently, the most interesting and important part of validation is the opportunity to correct diagnosed errors, provided mainly by local as opposed to global criteria, and available to you throughout the crystallographic process. Elsewhere in this book, you will hear about up-to-date methods in the data and model-to-data aspects of validation. This chapter addresses model validation and model improvement, first about current best-practice methodology (as done on the MolProbity website and elsewhere), and second about some important developments to anticipate in the near future.

Model validation has three primary parts: (a) geometry (bond lengths and angles, planarity, chirality), (b) conformation (rotamers, Ramachandran, ring pucker, RNA backbone conformers), (c) and sterics (clashes, H-bonds, packing). All of these both enhance and must be considered along with the information from electron density. The model criteria are primarily local, but their rate of occurrence can also be summarized as a global score.

Keywords Model validation • Model improvement • All-atom contacts • MolProbity • RNA backbone

1.1 Geometry Validation

Geometry differences from standard values ([7]; Fig. 1.1 shows how geometry outliers are represented in MolProbity graphics) mostly reflect non-ideal refinement strategy, but there are some circumstances where they help flag model-fitting errors.

J.S. Richardson (✉) • D.C. Richardson
Department of Biochemistry, Duke University Medical Center, 132 Nanaline Duke Building, Durham, NC 27710, USA
e-mail: jsr@kinemage.biochem.duke.edu

R. Read et al. (eds.), *Advancing Methods for Biomolecular Crystallography*, NATO Science for Peace and Security Series A: Chemistry and Biology,
DOI 10.1007/978-94-007-6232-9_1, © Springer Science+Business Media Dordrecht 2013

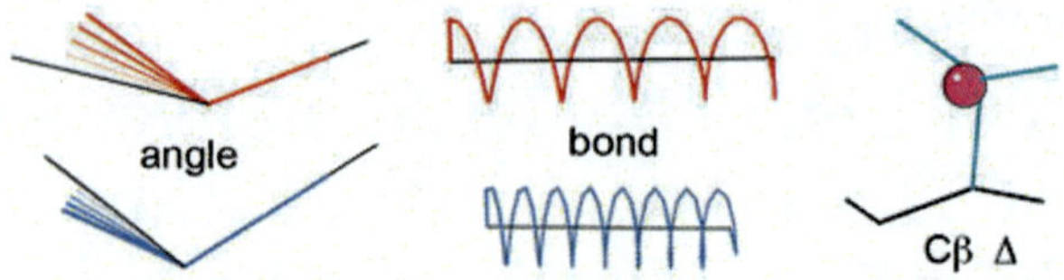

Fig. 1.1 Graphical icons in MolProbity for geometry outliers >4 σ in bond angles, bond lengths, and Cβ deviations

At high resolution, sometimes geometry is weighted very low or even turned off, which can produce truly dire results in high-B regions such as chain termini – be sure to look at the residues with the biggest outliers. Bad bond angles (or chirality or Cβ deviations, which are combinations of angles) are often symptoms of a sidechain or peptide turned around the wrong way [13], especially if there are also steric clashes or conformational outliers in the same residue. Another use of geometry is provided in WhatCheck [9], where overall deviations in bond length diagnose errors in cell dimensions.

1.2 Conformational Validation

1.2.1 Sidechain Rotamers

Conformational validation is most powerful when done on combinations of torsion angles, which see different, stronger constraints than the product of their individual preferences (as shown for lysine in Fig. 1.2). Initial fitting with sidechain rotamers is good strategy, later allowing poor ones only where clearly required by the density and stabilized by local interactions that can hold an unfavorable conformation in place (e.g. 2 or 3 H-bonds for an eclipsed χ angle). This avoids getting caught in a "decoy" rotamer (such as a doubly-eclipsed Thr or Leu, or an Arg with upside-down guanidinium) that approximately fits the electron density but is never energetically allowed, while there is a favorable alternative rotamer that will fit the density even a bit better [12].

1.2.2 RNA Backbone

For RNA, the analog of protein sidechain rotamers is RNA backbone conformers. There are more of them (over 50) because they have many more torsion angles per residue, and they are best defined for the "suite" unit from sugar to sugar rather than for the traditional nucleotide unit between phosphates [17]. They are diagnosed by MolProbity [4], either on the web site or in Phenix [1], and can be rebuilt in KiNG [3] or in Coot [6] with the RCrane feature [10].

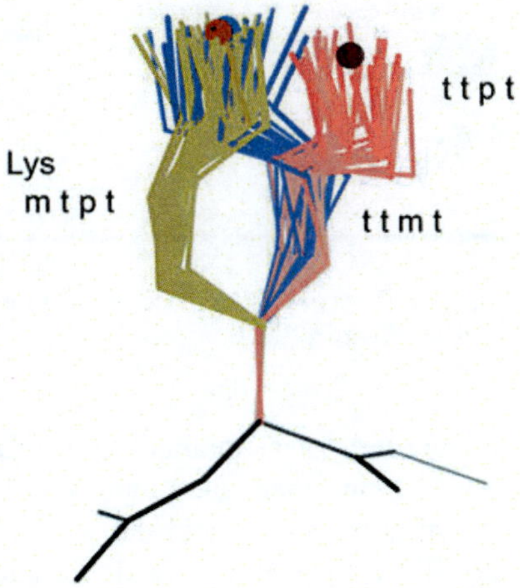

Fig. 1.2 Examples for three rotamers of Lys, showing tight clustering. χ angle naming: m = −60°, t = 180°, p = +60°

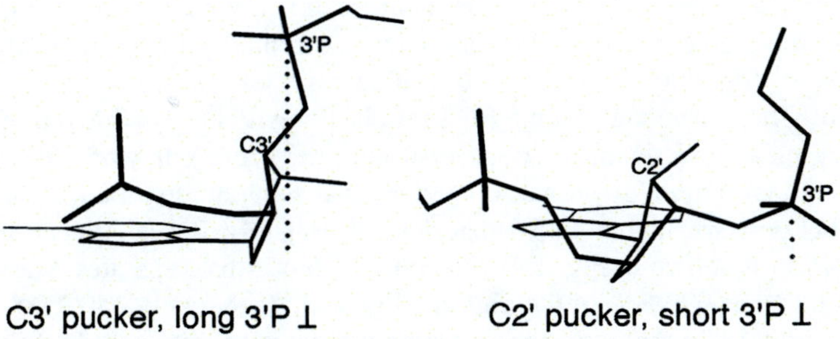

Fig. 1.3 Diagnosing ribose pucker by 3′P position relative to glycosidic bond

The most generic and powerful tool for diagnosing and correcting RNA backbone is a test for ribose pucker that measures the perpendicular distance from the 3′ phosphate to the glycosidic bond that joins ribose and base: if that distance is >2.9 Å the pucker is C3′-endo and if <2.9 Å it is C2′-endo (see Fig. 1.3). This is a rather cleanly bimodal distribution, reliable even at resolutions where it is hopeless to see the pucker directly, because the phosphate and base are the best positioned features in RNA models [17]. Historically, many C2′-endo puckers are misfit as the commoner C3′-endo, a mistake preventable by this diagnostic. In Phenix, this test is used to apply pucker-specific target parameters [1], which can keep the conformation correct and help avoid bad geometry in the ribose ring.

1.2.3 Protein Backbone

An especially significant part of conformational validation is the Ramachandran plot, pioneered by ProCheck [15] with the consideration that ϕ, ψ values are very

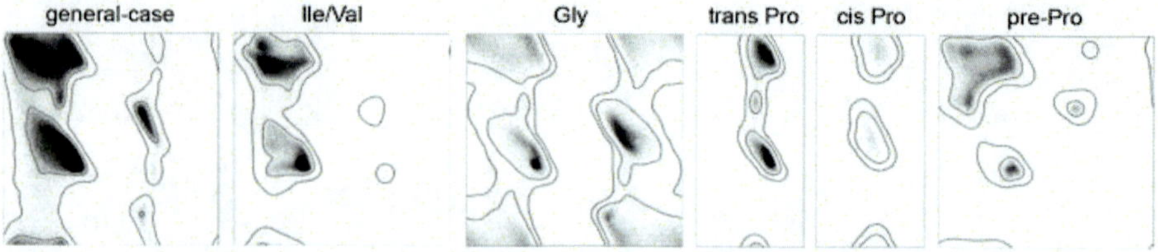

Fig. 1.4 Plots of the reference data for the six classes of Ramachandran plot recommended by the wwPDB X-ray Validation Task Force

seldom part of refinement target functions and therefore provide an independent check. Since then, the quantity and quality of data in the Protein Data Bank has grown enormously, and the accuracy and specificity of Ramachandran plots has steadily improved [11, 13]. Ramachandran and rotamer distributions meant for prediction or design [19] emphasize the favorable regions and amino-acid diversity, while for validation purposes the dominant issue is the outer contour that divides what is unfavorable but possible from what is essentially impossible. MolProbity now uses a "Top8000" dataset of 1.6 million residues with backbone $B < 30$, 1.3 million of which have maps at the EDS for checking up on possible outliers. These give very clean and definitive ϕ, ψ distributions. From our work for the X-ray Validation Task Force [16] it was concluded that separate distributions are needed for six, but only six, amino-acid categories: general, Gly, Ile/Val, trans Pro, cis Pro, and pre-Pro (as shown in Fig. 1.4). The outer contour that separates Allowed from Outliers contains 99.95 % of the high-quality data, so that only 1 in 2,000 residues should validly be in the outlier region even though it covers about half the area of the ϕ, ψ plot. It is well worth examining every outlier and either correcting it if possible, giving up gracefully if it really can't be improved (more often true at low resolution), or celebrating the significance of why it is being held in an unfavorable conformation.

1.3 Validation of Sterics

Validation of sterics includes the specific non-covalent interactions of hydrogen bonding, clashes (repulsive overlaps) and attractive van der Waals contacts (graphical icons for the three types are shown in Fig. 1.5), plus overall criteria of packing density and arrangement. The best measure so far available for evaluating underpacking is RosettaHoles [18]. Overpacking is taken care of by the all-atom clashscore (see below). A related issue is profile analysis, used mainly for predictive "threading" of a sequence into a potential 3D structure, but also useful for diagnosing an incorrect chain tracing [14].

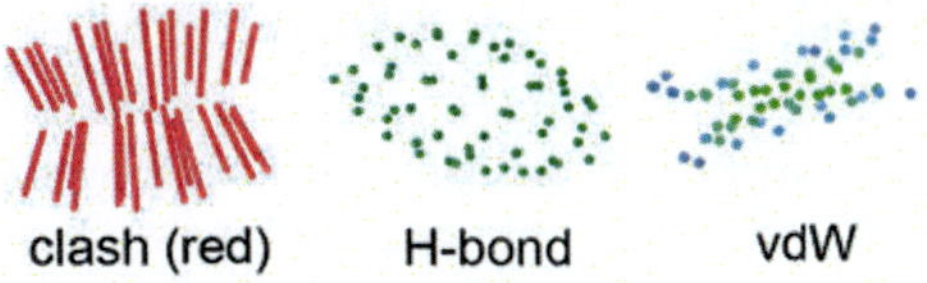

Fig. 1.5 Graphical icons in MolProbity for steric clashes >0.4 Å, for H-bonds, and for vdW contacts

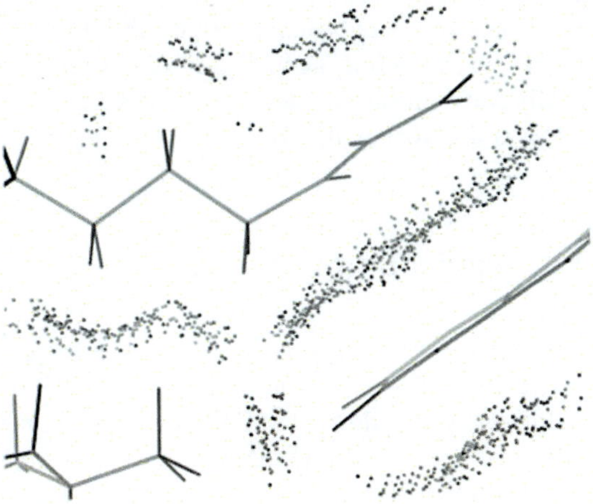

Fig. 1.6 All-atom contacts for excellent local packing of an Arg, Trp, and Leu in 3LZM at 1.7 Å, with explicit hydrogens included

1.3.1 All-Atom Contacts

Meaningful evaluation of specific atomic contacts requires the use of all explicit hydrogens, because H atoms are half of the total atoms and about three-fourths of all contacts involve at least one hydrogen. Therefore MolProbity validation uses the Reduce program [21] to add all H atoms, optimizing local H-bond networks and the 180° "flip" state of Asn/Gln/His sidechains. The Probe program [20] is then used to calculate "all-atom contacts", which constitute the most distinctive aspect introduced by MolProbity, now available in Phenix and Coot as well. Note that the flips can be done without reference to the diffraction data, since the density difference between N and O or N and C is so small. Asn/Gln flips change H-bonding, while His flips also frequently affect protonation assignments.

All-atom clashes are a very sensitive indicator of fitting problems, since H atom contacts are seldom refined against. If the heavier atoms are accurately placed at high resolution, then the H's added by Reduce almost never have serious clashes (defined as a non-H-bonding overlap ≥0.4 Å) because the hydrogens are really there in the molecule, helping to position those heavier atoms, as shown for a well-determined piece of structure in Fig. 1.6. The MolProbity "clashscore" is the number of serious clashes per thousand atoms, giving a global quality score strongly correlated with resolution. Individual all-atom clashes, or clusters of them, have proven very useful to guide rebuilding, since they are directional as well as local.

1.3.2 MolProbity Score

To satisfy user demand for a single number, there is a "MolProbity score" that combines clash, rotamer, and Ramachandran measures to give an overall measure of validation quality; it gives the approximate resolution at which that combination of scores would be typical. Both clashscore and MolProbity score are also reported as percentile scores relative to the similar-resolution cohort of PDB structures; working at corrections usually makes it possible to achieve percentiles in the 90's, at least for resolutions up to about 2.5 Å [2]. Satisfyingly, since the 2002 introduction of MolProbity, clashscore and Asn/Gln/His flips have decreased by over 30 % in PDB depositions worldwide [4].

1.4 Model Improvement

Correction of the diagnosed errors is a major goal of model validation, as was practiced in the Erice tutorials, using MolProbity and Phenix validation reports followed by rebuilding in KiNG or Coot. Some background is provided here, for aspects that may be unfamiliar. One such is the set of easy and satisfying fixes that come from recognizing systematic errors such as "decoy" rotamers. The electron density at tetrahedral branches often looks more straight across than boomerang-shaped, so it's easy for either people or programs to fit that group 180° rotated from the correct position (as shown for Thr in Fig. 1.7). The χ angle is then close to eclipsed rather than close to staggered, which is essentially never the right answer in these cases. These decoy fits of course have a bad rotamer, but they are usually flagged also by clashes, bond-angle outliers, or Cβ deviations [13].

An essential tool for rebuilding is the "backrub", a subtle dipeptide backbone shift with leverage on the Cα-Cβ direction that enables much larger two-state

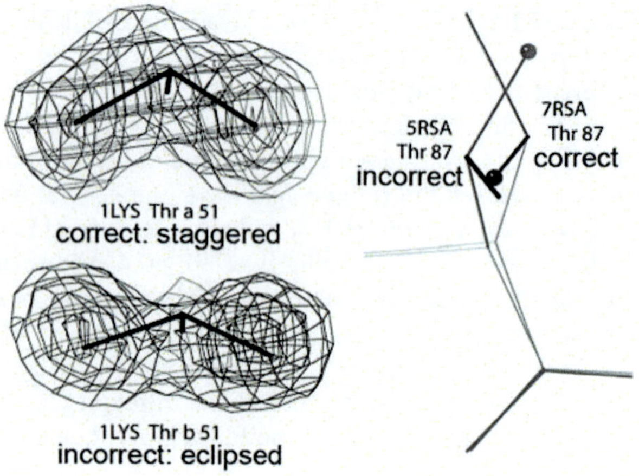

Fig. 1.7 Comparison of correct vs 180° backward "decoy" fitting of Thr sidechains into ambiguous density

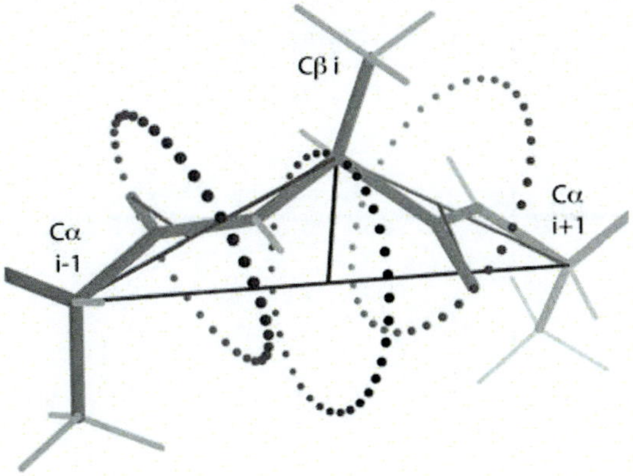

Fig. 1.8 Schematic of the "Backrub" motion that accomodates sidechain changes by subtle rotation of the backbone dipeptide

changes in the central sidechain. As shown in Fig. 1.8, the backrub is a rotation around an axis through the $i-1$ and $i+1$ Cα atoms, plus small compensating rotation of the individual peptides. It was shown to be the most prevalent backbone change between single-residue alternate conformations at high resolution [5]. That same motion accommodates misfit sidechains, and needs to be moved back to allow their correction. This is a fundamental feature in KiNG [3], is a possible move in Coot, and is being written into various software for protein design and refinement.

A major reason for making corrections is to improve the signal-to-noise for finding the few cases where the molecule has chosen to spend energy stabilizing an unfavorable conformation. These are apt to be significant features at functional sites.

1.5 Future Developments

1.5.1 wwPDB Validation Task Force

New developments are coming that will change the future of model validation and improvement. One important direction is the wwPDB Validation Task Force committees for the major experimental techniques. The X-ray VTF has made its report [16], and the wwPDB is working to implement many of those recommendations by the end of this year [8]. Summary validation reports will be available for referees, which the IUCr journals and JBC have already mandated for submission with structure manuscripts. There will be a brief graphical and numerical summary on the main PDB page for each structure, with more detail available on both global and per-residue statistics. The key scores will be reported as percentiles, both relative to the resolution cohort and relative to all PDB crystal structures; a plot of such scores is shown for clashscore in Fig. 1.9. The software to do these things will be part of the deposition process and also available independently, for easy and secure runs in trial mode.

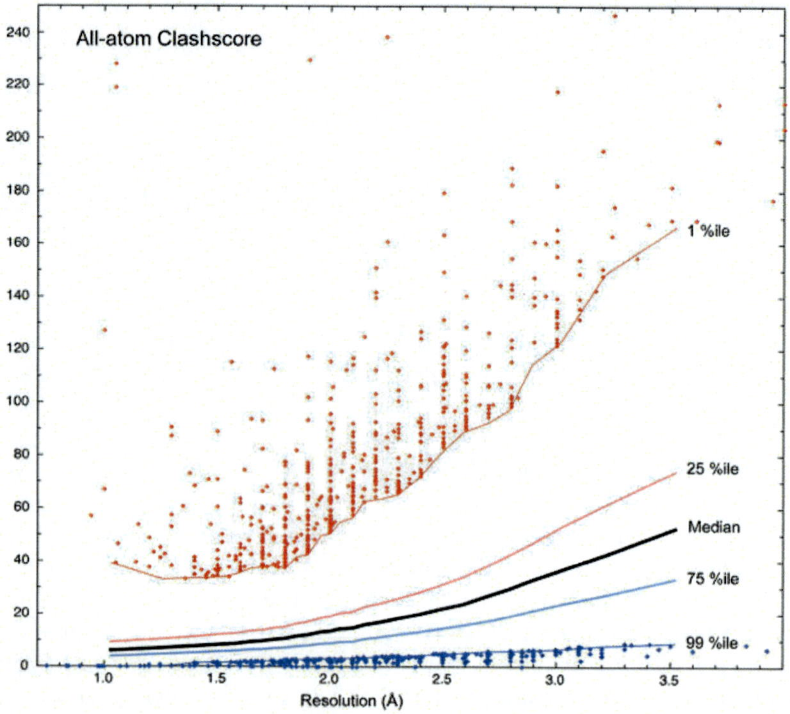

Fig. 1.9 wwPDB X-ray VTF percentiles for MolProbity clashscore as a function of resolution

1.5.2 More Help for the Hard Cases

The level of integrated automated for validation and correction is increasing rapidly. Many groups are working on better ways to deal with the still-difficult parts of crystallography such as low resolution, big mobile complexes, membrane proteins, RNA, etc. Our own lab is developing more consistent ways to model multiple conformations at high resolution, ways to build correct RNA backbone in full detail, and new techniques, including a new diagnostic parameter space, for better accuracy at low-resolution.

1.6 Conclusion

The final, most important point about model validation and improvement is summarized by the embarrassing example in Fig. 1.10, and by the precept that at least one person should look at the map!

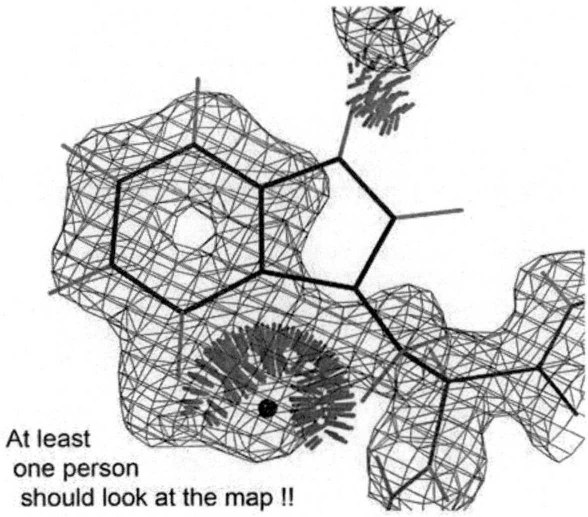

Fig. 1.10 Don't let this kind of no-brainer problem make it into the PDB for your structures

Acknowledgments Our work on model validation and improvement is supported by NIH grants: R01GM073919 for the *MolProbity* service, P01GM063210 for validation in *Phenix*, R01GM073930 for RNA validation, and R01GM088674 for improvement at low resolution.

References

1. Adams PD, Afonine PV, Bunkóczi G, Chen VB, Davis IW, Echols N, Headd JJ, Hung L-W, Kapral GJ, Grosse-Kunstleve RW, McCoy AJ, Moriarty NW, Oeffner R, Read RJ, Richardson DC, Richardson JS, Terwilliger TC, Zwart PH (2010) PHENIX: a comprehensive Python-based system for macromolecular structure solution. Acta Crystallogr D 66:213–221 (open access)
2. Arendall WB III, Tempel W, Richardson JS, Zhou W, Wang S, Davis IW, Liu Z-J, Rose JP, Carson WM, Luo M, Richardson DC, Wang B-C (2005) A test of enhancing model accuracy in high-throughput crystallography. J Struct Funct Genomics 6:1–11
3. Chen VB, Davis IW, Richardson DC (2009) KiNG (Kinemage, Next Generation): a versatile interactive molecular and scientific visualization program. Protein Sci 18:2403–2409
4. Chen VB, Arendall WB III, Headd JJ, Keedy DA, Immormino RM, Kapral GJ, Murray LW, Richardson JS, Richardson DC (2010) MolProbity: all-atom structure validation for macromolecular crystallography. Acta Crystallogr D 66:12–21 (open access)
5. Davis IW, Arendall WB III, Richardson JS, Richardson DC (2006) The backrub motion: how protein backbone shrugs when a sidechain dances. Structure 14:265–274
6. Emsley P, Lohkamp B, Scott WG, Cowtan K (2010) Features and development of *Coot*. Acta Crystallogr D 66:486–501
7. Engh RA, Huber R (1991) Accurate bond and angle parameters for x-ray protein structure refinement. Acta Crystallogr A 47:392–400
8. Gore S, Velankar S, Kleywegt GJ (2012) Implementing an X-ray validation pipeline for the Protein Data Bank. Acta Crystallogr D 68:478–483 (open access)
9. Hooft RWW, Vriend G, Sander C, Abola EE (1996) Errors in protein structures. Nature 381:272

10. Keating K, Pyle AM (2012) RCrane: semi-automated RNA model building. Acta Crystallogr D 68:985–995
11. Kleywegt GJ, Jones TA (1996) Phi/psi-chology: Ramachandran revisited. Structure 4: 1395–1400
12. Lovell SC, Word JM, Richardson JS, Richardson DC (2000) The penultimate rotamer library. Proteins: Struct Funct Genetics 40:389–408
13. Lovell SC, Davis IW, Arendall WB III, de Bakker PIW, Word JM, Prisant MG, Richardson JS, Richardson DC (2003) Structure validation by Ca geometry: ϕ, ψ and Cβ deviation. Proteins: Struct Funct Genetics 50:437–450
14. Luethy R, Bowie JU, Eisenberg D (1992) Assessment of protein models with three-dimensional profiles. Nature 356:83–85
15. Morris AL, MacArthur MW, Hutchinson EG, Thornton JM (1992) Stereochemical quality of protein structure coordinates. Proteins 12:345–364
16. Read RJ, Adams PD, Arendall WB III, Brunger AT, Emsley P, Joosten RP, Kleywegt GJ, Krissinel EB, Lütteke T OZ, Perrakis A, Richardson JS, Sheffler WH, Smith JL, Tickle IJ, Vriend G, Zwart PH (2011) A new generation of crystallographic validation tools for the protein data bank. Structure 19:1395–1412, (open access); Faculty of 1000 "Must Read"
17. Richardson JS, Schneider B, Murray LW, Kapral GJ, Immormino RM, Headd JJ, Richardson DC, Ham D, Hershkovits E, Williams LD, Keating KS, Pyle AM, Micallef D, Westbrook J, Berman HM (2008) RNA backbone: consensus all-angle conformers and modular string nomenclature. RNA 14:465–481 (open access)
18. Scheffler W, Baker D (2010) RosettaHoles2: a volumetric packing measure for protein structure refinement and validation. Protein Sci 19:1991–1995
19. Ting D, Wang G, Shapovalov M, Mitra R, Jordan MI, Dunbrack RL Jr (2010) Neighbor-dependent Ramachandran probability distributions of amino acids developed from a hierarchical dirichlet process model. PLoS Comp Biol 6:e1000763 (open access)
20. Word JM, Lovell SC, LaBean TH, Zalis ME, Presley BK, Richardson JS, Richardson DC (1999) Visualizing and quantitating molecular goodness-of-fit: small-probe contact dots with explicit hydrogen atoms. J Mol Biol 285:1711–1733
21. Word JM, Lovell SC, Richardson JS, Richardson DC (1999) Asparagine and glutamine: using hydrogen atom contacts in the choice of side-chain amide orientation. J Mol Biol 285:1735–1747

Web Sites

- EDS (Electron Density Server): http://eds.bmc.uu.se/eds
- MolProbity service: http://molprobity.biochem.duke.edu
- PDB (at RCSB) Validation Server: http://validate.rcsb.org
- PDBsum (ProCheck) is linked from the specific PDB page, eg for 1ubq: http://www.ebi.ac.uk/thornton-srv/databases/cgi-bin/pdbsum/GetPage.pl?template=main.html&o=PROCHECK&c=999&pdbcode=1UBQ
- Phenix: http://phenix-online.org
- ProSA-web: https://prosa.services.came.sbg.ac.at/prosa.php
- Richardson Lab for KiNG, Reduce, Suitename, Top8000, etc: http://kinemage.biochem.duke.edu
- Verify3D profile analysis: http://nihserver.mbi.ucla.edu/Verify_3D
- WhatCheck: http://swift.cmbi.ru.nl/gv/whatcheck
- Wikipedia article: https://en.wikipedia.org/wiki/Structure_validation
- wwPDB info about referee validation reports: http://www.wwpdb.org/validation.html
- wwPDB link to Xray Validation Task Force: http://www.wwpdb.org/workshop/2011

Chapter 2
Total Chemical Protein Synthesis for the Determination of Novel X-ray Structures by Racemic Protein Crystallography

Kalyaneswar Mandal and Stephen B.H. Kent

Abstract Total synthesis of proteins by modern chemical ligation methods enables the ready preparation of high purity protein molecules of typical size (up to ~300 amino acid residues). This in turn enables the preparation of mirror image D-protein molecules not found in nature. Use of a racemic protein mixture (i.e. D-protein + L-protein) greatly facilitates the formation of diffraction-quality crystals of otherwise recalcitrant proteins. Facilitated crystallization is also observed for quasi-racemic protein mixtures. Centrosymmetric crystals of racemic proteins diffract to high resolution and offer enhanced possibilities for structure solution by direct computational methods. Racemic protein crystallography has been successfully applied to a number of recalcitrant protein molecules, and has been used to determine the structure of a 35 kDa {L-protein target/D-protein ligand} complex.

Keywords Chemical protein synthesis • X-ray crystallography • Racemic protein crystallization • Quasi-racemates • Direct methods

2.1 Introduction

Chirality, from the Greek 'cheir' = hand, is a feature of everyday experience. In each of your two hands, the fingers and thumbs have the same connectivity yet your hands are evidently not identical: to a good approximation they are mirror images

K. Mandal • S.B.H. Kent (✉)
Department of Biochemistry & Molecular Biology, University of Chicago, Chicago, IL 60637, USA

Department of Chemistry, University of Chicago, Chicago, IL 60637, USA
e-mail: kmandal@uchicago.edu; skent@uchicago.edu

R. Read et al. (eds.), *Advancing Methods for Biomolecular Crystallography*, NATO Science for Peace and Security Series A: Chemistry and Biology,
DOI 10.1007/978-94-007-6232-9_2,

of one another. In the mid-nineteenth century Louis Pasteur found that a similar phenomenon exists at the level of individual molecules [26]. He showed that certain molecules exist in mirror image forms that are not superimposable on one another, and thus can have distinct chemical properties under some circumstances. Such non-superimposable mirror image molecules are called 'enantiomers'.

All proteins found in nature contain 19 genetically-encoded amino acids of the same chirality (L-amino acids) and the genetically-encoded achiral amino acid glycine. A protein molecule consists of a linear polypeptide chain folded into a defined tertiary structure; it is this folded structure that gives rise to the specific biological function and other properties of a protein molecule [5]. Folded protein molecules are chiral entities: the mirror image D-protein form (i.e. enantiomer) of a protein molecule is not super-imposable on the natural L-protein.

For the past several decades, protein molecules have usually been prepared by expression in genetically engineered microorganisms [12]. These recombinant DNA methods have enabled the preparation of a wide range of proteins in good purity for structural studies by X-ray crystallography and by biomolecular NMR [4]. Additionally, site-directed mutagenesis [13] has become a powerful tool for 'protein engineering' [32] in which the molecular basis of protein function is studied by the site-selective replacement of amino acids in the protein's polypeptide chain with any of the 20 genetically-encoded proteinogenic amino acids. Recently, it has become possible to use genetic methods to incorporate a wide range of non-standard amino acids at any one site in a protein molecule's polypeptide chain [34].

Despite the power of recombinant methods for protein expression and the innumerable successes achieved by the application of these methods to the study of protein structure and function, there are still protein molecules that have proved resistant to study. In particular, for many proteins it has proved difficult or impossible to obtain diffraction quality crystals. Methods have been developed to help in the crystallization of such recalcitrant proteins. These include 'entropy reduction', the expression of truncated forms of the protein molecule, and the crystallization of the target protein complexed with engineered antibody molecules [8]. Despite the use of these methods, it has proved possible to crystallize only about 30 % of proteins that were obtained in soluble, purified form [6].

In 1995 Yeates and Wukovitz predicted on theoretical grounds that a racemic protein mixture – that is, a mixture of the D-protein and L-protein enantiomers of a target protein molecule – would crystallize much more readily than the natural L-protein alone [33]. The unnatural D-proteins that are necessary to implement racemic protein crystallization can only be prepared by total chemical synthesis. To be able to make D-proteins of useful size it was necessary to develop novel synthetic methods, based on the chemistries, based on the chemical ligation of unprotected peptides. Using these tools, we have applied racemic protein crystallization (Fig. 2.1) to the determination of a number of novel protein structures by X-ray crystallography.

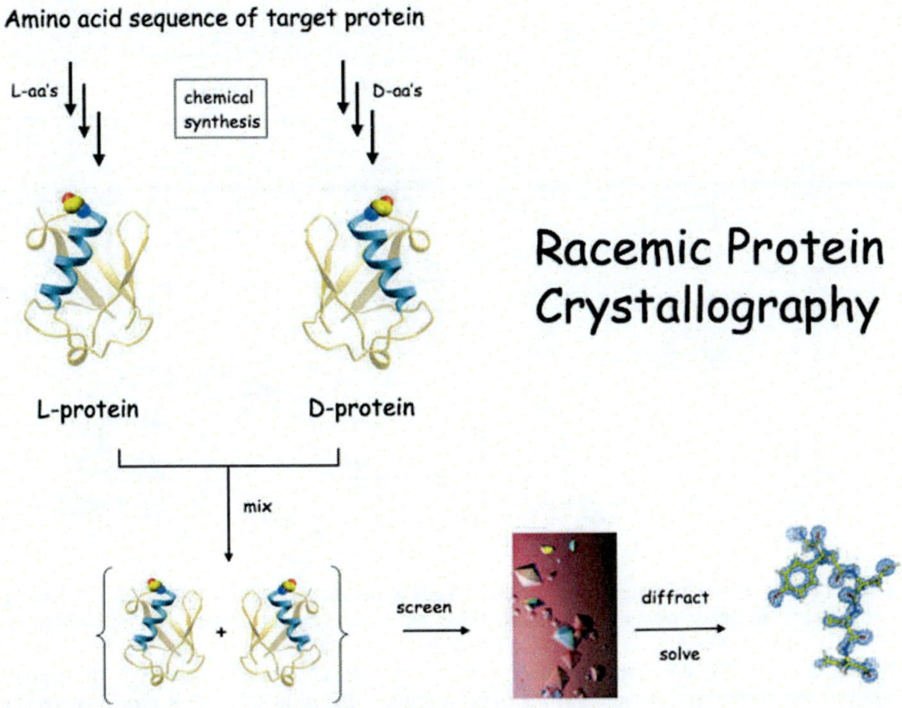

Fig. 2.1 Racemic protein crystallography. D-Protein and L-protein enantiomers are separately prepared by total chemical synthesis. A solution containing a 1;1 mixture of the D-protein + L-protein enantiomers is prepared and used in crystallization trials. After optimizatin of crystallization conditions, X-ray diffraction data are acquired, and the structure solved by MAD, molecular replacement, or direct methods [35]

2.2 Chemical Protein Synthesis

Anfinsen pointed out that a protein's polypeptide chain will spontaneously fold to form the defined tertiary structure of the functional protein molecule [1]. The building blocks of the natural protein world are 'domains', polypeptide chains of ~140 (±~30) amino acids that form autonomous folding units [5]. The typical protein found in nature has two such domains, in a single polypeptide chain of ~280 amino acids. From the early years of the twentieth century, chemists set out to make functional protein molecules by total synthesis. This endeavor was one of the 'grand challenges' of synthetic organic chemistry. Despite the invention and application of ingenious new chemistries based on novel physical principles, both solution chemistry and solid phase methods proved able to make only the smallest proteins [15]. By the mid-1970s, the largest protein reproducibly synthesized in chemically defined form was the insulin molecule (51 amino acids) [10]. It was clear that a new approach had to be found if the total chemical synthesis of protein molecules of typical size were to become a practical reality.

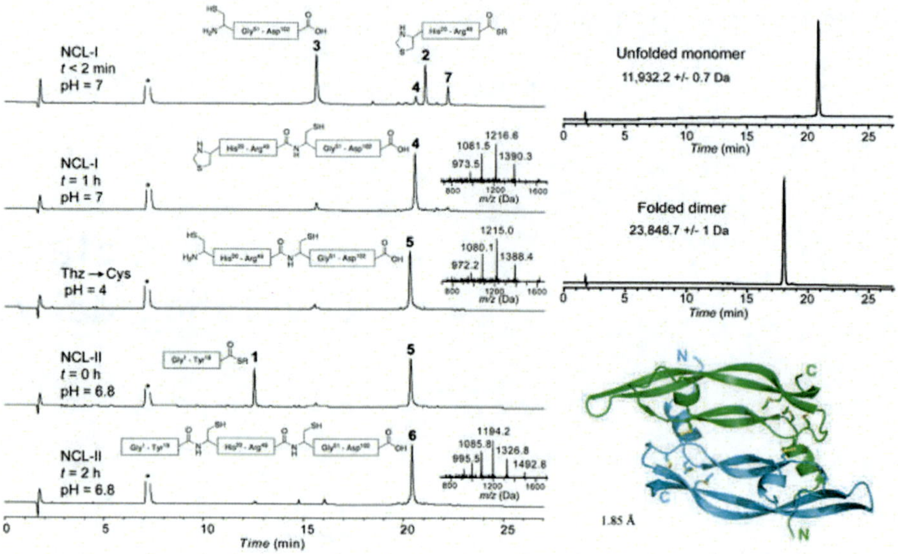

Fig. 2.2 Total chemical synthesis of VEGF-A (204 amino acid residues) [19]. (*Left*) Sequential native chemical ligations and other chemical manipulations were carried out without purification of intermediates. The *bottom* HPLC chromatogram shows the total crude products with a near-quantitative yield of the desired product **6**. (*Right*) LCMS characterization of the 102 residue synthetic polypeptide chain, and the folded covalent dimer 204 residue VEGF-A protein molecule. The X-ray structure of the synthetic protein was determined to a resolution of 1.85 Å

Coincidentally, at about the same time that Yeates and Woukovitz predicted the ease of racemic protein crystallization my laboratory developed a practical approach to the total chemical synthesis of proteins by means of chemoselective covalent condensation of unprotected synthetic peptides ('chemical ligation') [16]. Synthetic peptides up to ~50 amino acids are readily prepared in unprotected form by optimized solid phase peptide synthesis. The purity and covalent structure of the synthetic peptides are confirmed by high pressure liquid chromatography/electrospray mass spectrometry (LCMS).

The most effective ligation chemistry is 'native chemical ligation' [7], in which a peptide1-thioester is reacted with a Cys-peptide2 to give a product peptide1-Cys-peptide2 in which the two peptides are linked by a native amide bond. This chemical reaction is exquisitely precise. It is performed in aqueous solution at neutral pH and a chaotrope such as 6 M guanidine-HCl is used to insure the solubility of the reacting peptide segments. Quantitative yields of the desired ligation products are obtained within hours. LCMS is used to follow the progress of the ligation reactions and to confirm the identity of the products. Polypeptide chains of more than 200 amino acids can be prepared in straightforward fashion, and can routinely be folded in good yields to give high purity functional protein molecules with full biological activity. The total chemical synthesis of VEGF-A (204 amino acids) in this fashion is shown in Fig. 2.2.

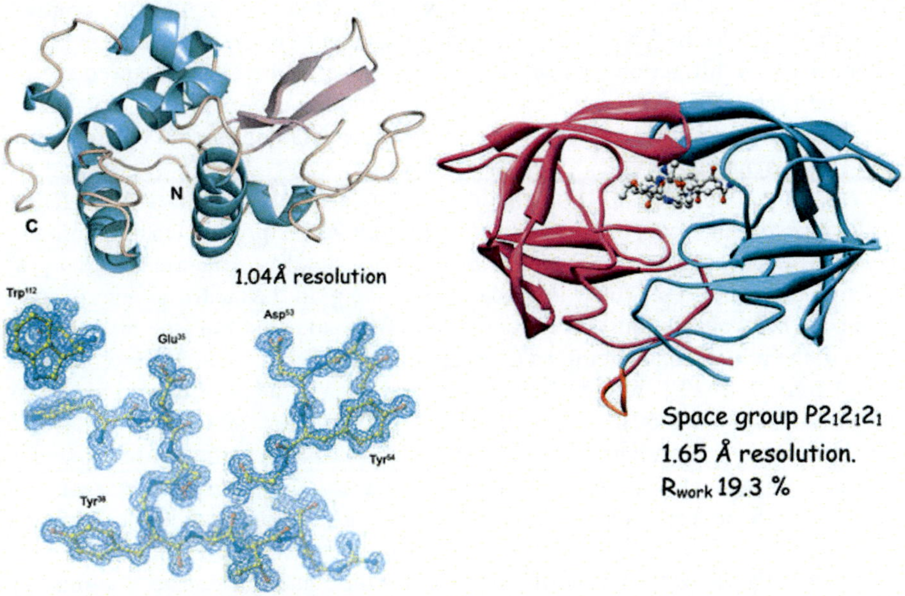

Fig. 2.3 X-ray structures of enzymes prepared by total chemical synthesis. (*Left*) Human lysozyme (130 amino acids) [9]. (*Right*) Covalent dimer of HIV-1 protease (203 amino acids synthetic polypeptide chain) [31]

The structure of the folded synthetic protein molecule can be confirmed by X-ray crystallography. Examples of enzymes prepared by total chemical synthesis are shown in Fig. 2.3.

2.3 Racemic Protein Crystallography

2.3.1 Snow Flea Antifreeze Protein

About 5 years ago, we set out to use total chemical synthesis to make a newly discovered protein. The snow flea antifreeze protein (sfAFP) was the smaller isoform of a novel thermal hysteresis protein, and its amino acid sequence had been predicted from cDNA sequencing [11]. At the time, data base searches found that sfAFP was unrelated to any known globular protein at the amino acid sequence level. The 81 residue polypeptide chain had a highly repetitive sequence and contained 37 glycine residues. The protein was believed to be thermally unstable and could not be expressed by recDNA techniques [11]. We used native chemical ligation to prepare the full-length sfAFP polypeptide chain from four synthetic peptide segments, and folded it with concomitant formation of two disulfide bonds to give synthetic sfAFP that was fully functional in an ice recrystallization assay [27]. In order to determine the structure of sfAFP, we attempted to crystallize the synthetic protein. After almost 9 months of exploring numerous conditions, we were able to obtain a 1.0 Å diffraction data set from a piece of a deliberately fractured crystal. Because no related proteins were known, we could not use molecular replacement methods to solve the structure, and we were not able to obtain crystals of a selenium-containing synthetic sfAFP.

We set out to use direct methods for structure solution. In 1989, Mackay had suggested that cocrystallization of D-protein and L-protein enantiomers in a centrosymmetric space group would considerably simplify structure solution [18], because the allowed phases would be quantized (e.g. in the centrosymmetric space group P1(bar), phases must be either 0 or π). We made D-sfAFP, the mirror image form of the protein, and attempted to crystallize sfAFP as a racemic mixture (i.e. L-sfAFP *plus* D-sfAFP]. To our pleasant surprise, where we had struggled for almost a year to get crystals of the L-protein form of sfAFP, now in a matter of days we obtained crystals from half of the conditions in a standard Hampton Index screen. It was at this point that we became aware of the 1995 prediction of Yeats and Wukovitz [33] that racemic proteins should crystallize more readily. The structure of sfAFP was solved by incorporating selenium into one sfAFP enantiomer only. Facilitated crystallization was still observed for the (quasi-)racemate, and multiple wavelength anomalous dispersion (MAD) was used to obtain phase values and solve the structure [28]. The X-ray structure of this unique globular protein molecule is shown in Fig. 2.4.

2.3.2 *Direct Methods*

Encouraged by our success with sfAFP, we set out to apply racemic protein crystallography to protein molecules for which there was no reported X-ray structure. Plectasin is an antibiotic protein isolated from the fungus *Pseudoplectania nigrella*. We made D-plectasin and L-plectasin by total chemical synthesis using native chemical ligation. Diffraction-quality crystals were obtained from both the L-protein alone, and from the racemic mixture of {D-plectasin + L-plectasin}. For the racemate, diffraction data was obtained to a resolution of 1.0 Å, and direct methods were used to solve the structure of the protein [20] (see Fig. 2.5). Another small protein, omwaprin, was first isolated from the venom of the inland taipan *Oxyuranus microlepidotus* [24]. We prepared D-omwaprin and L-omwaprin by total chemical synthesis using native chemical ligation. Crystals were obtained from the racemic mixture, but no crystals were obtained from L-omwaprin under the conditions used. Omwparin crystallized in space group $P2_1/c$ with 2 protein molecules in the asymmetric unit. Data was collected to a resolution of 1.33 Å, and the structure (see Fig. 2.5) was solved by direct methods [3]. Along similar lines, structures of two microprotein ion channel ligands (aka scorpion toxins), BmBKTx1 and kaliotoxin, which were difficult to crystallize as the L-proteins alone were solved from racemic crystals by direct methods [21, 29].

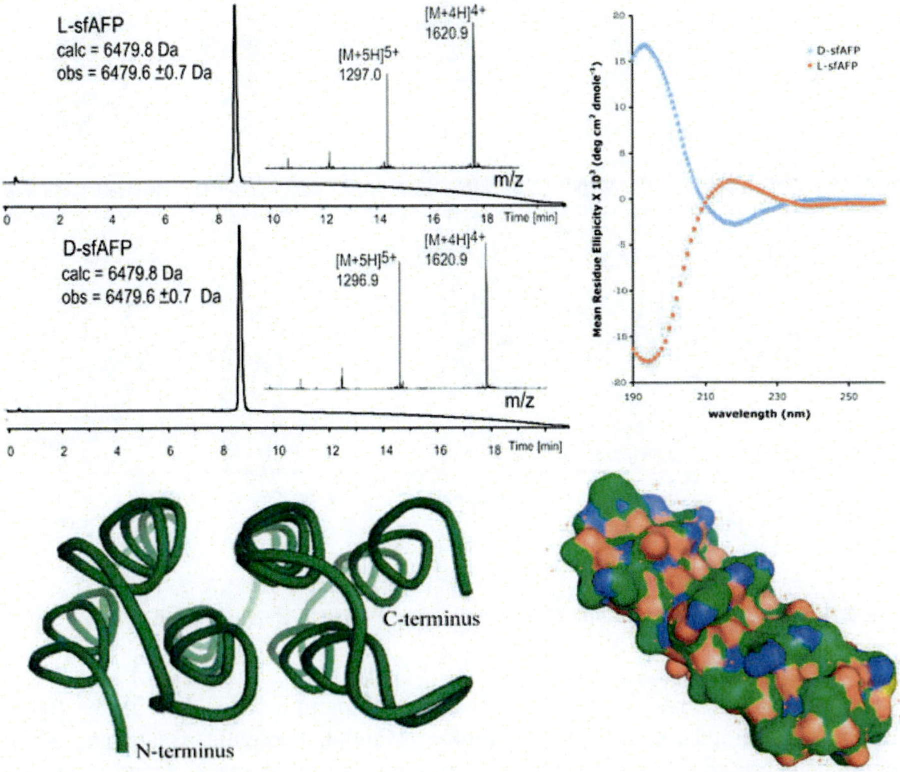

Fig. 2.4 Total chemical synthesis and X-ray structure of sfAFP [27, 28]. (*Top*) Characterization of D-sfAFP and L-sfAFP by LCMS and circular dichroism. (*Bottom*) *Left* – the secondary structure of sfAFP consists of two stacked sets of three antiparallel polyproline type II (*PPII*) helices. *Right* – the protein has a compact *brick-shaped* structure, with an upper polar surface, and an apolar lower surface (not shown)

2.3.3 Recalcitrant Proteins

We have applied racemic protein crystallization with considerable success to protein molecules that are resistant to crystallization by conventional methods. The predicted 94 residue Rv1738 protein from *Mycobacterium tuberculosis* is a notable example. Rv1738 is the most upregulated gene product when the *M. tuberculosis* bacterium enters persistent dormancy [30]. For that reason Rv1738 has been a major objective of an international consortium for structural genomics. However, in the Baker lab at Auckland University Rv1738 produced by recDNA expression was resistant to crystallization over a period of several years using a wide variety of methods to enhance crystallizability (EN Baker, personal communication). We prepared the D-protein and L-protein forms of Rv1738 using native chemical ligation, and obtained crystals from the racemic protein mixture. Diffraction data

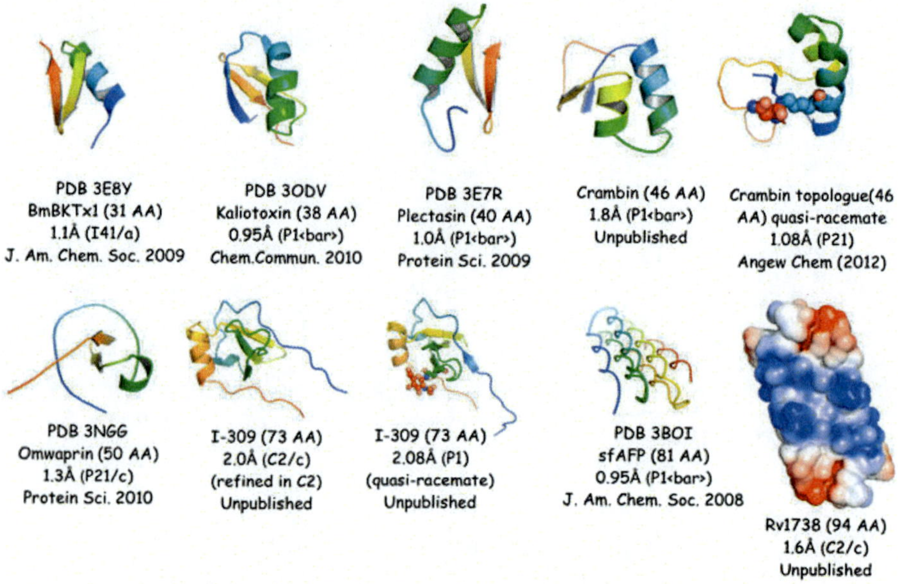

Fig. 2.5 Some structures determined by racemic protein crystallography in the Kent lab

were obtained to a resolution of 1.6 Å, and the structure was solved in the Baker lab by a modified direct method [Bunker, Mandal, Pentelute, Baker and Kent, unpublished data]. The structure suggests a critical biochemical function for the Rv1738 protein molecule as the bacterium enters persistent dormancy.

To date we have used racemic crystallization for about a dozen proteins that were known to be resistant to crystallization by conventional methods. In each case, the protein racemates gave diffraction quality crystals, with the single exception of the proinsulin protein molecule. Despite extensive efforts, we have been unable to crystallize proinsulin from a racemic protein mixture.

2.3.4 *Quasi-Racemate Crystallization*

The membrane-associated protein crambin (46 amino acid residues) has not been used for protein engineering because of its poor expression in microbial systems. Native crambin is isolated from plant seeds and has been used to determine the X-ray structure of this protein to a resolution of 0.54 Å [14]. We set out to use a covalent bond to replace an ion pair between the side chain of Arg^{10} and the alpha-carboxylate of the polypeptide chain at Asn^{46}. The resulting protein would have a novel linear-loop polypeptide chain topology. Successful synthesis of this protein topological analogue ('topologue') required us to develop novel synthetic chemistry for the condensation of unprotected peptides [2]. Ultimately, we were able to prepare

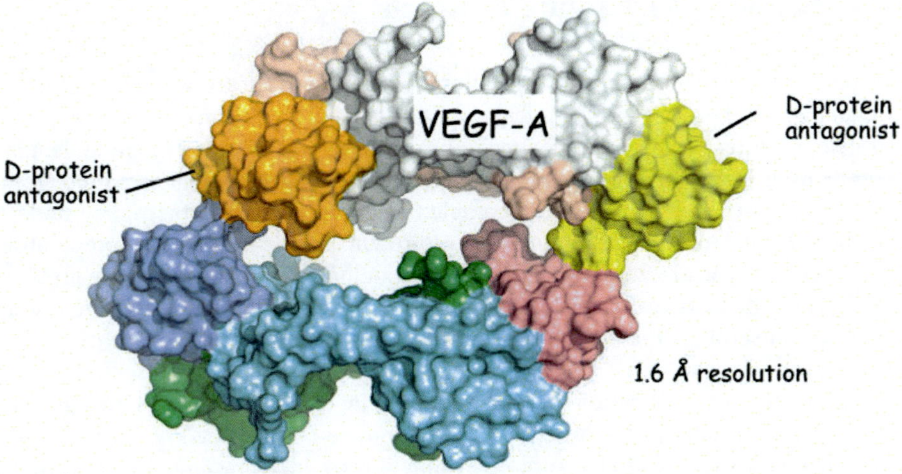

Fig. 2.6 X-ray structure determined by racemic crystallography of a heterochiral protein complex containing six chemically synthesized protein molecules; the complex has a structure weight of 73.2 kDa [23]

the topologue polypeptide chain by convergent synthesis of a branched peptide chain, followed by intramolecular native chemical ligation. The product 46 residue linear-loop polypeptide folded in quantitative yield to give a protein with three disulfide bonds. We were not able to obtain crystals from the L-topologue alone. To facilitate crystallization, we used a quasi-racemic mixture of {L-topologue + D-crambin}. Diffraction-quality crystals were readily obtained, and the X-ray structure was solved by molecular replacement [22]. The structure of the topological analogue protein is shown in Fig. 2.5. We have observed facilitated crystallization of quasi-racemic mixtures for a number of different protein molecules.

2.3.5 *Protein Complexes*

X-ray structures of interacting protein molecules enable the elucidation of the details of the interface between the protein molecules involved. This is of particular importance for the rational design of protein drugs. A D-protein that bound with good affinity to VEGF-A was developed by phage display of a library of designed variants of a novel small protein scaffold that was screened against the mirror image protein target D-VEGF-A [23]. In order to obtain the crystal structure of the D-protein bound to VEGF-A, we crystallized a mixture consisting of D-VEGF-A and L-VEGF-A with two equivalents of each of the D-protein binder and L-protein binder. The resulting crystals diffracted to 1.6 Å resolution, and the structure of the unique heterochiral complex was solved using molecular replacement. Each VEGF-A L-protein molecule was bound to two D-protein binder molecules, one at each end of the homodimeric VEGF-A protein (Fig. 2.6).

2.4 Summary and Future Prospects

In our hands, racemic protein crystallography has proven to be very useful for obtaining novel protein X-ray structures. Facilitated crystal formation from racemic protein mixtures has been almost invariably observed. What size protein molecules can be used for racemic protein crystallography? We have demonstrated the facile crystallization of a 35 kDa protein complex as the racemate. Furthermore, modern methods for the total chemical synthesis of proteins enable the practical preparation of proteins of typical size, up to ~30 kDa [16], including integral membrane proteins [17]. Thus, there is every reason to believe that racemic protein crystallography will prove useful for the determination of the structures of a wide range of protein molecules.

An important application of racemic protein crystallography is for the determination of the X-ray structures of natural glycoproteins. We have already demonstrated successful structure determination by quasi-racemic crystallization of a glycoprotein bearing a complex natural glycan with the corresponding non-glycosylated D-protein [25]. We are extending this approach to the determination of the X-ray structures of complex glycan moieties of natural glycoproteins by the use a of chemically synthesized L-neoglycan-D-protein crystallized as the quasi-racemate with the corresponding glycoprotein of natural configuration (i.e. D-neoglycan-L-protein).

The theory and practice of racemic protein crystallography are described in a current article in Annual Reviews of Biophysics [35].

Acknowledgments This research was supported by the Office of Science (BER), U.S. Department of Energy (grant no. DE-FG02 07ER64501 to S.B.H.K.) and by the National Institutes of Health (grant no. R01 GM075993 to S.B.H.K.). Use of NE-CAT beamline 24-ID at the Advanced Photon Source is supported by award RR-15301 from the National Center for Research Resources at the National Institutes of Health. Use of the Advanced Photon Source is supported by the U.S. Department of Energy, Office of Basic Energy Sciences, under contract no. DE-AC02-06CH11357.

References

1. Anfinsen CB (1973) Principles that govern the folding of protein chains. Science 181:223–230
2. Bang D, Pentelute BL, Kent SBH (2006) Kinetically-controlled ligation for the convergent chemical synthesis of proteins. Angew Chem Int Ed Eng 45:3985–3988
3. Banigan JR, Mandal K, Sawaya MR, Thammavongsa V, Hendrickx A, Schneewind O, Yeates TO, Kent SBH (2010) Determination of the X-ray structure of the snake venom protein Omwaprin by total chemical synthesis and racemic protein crystallography. Protein Sci 9:1840–1849
4. Berman HM, Westbrook J, Feng Z, Gilliland G, Bhat TN, Weissig H, Shindyalov IN, Bourne PE (2000) The protein data bank. Nucleic Acids Res 28:235–242
5. Branden C, Tooze J (1999) Introduction to protein structure, 2nd edn. Garland Science, New York
6. Chayen NE, Saridakis E (2008) Protein crystallization: from purified protein to diffraction-quality crystal. Nat Methods 5:147–153

7. Dawson PE, Muir TW, Clark-Lewis I, Kent SBH (1994) Synthesis of proteins by native chemical ligation. Science 266:776–779
8. Derewenda ZS (2004) Rational protein crystallization by mutational surface engineering. Structure 12:529–535
9. Durek T, Torbeev YV, Kent SBH (2007) Convergent chemical synthesis and high resolution X-ray structure of human lysozyme. Proc Natl Acad Sci USA 104:4846–4851
10. Eisler K, Kamber B, Riniker B, Rittel W, Sieber P, De Gasparo M, Marki F (1979) Synthesis and biological activity of five D-Cys analogs of human insulin. Bioorg Chem 8:443–450
11. Graham LA, Davies PL (2005) Glycine-rich antifreeze proteins from snow fleas. Science 310:461
12. Hannig G, Makrides S (1998) Strategies for optimizing heterologous protein expression in Escherichia coli. Trends Biotechnol 16:54–60
13. Hutchison CA III, Phillips S, Edgell MH, Gillam S, Jahnke P, Smith M (1978) Mutagenesis at a specific position in a DNA sequence. J Biol Chem 253:6551–6560
14. Jelsch C, Teeter MM, Lamzin V, Pichon-Pesme V, Blessing RH, Lecomte C (2000) Accurate protein crystallography at ultra-high resolution: valence electron distribution in crambin. Proc Natl Acad Sci USA 97:3171–3176
15. Kent SBH (1988) Chemical synthesis of peptides and proteins. Ann Rev Biochem 57:957–984
16. Kent SBH (2009) Total chemical synthesis of proteins. Chem Soc Rev 38:338–351
17. Kochendoerfer GG, Salom D, Lear JD, Kent SBH, DeGrado WF (1999) Total chemical synthesis of the integral membrane protein influenza A virus M2 proton channel: role of its cytoplasmic domain for pore assembly. Biochemistry 38:11905–11913
18. Mackay AL (1989) Crystal enigma. Nature 342:133
19. Mandal K, Kent SBH (2011) Total chemical synthesis of biologically active vascular endothelial growth factor. Angew Chem Int Ed 50:8029–8033
20. Mandal K, Pentelute BL, Tereshko V, Kossiakoff AA, Kent SBH (2009a) Racemic crystallography of synthetic protein enantiomers used to determine the X-ray structure of plectasin by direct methods. Protein Sci 18:1146–1154
21. Mandal K, Pentelute BL, Tereshko V, Kossiakoff AA, Kent SBH (2009b) X-ray structure of native scorpion toxin BmBKTx1 by racemic protein crystallography using direct methods. J Am Chem Soc 131:1362–1363
22. Mandal K, Pentelute BL, Bang D, Gates ZP, Torbeev VY, Kent SBH (2012a) Design, total chemical synthesis, and X-ray structure of a protein having a novel linear-loop polypeptide chain topology. Angew Chem Int Ed 51:1481–1486
23. Mandal K, Uppalapati M, Ault-Riché D, Kenney J, Lowitz J, Sidhu S, Kent SBH (2012b) Chemical synthesis and X-ray structure of a heterochiral {D-protein antagonist plus vascular endothelial growth factor} protein complex by racemic crystallography. Proc Natl Acad Sci USA 109:14779–14784
24. Nair DG, Fry BG, Alewood PF, Kumar PP, Kini RM (2007) Antimicrobial activity of omwaprin, a new member of the waprin family of snake venom proteins. Biochem J 402:93–104
25. Okamoto R, Kajihara Y, Kent SBH (2012), paper in preparation
26. Pasteur L (1848) Mémoire sur la relation qui peut exister entre la forme cristalline et la composition chimique, et sur la cause de la polarisation rotatoire. Comp Rend Paris 26:535–538
27. Pentelute BL, Gates ZP, Dashnau J, Vanderkooi JM, Kent SBH (2008a) Mirror image forms of snow flea antifreeze protein prepared by total chemical synthesis have identical antifreeze activities. J Am Chem Soc 130:9702–9707
28. Pentelute BL, Gates ZP, Tereshko V, Dashnau J, Vanderkooi JM, Kossiakoff AA, Kent SBH (2008b) X-ray structure of snow flea antifreeze protein determined by racemic crystallization of synthetic protein enantiomers. J Am Chem Soc 130:9695–9701
29. Pentelute BL, Mandal K, Gates ZP, Saway MR, Yeates TO, Kent SBH (2010) Total chemical synthesis and X-ray structure of kaliotoxin by racemic protein crystallography. Chem Commun 46:8174–8176
30. Sherman DR, Voskuil M, Schnappinger D, Liao R, Harrell MI, Schoolnik GK (2001) Regulation of the Mycobacterium tuberculosis hypoxic response gene encoding α-crystallin. Proc Natl Acad Sci USA 98:7534–7539

31. Torbeev VY, Kent SBH (2007) Convergent chemical synthesis and crystal structure of a 203 amino acid 'covalent dimer' HIV-1 protease enzyme molecule. Angew Chem Int Ed Eng 46:1667–1670
32. Wells JA, Estell DA (1988) Subtilisin – an enzyme designed to be engineered. Trends Biochem Sci 13:291
33. Wukovitz SW, Yeates TO (1995) Why protein crystals favor some space-groups over others. Nat Struct Biol 2:1062–1067
34. Xie J, Schultz PG (2005) Adding amino acids to the genetic repertoire. Curr Opin Chem Biol 9:548–554
35. Yeates TO, Kent SBH (2012) Racemic protein crystallography. Annu Rev Biophys 41:41–61

Chapter 3
Crystal Pathologies

Todd O. Yeates

Abstract Truly ideal crystals are rarely realized in macromolecular crystallography. The conformational complexity of protein molecules and the promiscuity of their chance interactions often conspire to give crystals in which the molecules are present in alternative configurations. When the alternative configurations occur randomly throughout the crystal, one is faced by a case of static disorder (often indistinguishable from thermal motion), leading to limited resolution and potential challenges in modeling the underlying structural variations. Despite those challenges, the case of random disorder is arguably the simplest to understand and interpret. A variety of more complex categories of crystal disorder occur when alternative molecular configurations, orientations, or positions are not random, but correlated to each other in one way or another throughout the crystal specimen.

Keywords Twinning • Crystal disorder • Intensity statistics

3.1 Twinning

Twinning describes a broad set of situations where a crystal specimen is composed of multiple domains, which individually behave like ideal crystals, but which are oriented differently relative to each other (Fig. 3.1). The subject of twinning in macromolecular crystals has been well-reviewed [3, 8, 11, 24, 25, 27, 28]. The misorientation of distinct crystal domains in a twinned specimen is made possible (or even probable) by the ability of molecular spacings to be matched at the interfaces between differently oriented twin domains. A fairly non-specific type of twinning, referred to as non-merohedral twinning, gives rise in a diffraction

T.O. Yeates (✉)
Department of Chemistry and Biochemistry, University of California,
Los Angeles, CA 90095, USA
e-mail: yeates@mbi.ucla.edu

R. Read et al. (eds.), *Advancing Methods for Biomolecular Crystallography*, NATO Science for Peace and Security Series A: Chemistry and Biology,
DOI 10.1007/978-94-007-6232-9_3,

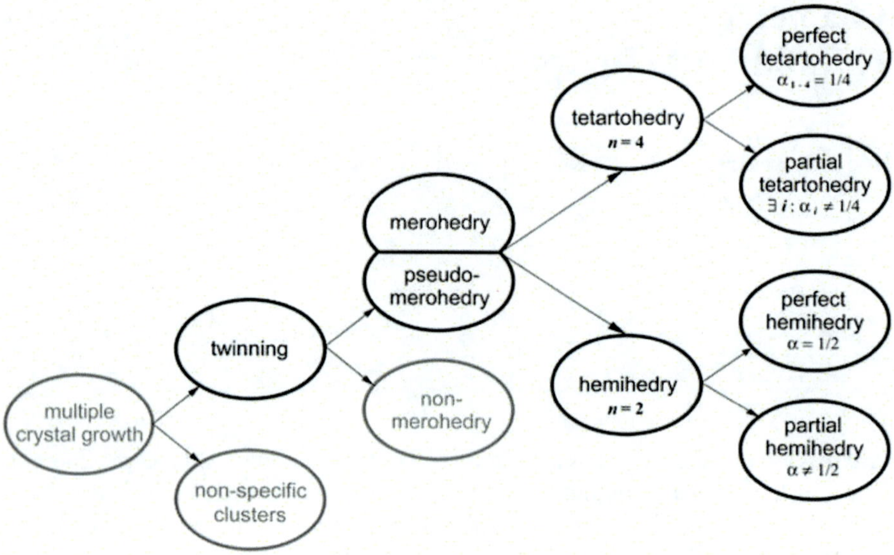

Fig. 3.1 Hierarchy of various types of twinning (Adapted from Yeates and Tsai [28])

experiment to a pattern composed of two (or more) independent, interpenetrating reciprocal lattices. This is usually easily recognized. Continual improvements in software have made it possible to deal effectively with diffraction patterns of this type, by integrating spots from distinct lattices separately, accounting for perfectly or closely overlapping reflections, etc.

Merohedral twinning is a more interesting, or at least more insidious, phenomenon. Here, the different twin domains of the specimen occur in (typically) two different orientations, related by an operation that is obeyed by the symmetry of the lattice (i.e. the holohedry) but which is not part of the crystal space group symmetry. This is possible whenever the lattice symmetry is higher than the space group symmetry. Figure 3.2 illustrates the case of space group P4; the underlying tetragonal lattice has extra rotational symmetry (422) not obeyed by the space group; the alternate twin domains are related by this extra operation. With merohedral twinning, the separate diffraction patterns arising from the multiple distinct twin domains are exactly superimposed, giving no visual indication that things are amiss.

The chief consequence of merohedral twinning is that the measured intensities are not really the true crystallographic intensities of individual reflections. Instead, each is a weighted sum of two twin-related but crystallographically independent reflections, $I(h_1)$ and $I(h_2)$, according to the value of the twin fraction, α

$$\mathrm{Iobs},1 = (1-\alpha)\,I(h_1) + \alpha\,I(h_2)$$

$$\mathrm{Iobs},2 = \alpha\,I(h_1) + (1-\alpha)\,I(h_2)$$

One challenge in recognizing and dealing with merohedral twinning is that the problem can manifest itself in different ways, depending on the twin fraction. When

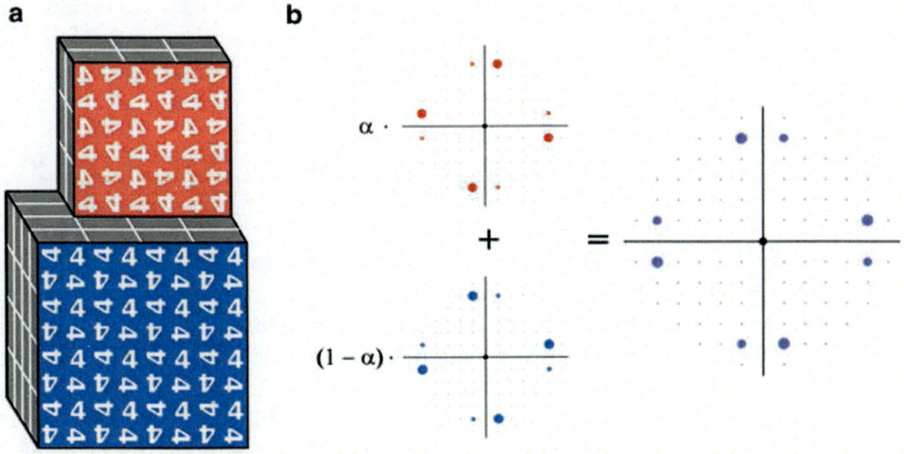

Fig. 3.2 A cartoon depicting partial merohedral twinning in space group P4. (**a**) Two twin domains growing together, related by a twofold twin-operation perpendicular to the fourfold symmetry axis. (**b**) The diffraction patterns of individual domains and their overlapping combination expected in a diffraction experiment (Adapted from Yeates and Fam [25])

α is equal to or very nearly equal to 1/2 (a situation often referred to as 'perfect twinning'), the outcome is erroneously high symmetry in the recorded X-ray data. For a successful structure determination, the crystallographer must come to realize that the true crystal space group symmetry is lower than it seems. When $\alpha < 1/2$ ('partial twinning'), the observed symmetry is correct, but one must realize and deal with the fact that the observed intensities do not reflect correct crystallographic quantities. Understanding which of these two scenarios is at play is essential in arriving at a correct interpretation in the end.

3.1.1 Testing for Twinning

It is possible to delineate two distinct effects that twinning has on intensity data. Each effect gives rise to various statistical tests for twinning. These are now broadly implemented in macromolecular software packages. Here again, understanding the distinction between the different effects of twinning and their respective tests, and how they relate to the dichotomy between perfect and partial twinning, is critical for a proper analysis.

First, twinning causes twin-related reflection pairs, which should be crystallographically independent, to have intensities more similar to each other than expected by chance; in the extreme case of $\alpha = 1/2$, they are exactly equal. The magnitude of the effect depends on α, and statistical tests based on a comparison of twin-pairs (sometimes referred to as tests for partial twinning) typically return an estimate for α. A variety of useful comparison metrics have been developed over the years

[5, 15, 23]. One of these gives an easy to remember interpretation. If H is defined to be the difference between two twin-related observed intensities divided by their sum, then the mean value of |H| over the data set should be equal to $(1-2\alpha)/2$ [23]. Rearrangement gives a quick estimate for α as $1/2 - <|H|>$. Methods for treating errors in estimating α have been developed [4, 5, 9, 15].

Two points of caution are called for in estimating the twin fraction by comparing potential twin-pairs. First, non-crystallographic symmetry (NCS) can cause the same effect as partial twinning – i.e. similarity between potentially twin-related reflections – so tests of this type, without further scrutiny, can lead to false conclusions of twinning. Second, tests of this type are of no utility in situations of perfect or near-perfect twinning; the equivalence between twin-related reflections would already be implicit from the apparent higher symmetry obtained during data reduction. And reducing data in a lower symmetry and then performing a test for partial twinning (i.e. comparing potentially twin-related reflections) can only lead to confusion; such a comparison would necessarily report near-equivalence of potential twin-pairs, which is consistent with perfect twinning, though no twinning may be present.

Various statistical measures are commonly employed to examine overall intensity distributions for evidence of perfect or very-high twinning [14, 18]. Perhaps the easiest to remember is $<I^2>/(<I>^2)$, which should be 2.0 for untwinned (acentric) data, and 1.5 for perfectly twinned (acentric) data. A more recent approach was designed to try to circumvent the obfuscating effects of anisotropy, whose presence along with other phenomena such as pseudo-translational symmetry can shift distributions in a way that masks the presence of twinning. In this more modern variation, the overall intensity distribution is not evaluated over individual reflections, but instead for reflection pairs nearby in reciprocal space (but not related by a potential twin operation). The local difference, $L = (I_A - I_B)/(I_A + I_B)$ obeys a simple distribution and has a simple expected mean value: $<|L|> = 1/2$ for untwinned (acentric) data and 3/8 for perfectly twinned (acentric) data [10]. This local test is generally more robust than the traditional approaches that date back to Wilson [22] (Fig. 3.3).

3.1.2 NCS

Non-crystallographic symmetry can confound attempts to analyze diffraction data for twinning. This situation is worsened by the observation that NCS very often occurs as an underlying feature in cases of twinning, typically with an NCS operator nearly parallel to a twin-operator. Although dissecting the two effects can be problematic, a generally useful approach can be to examine the behavior of various tests as a function of resolution; the effects of NCS typically break at higher resolution, whereas the effects of twinning persist across all resolution ranges.

Tests that give resolution-dependent results can be illuminating. For example, if an initial test for partial twinning (by comparing twin-pairs) suggests that twinning may be present, but repeating the test using only higher resolution data shows

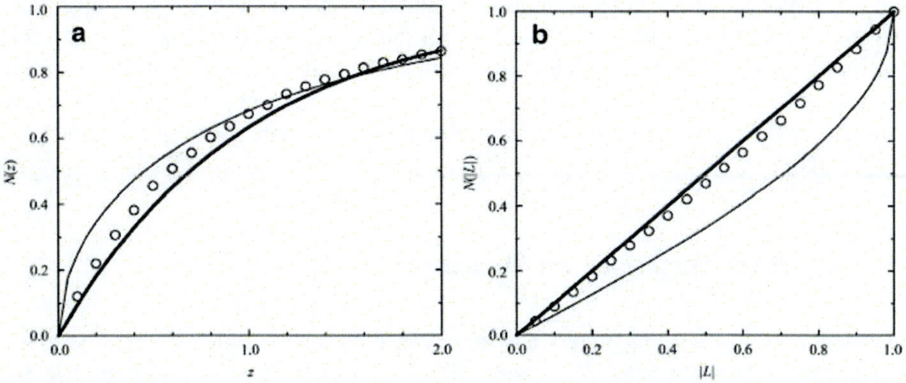

Fig. 3.3 Robustness of local intensity difference statistics in the presence of anisotropic scattering. Theoretical distributions for acentric data are shown by *bold curves*, while those for centric data are shown by the *thinner curves*. Distributions for observed acentric data are shown by *open circles*. The example (PDB code 1awu) illustrates a case where anisotropic scattering partially obscures the presence of twinning based on a traditional intensity distribution test (**a**), while the test of local differences, L, gives a clear indication of twinning, as seen in (**b**) (Adapted from Padilla and Yeates [10])

strongly reduced evidence for twinning, one might suspect that the situation results from simple NCS falsely mimicking twinning at lower resolution. Alternatively, one might examine the overall intensity statistics (in a test for perfect twinning) and find weak or ambiguous evidence for twinning at low to moderate resolution, but much stronger shifts in the intensity distribution based only on higher resolution data. This is consistent with true twinning nearly coincident with a nearly crystallographic NCS operation. At low resolution, the twin operator would mix together reflections whose intensities are already nearly equal to each other because of the NCS. Therefore, the intensity distribution might be almost normal at low resolution, and show strong evidence for twinning only at resolutions where the nearly crystallographic nature of the NCS breaks down.

3.1.3 *The End Game for Twinning*

If twinning is properly recognized, chances for successful structure determination are often good, especially by molecular replacement. Assuming that the true crystal symmetry has been correctly assigned, modern programs provide robust routines for refining structures against twinned data. How is this possible, given that the true intensities were never measured? In one type of approach, it is sometimes possible to effectively correct the observed data and estimate what the observed intensities should have been in the absence of twinning; this is referred to as 'detwinning'. In contrast, most approaches to structure determination and refinement take the reverse strategy, modeling the effects of twinning into the calculated intensities instead.

As a precautionary note, it should be understood that the averaging effects of twinning tend to produce lower R-values for purely statistical reasons not reflective of model quality. Therefore, obtaining a lower R-value in atomic refinement when twinning is invoked is, by itself, not evidence for the presence of twinning. More careful analyses of the type described above (and in more thorough reviews) are essential.

3.1.4 Other Variations on Twinning

The twinning situations noted above cover only the simpler types; there are numerous more complicated scenarios (Fig. 3.1). Pseudo-merohedral twinning can occur in space groups where twinning is not ordinarily expected, if a fortuitous unit cell geometry causes a low symmetry lattice to have nearly higher rotational symmetry. A rare situation known as reticular merohedral twinning can occur when only a subset of the reflections superimpose on each other; this can lead to strange diffraction patterns. Finally, twinning of higher order – i.e. with more than two distinct domain orientations – is possible. Several macromolecular cases of tetartohedral twinning ($n = 4$) have been reported in recent years [1, 6, 16, 29], and equations for handling such cases have been introduced [26].

3.2 Other Disorder Pathologies

Other kinds of disorder, distinct from twinning, have been reported in macromolecular crystals. One broad category includes cases where a single molecular configuration is maintained in a crystallographically ordered fashion in one layer (or row), but successive layers (or rows) might contain the molecule in an alternative configuration. When this occurs stochastically from layer to layer, the result has been described as an order-disorder (OD) phenomenon. Although cases are known where the distinction between alternate molecular configurations is a difference in orientation [12], most cases occur as a difference in relative position between molecules in different layers or rows [7, 17, 19–21, 30]. These cases are often described as lattice translocation disorders (LTD); their discovery dates to the case of methemoglobin in 1954, before the first crystal structures of proteins were determined [2].

LTD and other OD cases do not typically give the kinds of intensity distribution shifts seen in twinning; the short length scale of the stochastic variations between molecules causes structure factors to sum by interference in the usual way; complex F's add rather than intensities. LTD is therefore diagnosed in different ways. During the middle stages of structure determination, the presence of interpenetrating molecular density may provide a clue, echoing the presence of unmodeled molecular configurations that cannot exist simultaneously. Warning signs can often be seen before this stage. Intensity statistics can be 'hyper-centric' (shifted opposite from

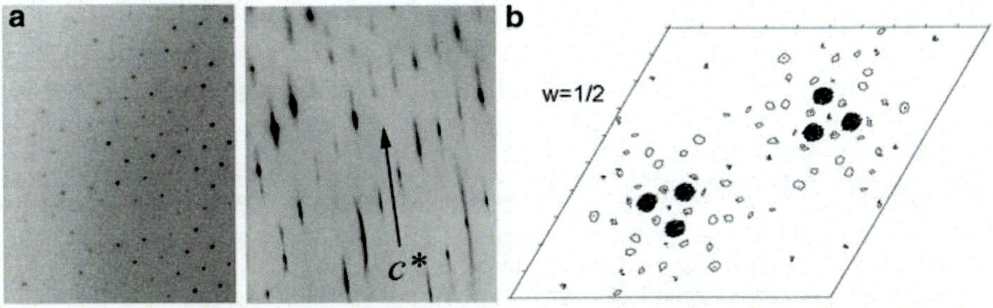

Fig. 3.4 A lattice translocation disorder in crystals of a bacterial microcompartment shell protein. (**a**) Prominent streaking is observed in certain directions, along c^* in this case. (**b**) Impossibly close packing peaks in a native Patterson map; the indicated molecular positions are not all simultaneously possible (Adapted from Tsai et al. [19])

the case of twinning) because of the modulating effects of translationally related molecules. However, hyper-centric intensity distributions are fairly common (e.g. whenever pseudo translational symmetry is present) even in crystals that do not suffer from disorder. Two features that appear to be common in LTD cases are systematic streaking of a subset of the reflections in a defined direction (Fig. 3.4a), and strong packing peaks in native Patterson maps at positions so close to the origin or to each other that they would imply impossibly close molecular packing if all the molecular positions were simultaneously occupied (Fig. 3.4b).

As with twinning, if a suitable model of the OD/LTD disorder can be developed, the structure can be determined correctly. As before, in some cases the observed intensities can be corrected by undoing the effects of having multiply shifted, partially occupied molecular positions. Alternatively, refinement can be performed in a way that incorporates the disorder into the model.

Beyond these kinds of disorders, others even more complex have been noted. Systematic off-Bragg peaks in crystals of a profilin-actin complex have been interpreted in terms of a complex modulated filament, whose period may not necessarily be commensurate with the lattice spacings in the crystal [13].

3.3 Concluding Remarks

Despite progress in identifying and dealing with disorder in macromolecular crystals, one thing that remains relatively clear is that the treatments employed are only approximations for what must be occurring in real crystals. The failures of final models to fully capture reality are especially evident in the cases of LTD treated so far. The models give reasonable approximations to the Bragg scattering (i.e. satisfactory R-values), but they do not account for the substantial scattering observed as streaking around Bragg peaks. To better understand and treat problems of disorder, renewed efforts are needed in the area of modeling diffraction from non-crystalline materials.

References

1. Barends TRM, de Jong RM, van Straaten KE, Thunnissen AWH, Dijkstra BW (2005) *Escherichia coli* MltA: MAD phasing and refinement of a tetartohedrally twinned protein crystal structure. Acta Crystallogr D 61:613–621
2. Bragg WL, Howells ER (1954) X-ray diffraction by imidazole methaemoglobin. Acta Crystallogr 7:409–411
3. Dauter Z (2003) Twinned crystals and anomalous phasing. Acta Crystallogr D 59:2004–2016
4. Dumas P, Ennifar E, Walter P (1999) Detection and treatment of twinning: an improvement and new results. Acta Crystallogr D 55:1179–1187
5. Fisher RG, Sweet RM (1980) Treatment of diffraction data from protein crystals twinned by merohedry. Acta Crystallogr A 36:755–760
6. Gayathri P, Banerjee M, Vijayalakshmi A, Azeez S, Balaram H, Balaram P, Murthy MRN (2007) Structure of triophosphate isomerase (TIM) from *Methanocaldococcus jannaschii*. Acta Crystallogr D 63:206–220
7. Hare S, Cherepanov P, Wang J (2009) Application of general formulas for the correction of a lattice-translocation defect in crystals of a lentiviral integrase in complex with LEDGF. Acta Crystallogr D 65:966–973
8. Helliwell JR (2008) Macromolecular crystal twinning, lattice disorders and multiple crystals. Crystallogr Rev 14:189–250
9. Lunin VY, Lunina NL, Baumstark MW (2007) Estimates of the twinning fraction for macromolecular crystals using statistical models accounting for experimental errors. Acta Crystallogr D 63:1129–1138
10. Padilla JE, Yeates TO (2003) A statistic for local intensity differences: robustness to anisotropy and pseudo-centering and utility for detecting twinning. Acta Crystallogr D 59:1124–1130
11. Parsons S (2003) Introduction to twinning. Acta Crystallogr D 59:1995–2003
12. Pletnev S, Morozova KS, Verkhusha VV, Dauter Z (2009) Rotational order-disorder structure of fluorescent protein FP480. Acta Crystallogr D 65:906–912
13. Porta J, Lovelace JJ, Schreurs AM, Kroon-Batenburg LM, Borgstahl GE (2011) Processing incommensurately modulated protein diffraction data with Eval15. Acta Crystallogr D 67: 628–638
14. Rees DC (1980) The influence of twinning by merohedry on intensity statistics. Acta Crystallogr A 36:578–581
15. Rees DC (1982) A general theory of x-ray intensity statistics for twins by merohedry. Acta Crystallogr A 38:201–207
16. Rosendal KR, Sinning I, Wild K (2004) Crystallization of the crenarchaeal SRP core. Acta Crystallogr D 60:140–143
17. Roversi P, Blanc E, Johnson S, Lea SM (2012) Tetartohedral twinning could happen to you. Acta Crystallogr D 68:418–424
18. Stanley E (1972) The identification of twins from intensity statistics. J Appl Crystallogr 5: 191–194
19. Tsai Y, Sawaya MR, Yeates TO (2009) Analysis of lattice-translocation disorder in the layered hexagonal structure of carboxysome shell protein CsoS1C. Acta Crystallogr D 65:980–988
20. Wang J, Kamtekar S, Berman AJ, Steitz TA (2005a) Correction of x-ray intensities from single crystals containing lattice-translocation defects. Acta Crystallogr D 61:67–74
21. Wang J, Rho S, Park HH, Eom SH (2005b) Correction of x-ray intensities from an HslV-HslU co-crystal containing lattice-translocation defects. Acta Crystallogr D 61:932–941
22. Wilson AJC (1949) The probability distribution of x-ray intensities. Acta Crystallogr 2: 318–321
23. Yeates TO (1988) Simple statistics for intensity data from twinned specimens. Acta Crystallogr A 44:142–144
24. Yeates TO (1997) Detecting and overcoming crystal twinning. Methods Enzymol 276:344–358
25. Yeates TO, Fam BC (1999) Protein crystals and their evil twins. Structure 7:R25–R29

26. Yeates TO, Yu F (2008) Equations for determining tetartohedral twin fractions. Acta Crystallogr D 64:1158–1164
27. Yeates TO, Sawaya MR (2011) Structure determination in the presence of twinning by merohedry. In: Arnold E, Himmel DM, Rossmann MG (eds) International tables for crystallography. Vol. F. Oxford, Wiley-Blackwell, pp 548–551
28. Yeates TO, Tsai Y (2011) Detecting twinning by merohedry. In: Arnold E, Himmel DM, Rossmann MG (eds) International tables for crystallography, vol F. Wiley-Blackwell, Oxford, pp 311–316
29. Yu F, Song A, Xu C, Sun L, Li J, Tang L, Yu M, Yeates TO, Hu H, He J (2009) Determining the DUF55-domain structure of human thymocyte nuclear protein 1 from crystals partially twinned by tetartohedry. Acta Crystallogr D 65:212–219
30. Zhu X, Xu X, Wilson IA (2008) Structure determination of the 1918 H1N1 neuraminidase from a crystal with lattice-translocation defects. Acta Crystallogr D 64:843–850

Chapter 4
Crystallizing Membrane Proteins for Structure-Function Studies Using Lipidic Mesophases

Martin Caffrey

Abstract The lipidic cubic mesophase or *in meso* method for crystallizing membrane proteins has posted some high profile successes recently. This is especially true in the area of G protein-coupled receptors with over a dozen new crystallographic structures emerging in the past 5 years. Slowly, it is becoming an accepted method with a proven record and convincing generality. However, it is not a method that is used in every membrane structural biology laboratory and that is unfortunate. The reluctance in adopting it is attributable, in part, to the anticipated difficulties associated with handling the sticky, viscous cubic mesophase in which crystals grow. Harvesting and collecting diffraction data with the mesophase-grown crystals is also viewed with some trepidation. It is acknowledged that there are challenges associated with the method. However, over the years we have worked to make the method user-friendly. To this end, tools for handling the mesophase in the pico- to nanolitre volume range have been developed for efficient crystallization screening in manual and robotic modes. Glass crystallization plates have been built that provide unparalleled optical quality and sensitivity to nascent crystals. Lipid and precipitant screens have been implemented for a more rational approach to crystallogenesis such that the method can now be applied to a wide variety of membrane protein types and sizes. These assorted advances are outlined here along with a summary of the membrane proteins that have yielded to the method. The challenges that must be overcome to further develop the method are described.

Parts of this article have been adapted from Caffrey [6] and Caffrey et al. [9].

M. Caffrey (✉)
Membrane Structural and Functional Biology Group, School of Biochemistry and Immunology, and School of Medicine, Trinity College Dublin, Dublin, Ireland
e-mail: martin.caffrey@tcd.ie

R. Read et al. (eds.), *Advancing Methods for Biomolecular Crystallography*, NATO Science for Peace and Security Series A: Chemistry and Biology,
DOI 10.1007/978-94-007-6232-9_4, © Springer Science+Business Media Dordrecht 2013

Keywords Alpha-helical membrane protein structure • Beta-barrel membrane protein • G protein-coupled receptor • GPCR • in meso • Integral membrane protein • Lipid cubic phase • Mesophase • Monoacylglycerol • Monoolein • Peptide • Precipitant • Rational design • Rastering • Remote data collection • Robotics • Screen • Sponge phase • Video demonstration • Workshop • X-ray crystallography

4.1 A Model for *In Meso* Crystallisation

A proposal has been advanced for how in meso crystallogenesis takes place at the molecular level [2–5]. It begins typically with an isolated biological membrane that is treated with detergent to solubilize the target protein. The protein-detergent complex is purified by standard wet-lab biochemical methods. Homogenizing with a monoacylglycerol (MAG) effects a uniform reconstitution of the purified protein into the bilayer of the cubic phase. The latter is bicontinuous in the sense that both the aqueous and bilayer compartments are continuous in three-dimensional space. Upon reconstitution, the protein ideally retains its native conformation and activity and has partial or complete mobility within the plane of the cubic phase bilayer. A precipitant is added to the mesophase, which triggers a local alteration in mesophase properties that include phase identity, microstructure, long-range order and phase separation. Under conditions leading to crystallization, one of the separated phases is enriched in protein, which nucleates and develops into a bulk crystal. The hypothesis envisions a local lamellar phase that acts as a medium in which nucleation and three-dimensional crystal growth occur. Molecular dynamics simulations highlight the hydrophobic/hydrophilic mismatch between the protein and the surrounding bilayer in the lamellar phase as a driving force for oligomerization in the membrane plane [19]. The local lamellar phase also serves as a conduit or portal for proteins on their way from the cubic phase reservoir to the growing face of the crystal. Initially at least, the proteins leave the lamellar conduit and ratchet into the developing crystal to generate a layered-type (Type I) packing of protein molecules. Given that proteins reconstitute across the bilayer of the cubic phase with no preferred orientation and the three-dimensional continuity of the mesophase, it is possible for the resulting crystals to be polar or nonpolar. These correspond to situations in which adjacent proteins in a layer have their long-axis director oriented in the same or in the opposite directions.

The proposal for how nucleation and crystal growth occur *in meso* relies absolutely on the three-dimensional continuity of the mesophase. Under the assumption that the sample exists as a single liquid crystallite or mono-domain, continuity ensures that the mesophase acts essentially as an infinite reservoir from which all protein molecules in the sample can end up in a bulk crystal. Neither the lamellar liquid crystal (L_α) nor the inverted hexagonal (H_{II}) phases, both of which are accessible mesophases in lipidic systems, have three-dimensional continuity and, alone, are unlikely to support membrane protein crystallogenesis by the *in meso* method.

However, it is possible to envision crystal growth that occurs by way of a local inverted hexagonal (H_{II}) phase. Indeed, there are several crystallization conditions, such as high salt, that favour this mesophase and that support crystal growth. As is the case with the cubic and lamellar phases, the cubic and H_{II} phases can and do coexist. Transitions between the two can be envisioned to involve inter-mesophase continuity. Since bitopic and polytopic membrane proteins span the bilayer at least once, the need to remain integral to the bilayer prevails also in the H_{II} phase. Indeed, locations where this can happen exist throughout the H_{II} phase, specifically at points of closest contact between lipid coated, water-filled rods. At such locations proteins can diffuse one-dimensionally along the length of the H_{II} phase rods to associate with one another – along and between rods – in nuclei first that, in time, evolve into macroscopic crystals. As with the lamellar phase model, the cubic phase will act as a reservoir to provide a continuous supply of proteins to the growing face of the crystal. A consequence of this growth mechanism type is that crystal packing, initially at least, will be hexagonal as opposed to layered or Type I.

That the *in meso* method works with bitopic and polytopic proteins, having one to several membrane crossings, respectively, is well proven. It has also been shown to support the crystallization of water-soluble proteins that include lysozyme and thaumatin. While there are no examples of the method working with monotopic or anchored proteins, we can anticipate these emerging in the not too distant future. A simple mechanism for crystallization would involve a form of inter-digitation. Here, the acyl chains of monolayers with which the protein are associated, inter-penetrate across the bilayer mid-plane. This would enable contact between proteins in and orthogonal to the membrane plane facilitating 3-D nucleation and crystal growth.

Because of the proposed need for the diffusion of proteins in the bilayer and of precipitant components in the aqueous channels of the mesophase, the expectation is that crystal growth rates might be tardy *in meso*. However, crystals have been seen to form within an hour, which suggests that the slowness associated with restricted diffusion can be compensated for by a reduction in dimensionality. The latter is a result of the protein being confined to a lipid bilayer with its long axis oriented perpendicular to the membrane plane. Thus, the number of orientations that must be sampled to effect nucleation and crystal growth is few *in meso* compared with its *in surfo* counterpart, in which all of three-dimensional space is accessible.

That crystal growth takes place in a mesophase implies it is happening in a convection-free environment. This is analogous to growth under conditions of microgravity or in a gel, which offers the advantage of a stable zone of depletion around the growing crystal and thus a slower and more orderly growth. Settling of crystals and subsequent growth into one another are also avoided under these conditions, as is the likelihood that impurities are wafted in from the surrounding solution to poison the face of the crystal and limit growth. For all these reasons *in meso* crystallogenesis is similar to crystallization in space with the prospect of producing high-quality, structure-grade crystals.

4.2 The *In Meso* Method. Practicalities and the Challenges It Presents

Setting up an *in meso* crystallization trial is straightforward. Typically, it involves combining two parts protein solution with three parts lipid at 20 °C. As noted, the lipid most commonly used is monoolein. According to the monoolein/water phase diagram [27], and assuming there is no major influence on phase behaviour of the protein solution components, this mixing process should generate, by spontaneous self-assembly, the cubic mesophase at or close to full hydration. The original method for mixing lipid and protein solution involved multiple cumbersome centrifugations in small glass tubes. Harvesting crystals required cutting the tubes and searching for small crystals through curved glass which was not simple, and required experience, time and patience.

The cubic phase is sticky and viscous in the manner of thick toothpaste. As such, it is not easy to handle. In the course of our earlier lipid phase science work we had developed tools and procedures for manipulating such refractory materials. One of these, the coupled-syringe mixing device, was ideally suited to the task of combining micro-litre volumes of monoolein with membrane protein solution in a way that produces protein-laden mesophase for direct use in crystallization trials with minimal waste [10]. The mixer consists of two Hamilton micro-syringes connected by a narrow bore coupler. Lipid is placed in one syringe, protein solution in the other. Mixing is achieved by repeatedly moving the contents of the two syringes back and forth through the coupler. The coupler is replaced by a needle for convenient dispensing of the homogenous mesophase into wells of custom-designed glass sandwich crystallization plates. Precipitant solutions of varying compositions are placed over the mesophase and the wells are sealed with a cover-glass. The plates are incubated at 20 °C and monitored for crystal growth. Optical quality is the best it can be given that the mesophase is held between two glass plates and the mesophase itself is transparent. This means that crystals, just a few micrometres in size, can be seen readily by microscope, whether or not the proteins are coloured. The use of cross-polarizers enhances the visibility of small crystals which usually appear birefringent in a dark background; the cubic phase itself is optically isotropic and non-birefringent. An added feature of the glass sandwich plates is that the double sided tape used to create the wells provides almost hermetic sealing ensuring minimal change in well composition during the course of trials that can last for months. Step-by-step instructions, complete with an open access, on-line video demonstration of the entire *in meso* crystallization process just described, have been published [7, 8, 24].

4.3 *In Meso* Robot

The protocol just described refers to the manual mode of setting up crystallization trials. Accurate and precise delivery of the protein-laden mesophase in volumes that range from pico- to micro-litres was made possible by use of an inexpensive repeat dispenser in combination with differently sized micro-syringes [8, 11, 12]. The smaller volumes means that the *in meso* method works with miniscule quantities of target protein. Thus, extensive crystallization trials can be set up with just a few micrograms of valuable membrane protein making the *in meso* method one of the most efficient in terms of protein requirement.

Whilst the repeat dispenser greatly facilitated the *in meso* method it was still a manual set-up with limits to the numbers of trials that any one person could comfortably set up at a sitting. The need to automate the process was obvious which led to the building of an *in meso* robot [15, 23]. The robot has two arms programmed to move simultaneously. One dispenses the viscous protein-laden mesophase while the other dispenses precipitant. Typical volumes used are 50 nL mesophase (consisting of 20 nL protein solution and 30 nL monoolein) and 800 nL precipitant solution. Custom, 96-well glass sandwich plates were designed which take about 5 min to fill using an 8-tip robot. The robot enables the precise and accurate setting up of *in meso* crystallization trials in high-throughput mode and, if required, under challenging conditions of reduced temperature and controlled lighting. Given the *in meso* robot's success, several are currently in use in labs throughout the world.

With the success that the *in meso* method has had it perhaps is not unexpected to find products appearing on the market in support of this novel crystallogenesis approach. In addition to the *in meso* robots, these include a number of precipitant screen kits, glass and plastic sandwich plates, and a plate that comes complete with lipid-coated wells. The vendors indicate that the latter can be used with a liquid dispensing robot for protein solution delivery first and precipitant post-swelling.

4.4 Mesophase Compatibility with Protein Solution Components

As alluded to above, what happens during *in meso* crystallization is intimately tied up with mesophase behaviour. The working hypothesis for how nucleation comes about begins with the protein reconstituting into the continuous bilayer of the cubic phase. Precipitant is added which triggers local formation of a lamellar phase into which the protein preferentially partitions and concentrates in a process that leads to nucleation and crystal growth. Experimental evidence in support of aspects of this model has been reported [9].

Experience built up over the years working with the *in meso* method suggests that the mesophase behaviour observed during the course of crystallization mimics that of the monoolein/water system [4]. The implication therefore is that the protein solution has minimal effect on the phase behaviour of the hosting lipidic mesophase into which the protein is reconstituted. That solution, along with the target protein, typically includes lipid, detergent, buffers, and salt at a minimum. Other components such as glycerol, sulphydryl reagents, denaturants, etc., are not uncommon. Each of these can impact on phase behaviour and, by extension, on the outcome of a crystallization trial. In the interests of learning about component compatibility, the sensitivity of the monoolein/water cubic phase system to their inclusion has been evaluated. Our findings indicate that the default cubic mesophase is remarkably resilient and retains its phase identity in the presence of a vast array of different additives. These include glycerolipids, cholesterol, free fatty acids, detergents, denaturants, glycerol and sulphydryl reagents, among others. Of course, for each there is a concentration beyond which the cubic phase is no longer stable. In most cases, these limits have been identified [14].

Occasionally, the concentration of a protein solution component is not known exactly. Detergent is a case in point. This poses a problem because if there is too much detergent the bulk lamellar phase may form, but it alone will not support crystallization. It may also be that a new detergent is being used who's compatibility with the cubic phase is not known. In this case, a small amount of the buffer used to solubilise the protein or the protein solution itself can be used to prepare mesophase. The physical texture, appearance between crossed polarizers, or small-angle X-ray scattering (SAXS) behaviour of the mesophase will indicate which phase has been accessed. If, for example, it is a lamellar phase that forms suggesting too much detergent then another purification step where its concentration in the final protein solution is reduced may be enough to solve the problem. We have encountered situations with bacteriorhodopsin where the particular preparation ended up having an excess of detergent. The mesophase first formed was lamellar but when it was used in combination with certain precipitants a transition back to the cubic phase was induced which went on to support crystal growth. This highlights the importance of understanding mesophase behaviour for more rational crystallisation.

4.5 When Protein Concentration Is Low

The driving force for nucleation is greater the more supersaturated is the system. Thus, a common strategy in the area of crystallisation is to work at the highest possible protein concentration to favor nucleation and to lower its concentration subsequently to just above the solubility limit for slow, orderly growth of a few good quality crystals. It is likely that the same principles apply to crystallization *in meso* where initially, the highest possible protein concentration should be used in support of nucleation. There are at least two issues that must be dealt with in this context that apply to membrane proteins. Firstly, most membrane proteins are prepared and purified in combination with detergents. Thus, the detergent is carried along with the protein into the crystallization mix. It follows then that as the protein concentration increases, the detergent concentration will rise in parallel. This may work against crystallization because high levels of detergent destabilize the hosting mesophase. Of course, the sensitivity to added detergent will depend, among other things, on the identities of the hosting lipid and detergent. Completely removing the detergent before folding the protein into the crystallization mix is usually not an option because it is commonly required to keep the protein soluble as a mixed micelle. One alternative is to reduce the detergent load to an acceptable level before combining the protein with the hosting lipid. This can be done with BioBeads or by eluting the protein in a highly concentrated form from an affinity column. Using detergents with low critical micelle concentration values, such as lauryl maltose neopentyl glycol (MNG-DDM) is also worth investigating.

Raising the protein concentration in the solution used to make the mesophase without simultaneously elevating detergent concentration is an important goal to work toward. This can be approached by selecting only the peak fractions from a final polishing gel filtration column and using the largest workable molecular weight cut-off filters for protein concentration. Glycerol can raise protein solubility and is compatible with the cubic phase. If this approach is used however the glycerol should be removed or its concentration dramatically reduced prior to running *in meso* crystallization trials. Simply equilibrating the mesophase so formed with excess precipitant under standard crystallization conditions (50 nL mesophase + 800 nL precipitant) will reduce glycerol concentration by ~40-fold. Further reductions are possible following the protocol described in the next paragraph.

The second issue has to do with raising the concentration of protein in the lipid bilayer of the cubic phase to facilitate nucleation. Two approaches can be tried that are quite different but that achieve the same end. The first exploits the water-carrying capacity of the cubic phase, a property that varies with lipid identity. Thus, the reconstituted protein will be more concentrated in the bilayer of the cubic phase prepared with a lipid of high water-carrying capacity than would obtain for a less hydrating lipid. The second approach involves sequential reconstitutions where the protein concentration in the bilayer rises with each round. The membrane protein preferentially partitions from the aqueous solution into the bilayer of the cubic phase. If the reconstitution step is repeated using a single mesophase bolus and with a series of solutions of protein at low concentration, the protein load in the bilayer of the mesophase will increase with each reconstitution round leaving excess aqueous solution depleted of protein. This protein-depleted solution is removed before the next round of reconstitution commences.

4.6 Screen Solution Compatibility

As noted, *in meso* crystallization relies upon a bicontinuous mesophase which acts as a reservoir to feed protein into nucleation sites and for crystal growth. The crystallization screening process requires that chemical space be interrogated over wide limits to find conditions that support crystal growth. In the screening process therefore, the protein-laden mesophase is typically exposed to precipitant solutions that encompass hundreds, perhaps even thousands of different chemical compositions. Screen solution components typically include buffers that cover a wide pH range, polymers, salts, small organics, detergents, apolar solvents, amphiphiles, etc., and all at different concentrations. Each component can potentially destabilize the mesophase. In a separate study using SAXS, we examined the compatibility of the default monoolein/water cubic phase with various commonly used precipitant screen solutions [13]. What we found was hardly surprising. Compatibility was temperature dependent and the usual suspects, that included organic solvents, destroyed the cubic phase rendering these screen solutions effectively useless. A goal of the study was to design screens that were mesophase friendly. However, that goal was not pursued then instead we opted for the convenience of commercial screen kits mindful of the fact that certain conditions are not relevant. As a result, certain kits are simply not used because they contain too few conditions that are compatible with the cubic phase.

4.7 Sponge Phase

During the course of mesophase compatibility studies we noticed that some screen components caused the cubic phase to 'swell' and, under certain conditions, to form what is referred to as the sponge phase. The latter evolves from the cubic phase as a result of the 'spongifying' component lowering bilayer interfacial curvature thereby enabling the mesophase to imbibe more lyotrope (aqueous solution). This is revealed in the SAXS pattern where the lattice parameter of the cubic phase rises [16]. Eventually, the mesophase looses order and the low-angle diffraction pattern becomes diffuse. Fortunately, the sponge phase retains its bicontinuity and, as a result, can support *in meso* crystallogenesis. One advantage of the sponge phase is that its aqueous channels are dilated. Thus, proteins with large extra-membranal domains should be accommodated in and amenable to crystallogenesis from the sponge phase [25, 29]. Further, the reduced interfacial curvature is likely to facilitate more rapid and long range diffusion within the lipid bilayer. Since net movement of protein from the bulk mesophase reservoir to the nucleation and growth sites is a requirement for crystallization this effect alone should contribute to improved crystallization. Interestingly, many of the proteins that have yielded to the *in meso* method have been crystallized under conditions that favour sponge phase formation (www.mpdb.tcd.ie).

Reflecting the utility of the sponge phase for *in meso* crystallogenesis a number of commercial screening kits now include spongifiers such as polyethyleneglycol, jeffamine, butanediol, among others. Some of these provide a preformed sponge phase to which the protein solution is added directly. We continue to use the original method that involves an active protein reconstitution step and where the entire crystallization screen space is available for sampling.

4.8 Rational Lipid Design for Low Temperature Crystallogenesis

Over the years, we have devoted considerable time and effort to establishing the rules for rationally designing lipids with specific end uses. One such application concerned the development of a host lipid for use in *in meso* crystallogenesis at low temperatures. Certain proteins are labile and require handling at low temperatures. The problem with the *in meso* method, in the default mode at least, is that it relies upon monoolein as the hosting lipid. The cubic phase formed by monoolein is not thermodynamically stable below about 17 °C and performing crystallization trials in a cold room at 4–6 °C is risky. For this low temperature application therefore a *cis*-monounsaturated monoacylglycerol, 7.9 MAG, was designed, using the rules referred to above [26]. The target MAG was synthesized and purified in-house and its phase behaviour mapped out using SAXS. As designed, it produced the cubic phase stable in the range from 6 to 85 °C. 7.9 MAG has been used in the crystallization of a number of membrane proteins in the MS&FB group. The objective now is to make it, along with other synthetic MAGs (see below), available to the community by way of a commercial vendor.

The word 'risky' was used in the previous paragraph when referring to low temperature crystallization with monoolein as the hosting lipid. This reflects the fact that it is possible to do successful *in meso* work with monoolein, and indeed other MAGs, at 4 °C provided the system undercools. Fortunately, the cubic phase is noted for this capacity and regularly we perform successful crystallization trials with monoolein in the 4–17 °C range. As expected, occasionally under these metastable conditions the mesophase will convert to the lamellar crystalline or solid phase which is useless as far as crystallization is concerned.

4.9 Additive Lipid Screening

Early on in the development of the *in meso* method the author recognized that monoolein, as the lipid used to create the hosting mesophase, is a most uncommon membrane lipid. The sense was that this MAG might rightly be regarded as foreign by certain target proteins and cause them to destabilize. One possible solution was to use a natural membrane lipid that would form the requisite cubic phase at 20 °C. None was available. An alternative was to use monoolein as the hosting lipid and to augment it with typical membrane lipids thereby creating a more native like environment. Accordingly, the carrying capacity of the monoolein cubic phase for a number of different lipids was established using SAXS [14]. This amounted to about 20 mol % in the case of phosphatidylcholine, phosphadidylethanolamine, and cholesterol with lesser amounts of phosphatidylserine and cardiolipin being accommodated. The approach of using additive lipids has had spectacular successes in the GPCR field where cholesterol doping of the cubic phase was critical to the production of structure yielding crystals [9, 29, 30].

4.10 Host Lipid Screening

Monoolein was the first lipid used for *in meso* crystallogenesis. From the outset, it was recognized that this one lipid may not work with all target membrane proteins. These, in turn, come from a variety of native membranes which differ in lipid composition, surface charge and packing density, fluidity and polarity profile, bilayer thickness, intrinsic curvature, etc. Thus, having a range of MAGs that differed in acyl chain characteristics with which to screen was deemed important. Using principles of rational design a number of suitable MAGs were identified with the requirement that they form the cubic phase at 20 °C under conditions of full hydration. A number of lipids meeting this specification have been synthesised and characterized in-house [31]. They constitute a successful hosting lipid screen in the MS&FB group [21, 22]. With several targets, that include β-barrels, α-helical proteins and an integral peptide antibiotic, crystals have been grown by the *in meso* method using these alternative MAGs [18, 21, 22, 25, 29]. In a number of cases, monoolein either failed to produce crystals or the crystals it did produce were not of diffraction quality. It was only when MAGs from the hosting lipid screen were used that structure grade crystals were obtained.

4.11 *In Meso* Structures

As of this writing, the *in meso* method accounts over 100 records in the PDB that relate to integral membrane proteins and peptides ([9]; www.mpdb.tcd.ie). This corresponds to about 10 % of all published membrane protein structures representing 8 distinct membrane protein types. With successes that include bacterial and eukaryotic rhodopsins, light harvesting complex II, photosynthetic reaction centres, β-barrels, GPCRs and their complexes, cytochrome oxidase, an ion exchanger and an integral membrane peptide the method has a record of versatility and range. Each of these membrane protein types represents bigger families the members of which become suitable candidates for *in meso* crystallogenesis. The GPCR family is a case in point with almost 800 distinct GPCRs coded for in the human genome alone. The *in meso* method therefore, in combination with the necessary protein engineering and receptor stabilization strategies, is now poised to contribute to the generation of GPCR structures in, what amounts to, production line fashion. Evidence in support of this statement is the recent spate of receptor structures (26 to date, see [9]; www.mpdb.tcd.ie) courtesy of the *in meso* method. We can only hope for the same degree of success with other membrane protein families.

The further development of the *in meso* crystallogenesis approach is an important goal for members of the MS&FB group. One direction this has taken recently concerns the utility of the method with small membrane proteins. A separate analysis performed using a model cubic phase under restricted conditions indicated that suitable targets would need to include at least five transmembrane helices. Our experience with the sponge phase variant of the cubic phase suggested otherwise.

Accordingly, the utility of the method with a 'mini-protein,' the pentadecapeptide antibiotic, linear gramicidin, was investigated. It worked remarkably well providing a structure with a resolution better than 1.1 Å [18]. This result is significant because it highlights the utility of the method with proteins having small transmembrane domains which abound in Nature.

4.12 The Membrane Protein Data Bank, Statistics On-line

Further details regarding the structure and function of integral, anchored and peripheral membrane proteins are available online in a convenient and searchable database, the Membrane Protein Data Bank (MPDB, [28]; www.mpdb.tcd.ie). Whilst records in the MPDB are obtained from the PDB, the former only includes entries for membrane proteins. Statistical analyses on the contents of the database can be performed and viewed online. Examples include detergents used for membrane protein structure work, number of structures published annually by method, and the like.

4.13 Prospects

The *in meso* method burst on the scene a decade and a half ago. It was received with great anticipation for what it would deliver; perhaps it was to be the panacea. However, output in the early years was limited to naturally abundant, bacterial α-helical proteins bedecked with stabilizing and highly colored prosthetic groups (www.mpdb.tcd.ie). The perceived restricted range, coupled to the challenges associated with handling the sticky and viscous cubic mesophase, meant that subsequent interest in the method waned. This was countered to some degree with the introduction of the *in meso* robot, a growing understanding for how the method worked at a molecular level, and a continued demonstration of the method's general applicability. However, interest in the method has rocketed of late with the success it has had in the GPCR field.

Improvements are needed of course if the method is to have longevity. Critically, the specialized materials and supplies upon which the method relies must be made more generally available and the method itself must be made user-friendly and routine. New and improved *in meso* robots available on the market are tackling the user-friendliness issue. Workshops that involve hands-on demonstrations contribute to making the method more accessible. Open access video demonstrations of the method are available online [8, 23, 24]. Recently, a simple and robust manual protocol for producing crystals in the lipidic cubic phase in less than an hour has been introduced. It is designed to provide newcomers to the *in meso* method with experience of preparing, handling and growing crystals in the sticky and viscous lipidic mesophase [1].

Developments are needed in the area of crystal identification. Optical clarity is of the highest quality with the glass sandwich plates currently in use and this provides for ready detection of colourless, micrometre-sized crystals in normal light and between crossed polarizers. Detection by UV fluorescence is particularly powerful and convenient for tryptophan containing proteins. Fluorescence labelling is also a route worth considering for the sensitive detection of early hits. Second-order nonlinear optical imaging of chiral crystals (SONICC) is a novel approach. It has been shown to sensitively and selectively detect membrane protein crystals growing *in meso* [20].

Recovering crystals from the mesophase for data collection is a non-trivial undertaking [24]. This is especially true when harvesting is done directly from glass sandwich plates. Typically, a glass cutter is used to open the well and to expose the mesophase. Teasing out and harvesting the crystal for immediate cryo-cooling is most conveniently done with a cryo-loop. This is a slow, pains-taking and cumbersome process especially if it must be done in a cold room and/or in subdued light. This whole area of harvesting calls out for innovation.

Data collection at the synchrotron is not exactly straightforward either. Given that *in meso*-grown crystals tend to be small, a mini-synchrotron X-ray beam is required. Oftentimes, the crystal of interest is hidden from view in a bolus of mesophase on the cryo-loop. This means that locating the crystal and centring it requires rounds of diffraction rastering with a beam of progressively smaller size [17]. This same approach is used to advantage in finding the best diffracting part of a crystal. Locating crystals and centering based on X-ray fluorescence from heavy atoms in the sample is in development. Effective and efficient rastering is recognized now as an important feature of the latest MX beamlines at synchrotron facilities worldwide and steady improvements in the rastering process are being made. *In situ* screening and data collection are other areas under active investigation. Increased efforts should be devoted to implementing protocols for crystal manipulation and diffraction data collection, viewing and processing remotely from the experimenter's home lab with the same ease and speed as are available (remotely) at the synchrotron lab.

The structures determined using *in meso*-grown crystals have, until very recently,[1] relied on molecular replacement for phasing. Increasingly, experimental phasing will be required. In our hands, with poorly diffracting crystals, this has proved to be a challenge. Several targets have been tackled using seleno-methionine labeling and pre-labeling, cocrystallization, and soaking with heavy atoms with only limited success. Problems derive, in part, from a low anomalous signal to noise due to a combination of background low- and wide-angle scatter from adhering mesophase and the need to work with small and sometimes poorly diffracting,

---

[1]The recent successes in using experimental phasing for structure determination have occurred with channelrhodopsin from *C. reinhardtii* (PDB entry 3UG9; mercury-MAD), the Na^{+}-Ca^{2+} exchanger from *M. jannaschii* (PDB entry 3V5U; samarium-SAD), β-barrels from *E. coli* (PDB entry 4E1S; seleno-methionine-SAD), and *Y. pseudotuberculosis* (PDB entry 4E1T; seleno-methionine-SAD) and with a membrane kinase from *E. coli* (D. Li, J. Lyons, V. Pye, D. Aragao, and M. Caffrey, in preparation; seleno-methionine-SAD).

radiation-sensitive crystals. As often as not, data must be collected in angular wedges on different parts of a single crystal or on multiple crystals, and merging data satisfactorily is a challenge. This part of the *in meso* pipeline is in need of work.

Finally, the method should begin to be used with really big proteins and complexes. The sponge phase, with its open aqueous channels and flatter bilayer, should prove particularly useful in this regard. Using it in combination with novel hosting and additive lipid screens will go a long way toward producing crystals and ultimately high resolution structures where interactions that are integral to human health are revealed.

Acknowledgements There are many who contributed to this work and most are from the MS&FB Group, both past and present members. To all I extend my warmest thanks and appreciation. This work was supported in part by grants from Science Foundation Ireland (07/IN.1/B1836, 12/IA/1255), FP7 COST Action CM0902, Marie Curie Actions (PIEF-GA-2009-254103) and the National Institutes Health (GM75915, P50GM073210, U54GM094599).

References

1. Aherne M, Lyons JA, Caffrey M (2012) A fast, simple and robust protocol for growing crystals in the lipidic cubic phase. J Appl Crystallogr 45(6):1330–1333
2. Caffrey M (2000) A lipid's eye view of membrane protein crystallization in mesophases. Curr Opin Struct Biol 10:486–497
3. Caffrey M (2003) Membrane protein crystallization. J Struct Biol 142:108–132
4. Caffrey M (2008) On the mechanism of membrane protein crystallization in lipidic mesophases. Cryst Growth Des 8:4244–4254
5. Caffrey M (2009) Crystallizing membrane proteins for structure determination. Use of lipidic mesophases. Annu Rev Biophys 38:29–51
6. Caffrey M (2011) Crystallizing membrane proteins for structure-function studies using lipidic mesophases. Biochem Soc Trans 39:725–732
7. Caffrey M, Cherezov V (2009) Crystallizing membrane proteins in lipidic mesophases. Nat Protoc 4:706–731
8. Caffrey M, Porter C (2010) Crystallizing membrane proteins for structure determination using lipidic mesophases. J Vis Exp 45:e1712, www.jove.com/index/details.stp?id=1712
9. Caffrey M, Li D, Dukkipati A (2012) Membrane protein structure determination using crystallography and lipidic mesophases: recent advances and successes. Biochemistry 51:6266–6288
10. Cheng AH, Hummel B, Qiu H, Caffrey M (1998) A simple mechanical mixer for small viscous lipid-containing samples. Chem Phys Lipids 95:11–21
11. Cherezov V, Caffrey M (2005) A simple and inexpensive nanoliter-volume dispenser for highly viscous materials used in membrane protein crystallization. J Appl Crystallogr 38:398–400
12. Cherezov V, Caffrey M (2006) Picoliter-scale crystallization of membrane proteins. J Appl Crystallogr 39:604–609
13. Cherezov V, Fersi H, Caffrey M (2001) Crystallization screens: compatibility with the lipidic cubic phase for in meso crystallization of membrane proteins. Biophys J 81:225–242
14. Cherezov V, Clogston J, Misquitta Y, Abdel Gawad W, Caffrey M (2002) Membrane protein crystallization in meso. Lipid type-tailoring of the cubic phase. Biophys J 83:3393–3407
15. Cherezov V, Peddi A, Muthusubramaniam L, Zheng YF, Caffrey M (2004) A robotic system for crystallizing membrane and soluble proteins in lipidic mesophases. Acta Crystallogr D 60:1795–1807

16. Cherezov V, Clogston J, Papiz M, Caffrey M (2006) Room to move. Crystallizing membrane proteins in swollen lipidic mesophases. J Mol Biol 357:1605–1618
17. Hilgart MC, Sanishvili R, Ogata CM, Becker M, Venugopalan N, Stepanov S, Makarov O, Smith JL, Fischetti RF (2011) Automated sample-scanning methods for radiation damage mitigation and diffraction-based centering of macromolecular crystals. J Synchrotron Radiat 18:717–722
18. Hofer N, Aragao D, Lyons JA, Caffrey M (2011) Membrane protein crystallization in lipidic mesophases. Hosting lipid effects on the crystallization and structure of a transmembrane peptide. Cryst Growth Des 11:1182–1192
19. Khelashvili G, Albornoz P, Johner N, Mondal S, Caffrey M, Weinstein H (2012) Why GPCRs behave differently in cubic and lamellar lipidic mesophases. J Am Chem Soc 134:15858–15868
20. Kissick DJ, Gualtieri EJ, Simpson GJ, Cherezov V (2010) Nonlinear optical imaging of integral membrane protein crystals in lipidic mesophases. Anal Chem 82:491–497
21. Li D, Shah, Caffrey M, Host lipid and temperature as important screening variables for crystallizing integral membrane proteins in lipidic mesophases. Trials with diacylglycerol kinase. Cryst Growth Des (In press)
22. Li D, Lee J, Caffrey M (2011) Crystallizing membrane proteins in lipidic mesophases. A host lipid screen. Cryst Growth Des 11:530–537
23. Li D, Boland C, Walsh K, Caffrey M (2012a) Use of a robot for high-throughput crystallization of membrane proteins in lipidic mesophase. J Vis Exp 67:e4000, www.jove.com/index/details.stp?id=4000
24. Li D, Boland C, Aragao D, Walsh K, Caffrey M (2012b) Harvesting and cryo-cooling crystals of membrane proteins grown in lipidic mesophases for structure determination by macromolecular crystallography. J Vis Exp 67:e400, www.jove.com/index/details.stp?id = 4001
25. Lyons JA, Aragao D, Slattery O, Pisliakov AV, Soulimane T, Caffrey M (2012) Structural insights into electron transfer in caa3-type cytochrome oxidase. Nature 487:514–518
26. Misquitta Y, Cherezov V, Havas F, Patterson S, Mohan J et al (2004) Rational design of lipid for membrane protein crystallization. J Struct Biol 148:169–175
27. Qiu H, Caffrey M (2000) The phase diagram of the monoolein/water system: metastability and equilibrium aspects. Biomaterials 21:223–234
28. Raman P, Cherezov V, Caffrey M (2006) The membrane protein data bank. Cell Mol Life Sci 63:36–51, www.mpdb.tcd.ie
29. Rasmussen SGF, DeVree BT, Zou Y, Kruse AC, Chung KY, Kobilka TS, Thian FS, Chae PS, Pardon E, Calinski D, Mathiesen JM, Shah STA, Lyons JA, Caffrey M, Gellman SH, Steyaert J, Skiniotis G, Weis WI, Sunahara RK, Kobilka BK (2011) Crystal structure of the β2 adrenergic receptor–Gs protein complex. Nature 477:549–555
30. Rosenbaum DM, Zhang C, Lyons JA, Holl R, Aragao D, Arlow DH, Rasmussen SGFR, Choi H-J, DeVree BT, Sunahara RK, Chae PS, Gellman SH, Dror RO, Shaw DE, Weis WI, Caffrey M, Gmeiner P, Kobilka BK (2011) Structure and function of an irreversible agonist-b2 andrenoceptor complex. Nature 469:236–240
31. Yang D, Cwynar VA, Hart DJ, Madanmohan J, Lee J, Lyons J, Caffrey M (2012) Preparation of 1-monoacylglycerols via the Suzuki-Miyaura reaction: 2,3-dihydroxypropyl (Z)-tetradec-7-enoate. Org Synth 89:183–201

Chapter 5
Searching for Needles in Haystacks: Automation and the Task of Crystal Structure Determination

Seán McSweeney

Abstract Chambers dictionary defines the expression **look for a needle in a haystack** as *to undertake a hopeless search.* Crystallographic investigations seem on many occasions to fit this definition closely. The use of synchrotron radiation sources with automated methods for beam delivery and sample changing has revolutionised the process of finding those crystals that have the properties required to elucidate a crystal structure.

Keywords Structural Biology • Synchrotron Radiation • Automation • MX: Macromolecular Crystallography

5.1 Introduction

There has been a dramatic growth in the use of Macromolecular Crystallography (MX) over the last 15 years. This growth has been coupled with some spectacular scientific successes. These have been achieved across the whole spectrum of biology including (but not limited to) membrane proteins (G-protein coupled receptors [34], ABC transporters [12], ion channels [14], photosystems I and II [2, 16, 37], viruses (Blue tongue virus [17]) and more recently lipid bilayer containing bacteriophages [7]), molecular machines such as F1-ATPase, the bacterial ribosome [28], DNA polymerase II [10], the muscle contractile protein myosin [9], macromolecular complexes (the proteosome, tumour necrosis factor) and high throughput ligand screening for improved medicines (human cytochrome P450 [35, 36] and human phosphodiesterase structures [32]). The impact of these results has extended from an improvement of the basic understanding of Molecular Biology to wealth generation in the European Community.

S. McSweeney (✉)
European Synchrotron Radiation Facility, 6 rue Jules Horowitz, 38043 Grenoble, France
e-mail: seanmcs@esrf.fr

R. Read et al. (eds.), *Advancing Methods for Biomolecular Crystallography*,
NATO Science for Peace and Security Series A: Chemistry and Biology,
DOI 10.1007/978-94-007-6232-9_5, © Springer Science+Business Media Dordrecht 2013

Spurred on by the successes that have already been achieved, Structural Biologists are tackling ever more ambitious projects, for example more complex membrane proteins and larger macromolecular assemblies. Almost invariably, these projects pose significant challenges both in terms of specimen preparation and of obtaining sufficient material of the appropriate levels of purity and stability for crystallisation trials. These problems are particularly an issue for mammalian membrane proteins, which are very unstable in the detergents that are used for crystallisation, and for which the cost of obtaining sufficient quantities of sample is extremely high.

It is very common for crystals of large biological macromolecules, even when obtained under notionally identical conditions, to show considerable variation in the quality of their diffraction. This is almost certainly the result of the intrinsic physical properties of the systems being studied. The molecules involved are frequently very flexible, leading to conformational heterogeneity. In the case of complexes of several proteins, there is also the risk of compositional heterogeneity where some complexes have lost one or more components. When the crystals obtained are small (less than $\sim$50 μm) they are also often mechanically fragile, and therefore susceptible to damage during transfer (from the crystallisation trays to the sample holders) or essential cryo-protection steps.

Advances in technology have played a significant role in realising these achievements. For MX this includes improvements in protein expression and crystallisation and, crucially, the development of intense, tunable beamlines at third generation synchrotron sources. During the last 5 years, two further developments at synchrotrons have played important roles, namely high levels of automation of sample handling and access to microfocus beamlines that are optimised for MX applications.

Current data collection trends at the ESRF MX beamlines show clearly that automation has led to a fundamental change in the way that MX data are collected. With the introduction of the automatic sample changer [6] at the ESRF MX beamlines circa 2005, a dramatic increase in the number of samples that were tested for diffraction quality before any full data collections were carried out was observed (Fig. 5.1), where we define a dataset as the collection of 20 or more images in one sweep. As is clear from the figure we are reaching a point where fewer than one sample in 20 or 30 is deemed suitable for further investigation by data collection.

These examples are not atypical of many ongoing structural biology projects. Access to highly automated, high throughput screening facilities, with the possibility to then transfer the best crystals to another beamline optimised for data collection, will become an increasingly important element for success in Structural Biology. In practice, the ability to identify those "one in a thousand" samples that are well ordered will enable new science to be performed with new projects becoming viable. However providing turnkey systems that will be usable by scientists who are inexperienced remains challenging and describing our efforts in this direction is the subject of the remainder of this chapter.

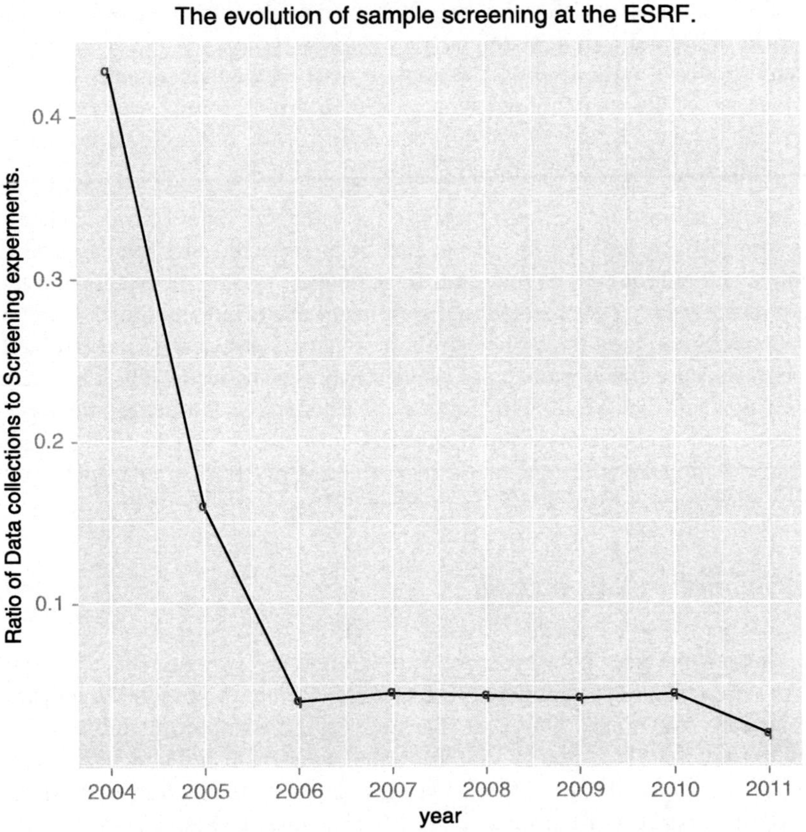

Fig. 5.1 The change of beamline use since the availability of a reliable MX sample changer described in the form of the ratio of data collection experiments to screening experiments. The rapid increase in the number of samples tested before a data collection is launched is obvious. Here we define Screening as the collection of four or fewer non-contiguous images, data collection being 20 or more contiguous images

5.2 Specification of an Automated MX System

Based upon the feedback from the ESRF User community and the use of automated systems at synchrotrons worldwide we identify some essential elements for an optimised automated system supporting MX. The principle components of this definition include the following elements:

- It is expected that the number and nature of crystals requiring screening will increase (intra-crystal, inter-crystal, micro-crystals, in plate screening, low resolution...).

- Rapid data collection after the screening even if the optimal experiment must be performed on a different beamline.
- Remote access to the beamline as a major necessary enhancement.
- It is expected that an increased synergy with different, complementary techniques including small angle scattering, time resolved MX, Cryo-electron-microscopy, imaging, micro-specrophotometry is highly desirable.

Primary among these requirements is the need for reliable and accurate data collection. To achieve this on all samples it is expected that increased levels of beamline automation will be required. To achieve a measure of success in these two requirements it is to be expected that significant effort in technology development will be required. Inevitably the number of partial datasets to be handled will increase. Because radiation damage plays a major causative role in this situation it is likely that new data handling methods will be necessary. Ultimately we expect that resolution of the requirements outlined above need also be matched with a beamline capable of supporting the concept of the evaluation of "1000 crystals per day".

5.3 Beamline Automation

A concern to the user of a synchrotron beamline are two sources of automation that must be taken in to account before the task of determining the best diffraction strategy can take place. These are the delivery of an appropriately conditioned X-ray beam with which one may perform the experiment and a mechanism for the changing of samples that is both reliable and convenient to use. Space does not allow for a full exploration of all the details of the methods so a summary of the major themes is provided.

5.3.1 X-ray Beam Delivery

When thinking about the automated provision of X-ray beams it is important to keep in mind that the task is to provide an intense X-ray source, RMS stability to better than 1 % on intensity and with a positional stability of order 2 μm whilst handling distances from X-ray source to sample of 60 m or more. It is not surprising then that the "simple" task of delivering the X-ray beam has taken many years of effort to achieve the level of performance we experience today.

In the alignment of a beamline an important first step if the definition of the X-ray beam emerging from the storage ring. This step is made by aligning (typically water cooled) apertures thereby providing a defined and characterised starting point for subsequent alignment. Due to the high power density generated by third generation sources most beamlines use liquid-nitrogen-cooled silicon monochromators. At the ESRF we have chosen to standardise on the use of channel cut crystals, although other designs have been used with success elsewhere. In our experience with this system it is possible to provide reliable automatic energy variation on request with good positional reproducibility and error in the photon energy of less than 0.2 eV in the range 6 and 25 keV.

Automated beam focusing has two objectives: it must optimize the X-ray flux at the sample position and ensure a smooth beam cross section. This latter point is crucial when dealing with highly collimated beams and micrometre-sized crystals. The automatic-focusing procedure we use performs an alignment of the mirror to the beam (necessary in our case because of the properties of the monochromator) and followed by an optimisation of the focal spot at the sample or at the detector front face. A principle objective in this process is to be able to match the beam dimensions to those of the crystal.

By employing these techniques we are able to allow complete control of the beamline configuration to the scientists using the beamline. They are assured that the X-rays are as they requested, we are reassured that no damage or mishap can come to the expensive optical components of the beamline.

5.3.2 *Automatic Sample Changers*

A major bottleneck encountered by a user at a synchrotron MX beamline has been the time required to get the sample of interest mounted/centred in the X-ray beam and ready for data acquisition. Analysis of manual method of sample changing revealed that the process was amenable to automation and a number of groups have successfully developed methods for automatic sample handling [6, 8, 20, 29, 30, 33].

Automation of sample handling procedures allows one to streamline the data-collection process and ideally reduces the error rate encountered during manual mounting and dismounting of samples. When initially deployed at the ESRF we envisaged that the provision of an intelligent automated mounting system would ultimately allow for the tracking of crystal samples throughout the entire experimental process. It is perhaps obvious that when combined with a suitable data management system we have paved the way for the automation of screening procedures for the purpose of identifying the best available sample for subsequent careful data acquisition.

From the initial soundings undertaken with the User community it was clear that if the goal of routine sample exchange was to be provided a preliminary first step was required. Consequently the development of a standard sample holder for use with robotic sample changers was undertaken. By adapting existing models (available in 2004) it was possible to provide a solution that allowed synchrotron users to use several different sample changers without having to resort to specialised holders for each robot. Therefore, although the definition of the SPINE standard sample holder [6] was a decisive step for moving from manual to automatic sample handling, the design and size are nevertheless based on a concept initially developed for manual use. In the future, further optimization of sample throughput will probably require the design of a sample holder specifically optimized for handling by robotic systems. Despite the compromises made this system has nevertheless proved to be an extremely useful – enabling the loading in excess of 120,000 samples per year since 2006 at the ESRF.

5.4 Sample Evaluation and Screening

With automated systems in place for delivering X-rays and exchanging samples, the chief problem that remains is finding and exploiting to the full the diffraction potential of "good" crystal samples. We have attempted to produce software infrastructures to enable automated data evaluation this is most visible in the EDNA project [19]. In addition we attempt to provide advanced tools for the evaluation of samples and in particular understanding the diffraction potential of the crystal sample throughout the crystal volume.

5.4.1 EDNA

Typical data collection in macromolecular crystallography (MX) using the rotation method [1] involves preliminary steps: measuring a few diffraction patterns (reference images) of a crystal under investigation, and their evaluation. The procedure aims to determine the basic crystallographic parameters (unit-cell dimensions and putative Laue class, crystal orientation, mosaicity, diffraction spot shapes) and to evaluate the signal-to-noise ratio in the data. On the basis of these processing results, the data collection parameters (data collection resolution, rotation range, oscillation width and exposure time, jointly called a 'data collection strategy') are determined, along with the decision on suitability of a sample for a particular crystallographic problem. An additional consideration that must be taken into account in a modern data collection procedure at third-generation beamlines is an anticipated sustainable radiation dose for the crystal [24, 26]. At weaker sources, the total time available for the experiment may form a significant restriction. Such procedures are essential for obtaining acceptable quality diffraction data and may not be replaced by, for example, a set of 'standard' data collection conditions that are satisfactory for most samples [11, 15]. The main goal of the EDNA MX application is to provide the software which implements the procedure described above in a fully automated way [19, 22]. Thus as deployed on a beamline the EDNA MX package will initiate the collection of indexing images, determine unit cell parameters using both mosflm [21], and Labelit [27]. Following success at this level the integrated images are passed to BEST [3, 25] where a data collection strategy (taking into account radiation damage) is provided.

5.4.2 *Diffraction Cartography*

As has been noted it is very common for crystals of biological macromolecules, even when obtained under apparently identical conditions, to show variation in the quality of their diffraction.

As sample evaluation, automation and microbeams have become more widely available, more advanced screening methods have evolved. These include locating the best region of a crystal on which to perform data collection [18], locating and evaluating the diffraction properties of crystals in crystallization drops [20, 23] and locating and evaluating the diffraction properties of very small crystals contained in large sample mounts [5]. These types of sample evaluation are already performed empirically by many crystallographers. However, we have sought to formalize and automate these procedures and thus allow them to become routine. To achieve this goal requires the combination of current 'line' and 'mesh' scans [31] with diffraction-quality characterization and experiment planning using EDNA and BEST.

Two typical cases are common to most crystallographers [4] "Lilliput" where a small beam is used to scan a support for small crystals – in this case almost any diffraction information is valuable and may be happily compared to the cases exemplified for GPCR proteins [5]. The other example being "Brobdinnag" where crystal dimensions are larger than the beam probing it, interestingly in this case we were able to demonstrate intra-crystal heterogenity that could be advantageously used in planning experiments [4]. Since the results of both these investigations is a type of map of the diffraction potential of the sample it will be possible to transfer this information with the sample as data collection advances or screening concludes.

5.4.3 *Screening for Samples*

The selection of samples of suitable quality is a task that comes naturally to an experienced crystallographer. Indeed the term (and the task) is used for the generation of crystals in the first place as well as the initial testing for the diffraction properties of crystallisation hits. In our terms the screening step means testing if the diffraction observed from a crystal (or set of crystals) corresponds to the requirements for the experiment. This assumes of course that it is possible to articulate the needs of the experiment in a way that is understandable. We have chosen to apply the constraint that we will strive to deliver the best possible datasets given the samples available rather than, for example, attempting to generate the greatest number of datasets per experimental session. Our approach is shown schematically in Fig. 5.2 where samples of the same type are sequentially tested

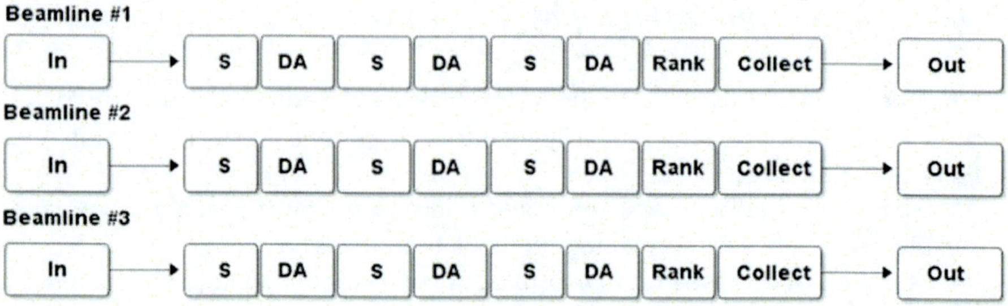

Fig. 5.2 With a conventional CCD detector the screening process (S) tends to involve more data analysis of the individual samples (DA step). When several samples have been observed the relative merits of each are evaluated (Rank) and a selection for data collection made

Fig. 5.3 With a high speed detector the screening strategy can shift. Here collection of data is so fast that it is often tempting to initiate data collection using the simpelest strategy possible as soon as the crystal is centered. Of course this act only postpones the inevitable need to check that the data are consistent with the desired out come

on one beamline. When all are checked a ranking is applied associated to a number of possible criteria : diffraction limits; minimum data collection time; lowest mosaicity; or combinations of these values. Data collection can then proceed on the most appropriate crystal. Of course other schemes are possible (and implemented) of which the most popular is the option to the collect the first sample that meets a certain minimum diffraction resolution.

The availability of fast read out detectors posses a problem for the screening scheme we favour. The initial data analysis is now probably longer than the likely data collection time for a simple data collection strategy (most people request 180° rotation). Thus provided diffraction is seen a dataset will be collected (Fig. 5.3). However appealing this approach is it simply delays the sorting process until later in the analysis process since not all datasets will be useful. Given the high cost of delivering the X-rays and maintaining the beamlines we find this approach unconvincing and prefer to enable an approach that will provide the best possible data from a wide variety of samples in a flexible manner.

The current focus of our efforts at the ESRF is to provide the next level of automation to allow for more rapid initial sample screening (Fig. 5.4). The techniques described will be backed by a data management system ISPyB [13] capable of recording the data and meta-data of the experiment and diffraction analysis. The key concept is that with an analysis of the crystallographic potential of each sample it will be possible to decide which of the beamlines has the characteristics to allow for a optimal experiment to be performed. This idea does require that the sample handling allows for redistribution of crystals between beamlines.

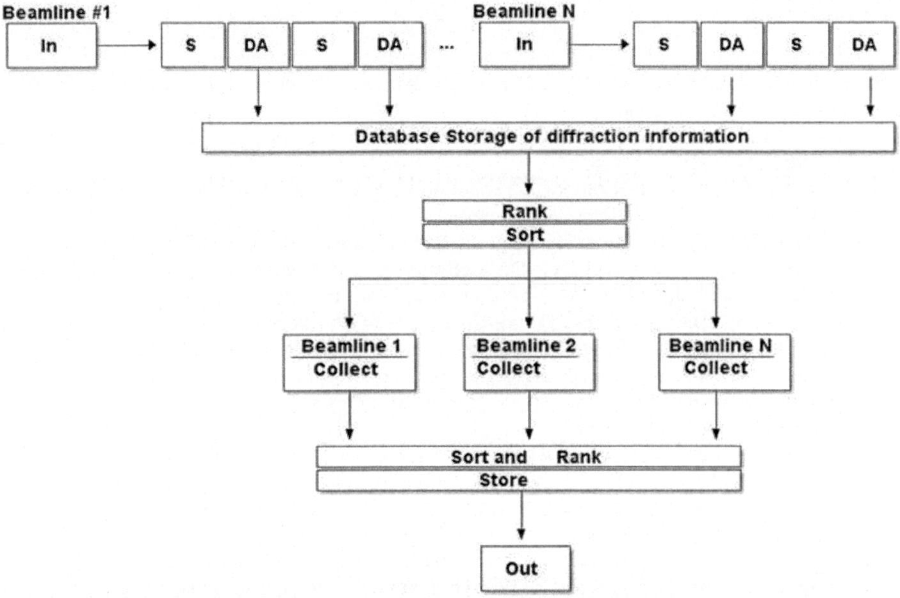

Fig. 5.4 As envisaged samples will be able to be screened before being reorganised for data collection on beamlines where the X-ray characteristics most closely match the needs of the experiment

5.5 Conclusion

Automation and advances in technology are the key elements in addressing the steadily increasing complexity of MX experiments. Much of this complexity is due to inter- and intra-crystal heterogeneity in diffraction quality often observed for crystals of multi-component macromolecular assemblies or membrane proteins. Such heterogeneity makes high-throughput sample evaluation an important and necessary tool for increasing the chances of a successful structure determination.

The screening approach described here raises the challenge of handling (and reporting) large quantities of diffraction meta-data and will require considerable work to understand how to use these data. Employing the tools available to find these "crystallographic needles" in the haystack of our data will open the possibility to elucidate ever more challenging crystallographic projects.

Acknowledgments The work described in this paper represents the efforts of many staff members at the ESRF and the EMBL synchrotron groups over more than a decade. Without their countless hours of toil none of the results from the ESRF would have been possible.

References

1. Arndt UW, Wonnacott AJ (1977) The rotation method in crystallography. North Holland, Amsterdam
2. Ben-Shem A, Frolow F, Nelson N (2003) Crystal structure of plant photosystem I. Nature 426(6967):630–635
3. Bourenkov GP, Popov AN (2006) A quantitative approach to data-collection strategies. Acta Crystallogr D Biol Crystallogr 62(Pt 1):58–64
4. Bowler MW, Guijarro M, Petitdemange S, Baker I, Svensson O, Burghammer M, Mueller-Dieckmann C, Gordon EJ, Flot D, McSweeney SM, Leonard GA (2010) Diffraction cartography: applying microbeams to macromolecular crystallography sample evaluation and data collection. Acta Crystallogr D Biol Crystallogr 66(Pt 8):855–864
5. Cherezov V, Hanson M, Griffith M, Hilgart M, Sanishvili R, Nagarajan V, Stepanov S, Fischetti R, Kuhn P, Stevens R (2009) Rastering strategy for screening and centring of microcrystal samples of human membrane proteins with a sub-10 mm size x-ray synchrotron beam. J R Soc Interface 6(Suppl 5):S587–S597
6. Cipriani F, Felisaz F, Launer L, Aksoy JS, Caserotto H, Cusack S, Dallery M, di Chiaro F, Guijarro M, Huet J, Larsen S, Lentini M, McCarthy J, McSweeney S, Ravelli R, Renier M, Taffut C, Thompson A, Leonard GA, Walsh MA (2006) Automation of sample mounting for macromolecular crystallography. Acta Crystallogr D Biol Crystallogr 62:1251–1259
7. Cockburn JJ, Abrescia NG, Grimes JM, Sutton GC, Diprose JM, Benevides JM, Thomas GJ, Bamford JK, Bamford DH, Stuart DI (2004) Membrane structure and interactions with protein and DNA in bacteriophage PRD1. Nature 432(7013):122–125
8. Cohen AE, Ellis PJ, Miller MD, Deacon AM, Phizackerley RP (2002) An automated system to mount cryo-cooled protein crystals on a synchrotron beamline, using compact sample cassettes and a small-scale robot. J Appl Crystallogr 35(6):720–726
9. Coureux PD, Wells AL, Ménétrey J, Yengo CM, Morris CA, Sweeney HL, Houdusse A (2003) A structural state of the myosin v motor without bound nucleotide. Nature 425(6956):419–423
10. Cramer P, Bushnell DA, Kornberg RD (2001) Structural basis of transcription: RNA polymerase II at 2.8 angstrom resolution. Science 292(5523):1863–1876
11. Dauter Z (1999) Data-collection strategies. Acta Crystallogr D Biol Crystallogr 55(Pt 10):1703–1717
12. Dawson RJ, Locher KP (2006) Structure of a bacterial multidrug ABC transporter. Nature 443(7108):180–185
13. Delagenière S, Brenchereau P, Launer L, Ashton AW, Leal R, Veyrier S, Gabadinho J, Gordon EJ, Jones SD, Levik KE, McSweeney SM, Monaco S, Nanao M, Spruce D, Svensson O, Walsh MA, Leonard GA (2011) ISPyB: an information management system for synchrotron macromolecular crystallography. Bioinformatics 27(22):3186–3192
14. Doyle DA, Morais Cabral J, Pfuetzner RA, Kuo A, Gulbis JM, Cohen SL, Chait BT, MacKinnon R (1998) The structure of the potassium channel: molecular basis of K+ conduction and selectivity. Science 280(5360):69–77
15. Evans PR (1999) Some notes on choices in data collection. Acta Crystallogr D Biol Crystallogr 55(Pt 10):1771–1772
16. Ferreira KN, Iverson TM, Maghlaoui K, Barber J, Iwata S (2004) Architecture of the photosynthetic oxygen-evolving center. Science 303(5665):1831–1838
17. Grimes JM, Burroughs JN, Gouet P, Diprose JM, Malby R, Ziéntara S, Mertens PP, Stuart DI (1998) The atomic structure of the bluetongue virus core. Nature 395(6701):470–478
18. Higuchi Y, Okamoto T, Yasuoka N (1996) The heterogeneity in a protein crystal revealed by synchrotron radiation. J Cryst Growth 168(1):99–105
19. Incardona MF, Bourenkov GP, Levik K, Pieritz RA, Popov AN, Svensson O (2009) *EDNA*: a framework for plugin-based applications applied to X-ray experiment online data analysis. J Synchrotron Radiat 16(6):872–879

20. Jacquamet L, Joly J, Bertoni A, Charrault P, Pirocchi M, Vernede X, Bouis F, Borel F, Périn JP, Denis T, Rechatin JL, Ferrer JL (2009) Upgrade of the CATS sample changer on FIPBM30A at the ESRF: towards a commercialized standard. J Synchrotron Radiat 16(1):14–21
21. Leslie AGW, Powell HR (2007) Processing diffraction data with mosflm. In: Read RJ, Sussman JL (eds) Evolving methods for macromolecular crystallography, vol 245, NATO Science series. Springer, Dordrecht, pp 41–51
22. Leslie AGW, Powell HR, Winter G, Svensson O, Spruce D, McSweeney S, Love D, Kinder S, Duke E, Nave C (2002) Automation of the collection and processing of X-ray diffraction data – a generic approach. Acta Crystallogr D Biol Crystallogr 58(11):1924–1928
23. Ohana J, Jacquamet L, Joly J, Bertoni A, Taunier P, Michel L, Charrault P, Pirocchi M, Carpentier P, Borel F, Kahn R, Ferrer JL (2004) CATS: a Cryogenic Automated Transfer System installed on the beamline FIP at ESRF. J Appl Crystallogr 37(1):72–77
24. Owen RL, Rudiño-Piñera E, Garman EF (2006) Experimental determination of the radiation dose limit for cryocooled protein crystals. Proc Natl Acad Sci USA 103(13):4912–4917
25. Popov AN, Bourenkov GP (2011) Features and development of BEST. Acta Crystallogr A 67(a1):C45
26. Ravelli RB, Garman EF (2006) Radiation damage in macromolecular cryocrystallography. Curr Opin Struct Biol 16(5):624–629
27. Sauter NK, Grosse-Kunstleve RW, Adams PD (2004) Robust indexing for automatic data collection. J Appl Crystallogr 37(Pt 3):399–409
28. Selmer M, Dunham CM, Murphy FV, Weixlbaumer A, Petry S, Kelley AC, Weir JR, Ramakrishnan V (2006) Structure of the 70s ribosome complexed with mRNA and tRNA. Science 313(5795):1935–1942
29. Snell G, Cork C, Nordmeyer R, Cornell E, Meigs G, Yegian D, Jaklevic J, Jin J, Stevens RC, Earnest T (2004) Automated sample mounting and alignment system for biological crystallography at a synchrotron source. Structure 12(4):537–545
30. Soltis SM, Cohen AE, Deacon A, Eriksson T, González A, McPhillips S, Chui H, Dunten P, Hollenbeck M, Mathews I, Miller M, Moorhead P, Phizackerley RP, Smith C, Song J, van dem Bedem H, Ellis P, Kuhn P, McPhillips T, Sauter N, Sharp K, Tsyba I, Wolf G (2008) New paradigm for macromolecular crystallography experiments at SSRL: automated crystal screening and remote data collection. Acta Crystallogr D Biol Crystallogr 64(Pt 12):1210–1221
31. Song J, Mathew D, Jacob S, Corbett L, Moorhead P, Soltis S (2007) Diffraction-based automated crystal centering. J Synchrotron Radiat 14(2):191–195
32. Sung BJ, Hwang KY, Jeon YH, Lee JI, Heo YS, Kim JH, Moon J, Yoon JM, Hyun YL, Kim E, Eum SJ, Park SY, Lee JO, Lee TG, Ro S, Cho JM (2003) Structure of the catalytic domain of human phosphodiesterase 5 with bound drug molecules. Nature 425(6953):98–102
33. Ueno G, Hirose R, Ida K, Kumasaka T, Yamamoto M (2004) Sample management system for a vast amount of frozen crystals at SPring-8. J Appl Crystallogr 37(6):867–873
34. Warne T, Serrano-Vega MJ, Baker JG, Moukhametzianov R, Edwards PC, Henderson R, Leslie AG, Tate CG, Schertler GF (2008) Structure of a beta1-adrenergic G-protein-coupled receptor. Nature 454(7203):486–491
35. Williams PA, Cosme J, Ward A, Angove HC, MatakVinković D, Jhoti H (2003) Crystal structure of human cytochrome p450 2c9 with bound warfarin. Nature 424(6947):464–468
36. Williams PA, Cosme J, Vinkovic DM, Ward A, Angove HC, Day PJ, Vonrhein C, Tickle IJ, Jhoti H (2004) Crystal structures of human cytochrome p450 3a4 bound to metyrapone and progesterone. Science 305(5684):683–686
37. Zouni A, Witt HT, Kern J, Fromme P, Krauss N, Saenger W, Orth P (2001) Crystal structure of photosystem II from synechococcus elongatus at 3.8 Å resolution. Nature 409(6821):739–743

Chapter 6
Data Processing: How Good Are My Data *Really*?

Kay Diederichs and P. Andrew Karplus

Abstract Since its inception more than 60 years ago, a "reliability index", later called "R-value", has been used to measure the agreement of model and averaged data, and a similar quantity, R_{merge}, has been defined to assess the quality of the averaged data.

However, a little known fact is that the two kinds of R-values have very different properties and asymptotic behaviors, and cannot be compared with each other. This is the reason that decisions concerning the high-resolution cutoff of data that are based on these R-values are questionable, and also helps explain why disagreements between journal authors and the manuscript reviewers have been so common.

Here, the authors will show that a different statistic can be used to overcome these deficiencies, and will establish a direct quantitative relation between data and model quality. This relation is important to judge the extent to which the data are useful, and also gives insight into the quality of the model that is derived from the data. The theoretical and practical consequences are at variance with several commonly employed crystallographic concepts and procedures.

Keywords Data quality • Model quality • Resolution • R-value • R_{merge} • R_{work} • R_{free} • Correlation coefficient

K. Diederichs (✉)
Department of Biology, University of Konstanz, Universitätsstr. 10, 78457 Constance, Germany
e-mail: Kay.Diederichs@uni-konstanz.de

P.A. Karplus
Department of Biochemistry and Biophysics, Oregon State University, Corvallis, OR 97331, USA
e-mail: karplusp@science.oregonstate.edu

R. Read et al. (eds.), *Advancing Methods for Biomolecular Crystallography*, NATO Science for Peace and Security Series A: Chemistry and Biology, DOI 10.1007/978-94-007-6232-9_6,

6.1 How to Measure Data Quality?

The quantity that is most often used to indicate the quality of data used to solve a crystal structure is R_{merge}. This quantity, named R_{sym} at the time, was invented by Arndt and coworkers [1] to characterize the data that were obtained from the first two-dimensional detector with electronic readout. Until the present, its overall and highest-shell value is an important part of both the "Table 1" of crystallographic papers, and of structures deposited in the PDB.

About 15 years ago, we pointed out [2] deficiencies of R_{merge} and suggested alternative formulas both for the characterization of data consistency (R_{meas} in place of R_{merge}), and data quality ($R_{mrgd\text{-}I}$) and suggested that only the latter of these is relevant to making a decision about the high resolution cutoff. Since then, we have been contacted many times by colleagues who reported to us that reviewers rejected their papers due to high values of R_{merge} in the highest-resolution shell. These reports confirm a view often held in the crystallographic community, namely that use of data that are characterized by high R_{merge} values has a detrimental influence on the quality of structural models, and that these data should therefore not be used for refinement. Based on the latter assertion, a high-resolution data cutoff is often employed such as to reduce R_{merge} in the highest-resolution shell to an "acceptable" value.

After enumerating data quality indicators in crystallography in Sect. 6.2, Sect. 6.3 is concerned with the question of selecting a suitable high-resolution cutoff for crystallographic data. It will be shown that the inclusion of weak high-resolution data improves the refined model, and that there are substantial problems with understanding the meaning of high-resolution R-values.

Section 6.4 introduces a new concept and indicator, which quantitatively links model and data quality in crystallography.

6.2 Crystallographic Statistics – Which Indicators Are Being Used?

To calculate aspects of data quality, the formulas given in Table 6.1 are in use; note that $R_{mrgd\text{-}I}$ [2] has been superseded by R_{pim}. For these "data R-values", the following relation holds:

$$R_{pim} < R_{merge} < R_{meas}$$

However, since neither R_{pim} nor R_{meas} are regularly used, they will not be discussed further.

In addition to R_{merge}, the mean signal-to-noise ratio ($<I/\sigma>$) of the merged data is often cited, both as an overall value and as a value for the highest-resolution shell.

Table 6.1 Data R-values that are in common use

Name	Formula	Reference
Merging R-value	$R_{merge} = \frac{\sum_{hkl} \sum_{i=1}^{n} \left\| I_i(hkl) - \bar{I}(hkl) \right\|}{\sum_{hkl} \sum_{i=1}^{n} I_i(hkl)}$	Arndt et al. [1]
Redundancy-independent merging R-value	$R_{meas} = \frac{\sum_{hkl} \sqrt{\frac{n}{n-1}} \sum_{i=1}^{n} \left\| I_i(hkl) - \bar{I}(hkl) \right\|}{\sum_{hkl} \sum_{i=1}^{n} I_i(hkl)}$	Diederichs and Karplus [2, 10]
Precision-indicating merging R-value	$R_{pim} = \frac{\sum_{hkl} \sqrt{1/n-1} \sum_{i=1}^{n} \left\| I_i(hkl) - \bar{I}(hkl) \right\|}{\sum_{hkl} \sum_{i=1}^{n} I_i(hkl)}$	Weiss et al. [11]

Model quality is most often also measured by R-values ("model R-values"); the formula is

$$R_{work/free} = \frac{\sum_{hkl} \left| F_{obs}(hkl) - F_{calc}(hkl) \right|}{\sum_{hkl} F_{obs}(hkl)}$$

where the summation extends over those reflections against which the refinement target is minimized (the "working set"), or (for R_{free}) the "free set" which is usually a small, randomly selected subset of all reflections and is only used for cross-validation of the refinement progress.

Unfortunately, X-ray crystallography is disconnected from mainstream statistics, when it comes to R-values: "The linear residual is actually an odd statistic; you will not find anything like an R-value in a "real" statistics textbook. The insufficiency of this basic linear residual becomes painfully obvious in the merging R-values for intensities ..." [5, p 331].

Concerning the relation of data and model R-values, this means that no proper theory exists: "Note that the meaning of the merging R-value for a shell with random noise data is entirely different from the maximum expected R-value on F for a random structure ($R_F = 0.59$)" [5, p 415].

6.3 What Is a Suitable High-Resolution Cutoff to Be Chosen for Refinement?

Obviously, higher resolution means better accuracy and more detailed maps. However, experience shows that a cutoff that is chosen at "too high" resolution yields high overall R_{work} and R_{free} values. The following questions must be answered:

1. Are any of the data R-values (overall, and by resolution) really a good predictor and indicator of model quality?

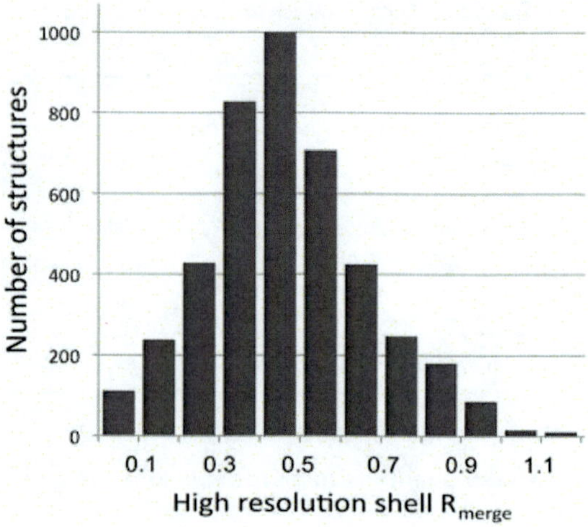

Fig. 6.1 Histogram of high resolution R_{merge} values for structures deposited in 2010 (Figure from Karplus and Diederichs [4])

2. To what extent does the data quality influence the refinement result?
3. What to refine, and when to stop refining?
4. How to discover and avoid overfitting?

It is enlightening to investigate what the high-resolution R_{merge} values are for structures deposited in the PDB in 2010 (Fig. 6.1). This distribution peaks around a cutoff of 50 %, consistent with published recommendations [13].

We believe that a choice of cutoff around 60 %, which we call "conservative cutoff" in the following, is not the best possible one. First, the decision for this cutoff is not based on values derived from first principles, or arising from statistical insight. In the absence of these, such a decision should be validated with refinement programs that are currently available. Second, a mechanism of self-fulfilling prophecy seems to be at work: the practical decisions of principal investigators are strongly influenced by rejections of their manuscripts by reviewers who favor the conservative cutoff. When these principal investigators are themselves reviewers, the request for a conservative cutoff that they were forced to choose sometimes seems to spread into their own reviews.

6.3.1 The Problem with R-Values at High Resolution

In the absence of a proper theory, most crystallographers expect a quantitative relationship to exist between data R-value and model R-value; in particular, they expect the data R-value to be a lower limit for the model R-value ("the model R-value cannot be lower than the data R-values since the data prevent it").

In case of the model R-value, it can be shown [12] that its value is limited, for a random or wrong model, to 0.584 and 0.828 for acentric and centric reflections, respectively. This explains why a maximum value of R_{merge} around 0.6 is often suggested and employed. Interestingly, even if this argumentation were correct (which it is not as shown below), since data R-values are calculated from intensities, a more suitable cutoff for data R-values would be 1.00 and 1.273 for acentric and centric reflections, respectively.

6.3.2 *The Asymptotic Behavior of Model and Data R-Values Is Different at High Resolution*

However, when the signal vanishes at high resolution, data R-values (R_{pim}, R_{merge}, R_{meas}) diverge to *infinity* since the numerator of their respective formula is then constant (being determined by variation of counts in the background), and the denominator approaches zero. The factor $1/\sqrt{n}$ present in the R_{pim} formula does not fundamentally change this asymptotic behaviour. This is in stark contrast to the behaviour of model R-values, which asymptotically approach a constant as mentioned above.

6.3.3 *Is There Information Beyond the Conservative High-Resolution Cutoff?*

To investigate the influence of the high-resolution cutoff on the quality of the model, we developed a novel procedure for the quality comparison of two models and employed it to data from cysteine dioxygenase (CDO; [7, 8]) collected previously. The dataset that we used was 15-fold weaker than the one that the model was refined against originally (Fig. 6.2).

For these data, a conservative resolution cutoff would be around 1.8 Å ($R_{merge} = 70$ %); a less conservative cutoff (using a $<I/\sigma> = 2$ cutoff) would be around 1.6 Å.

To judge the quality of different models, we developed the following "paired refinements technique":

- Refine at two different resolution cutoffs, e.g. at 2.0 Å and at 1.9 Å, using the same starting model and refinement parameters
- Since it is meaningless to compare the resulting R-values at different resolutions, calculate the overall R-values (R_{work} and R_{free}) of the higher-resolution (1.9 Å) model at the lower-resolution (2.0 Å) cutoff, without any change of the model.
- Calculate $\Delta R = R_{1.9}(2.0) - R_{2.0}(2.0)$ i.e. compare the R-values of the two models at the same (lower) resolution. This is meaningful because the same set of reflections is used for the comparison. If the R_{free}-value difference is negative, the data in the resolution shell between 2.0 Å and at 1.9 Å are beneficial for model quality.

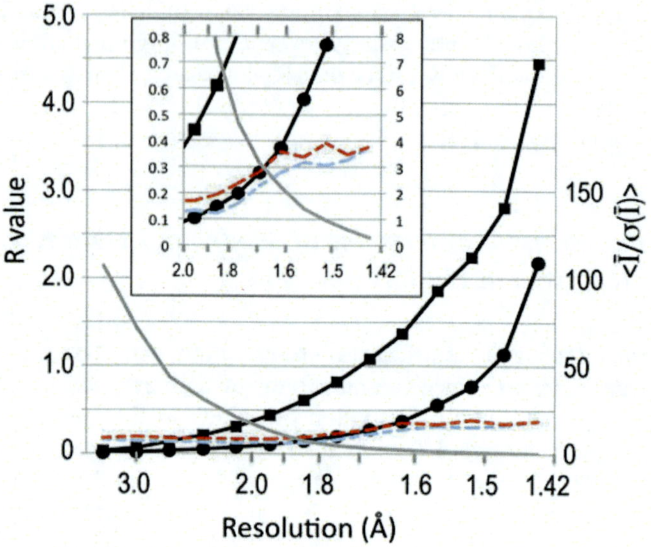

Fig. 6.2 Statistics for cysteine dioxygenase (*CDO*), PDB 3ELN, re-refined against 15-fold weaker data. Main canvas: *upper black curve (square markers)* is R_{merge}; *lower black curve (filled circle markers)* is R_{pim}; *upper dashed curve (red/dark grey)*: R_{free}; *lower dashed curve (blue/light grey)*: R_{work}; *continuous grey curve* is <I/σ>. *Inset*: close-up beyond 2 Å resolution (Figure from Karplus and Diederichs [4]) (Color figure online)

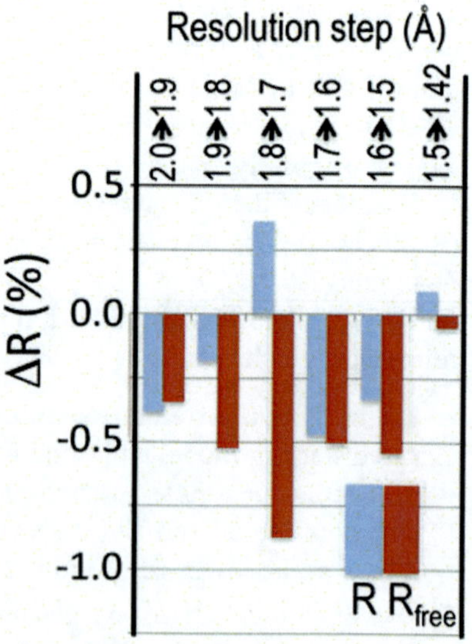

Fig. 6.3 Comparisons of R_{work} and R_{free} reveal that high-resolution data improve the model. For each resolution shell, the *left bar (blue/light grey)* indicates the difference in R_{work} and the *right (red/dark grey)* that in R_{free} (Figure from Karplus and Diederichs [4]) (Color figure online)

The results shown in Fig. 6.3 demonstrate clearly that R_{free} drops when including higher-resolution data, and that also $R_{work} - R_{free}$ drops. The latter is a measure of overfitting; a smaller $R_{work} - R_{free}$ difference indicates less overfitting. In this case, even the last shell of data out to 1.42 Å improves the results, although only marginally. This result demonstrates that the inclusion of weak high-resolution data, well beyond the conservative cutoff, improves the refinement result.

6.4 Relation of Data to Model Quality

6.4.1 *Beyond R-Values – Measuring Data Quality with a Correlation Coefficient*

In this chapter, we investigate a correlation coefficient (CC) as an alternative to using an R-value. Generally, a CC has clear meaning and well-known statistical properties: the significance of its value can be assessed by Student's t-test: e.g. $CC > 0.3$ is significant at $p = 0.01$ (i.e. may be obtained by chance in 1 % of the cases) for $n > 100$ pairs, and $CC > 0.08$ is significant at $p = 0.01$ for $n > 1{,}000$.

To obtain a meaningful quantity, we are guided by our work on random half-datasets [2] and by experience with a CC that has been shown to be useful for assessing the quality of the anomalous signal in a dataset. Two "random half-datasets" are obtained when all observations of each unique reflection from a full dataset are evenly but randomly assigned to two subsets. Each half-dataset subset is then merged individually and the desired signal (e.g. the anomalous difference) is extracted. For substructure solution, it was found that useful data can be identified by employing a CC_{anom} cutoff at around 0.3 [6].

We define $CC_{1/2}$ as the CC between intensities of crystallographic random half-datasets. This quantity has been available from the SCALA [3] logfile under the name CC_{Imean}. $CC_{1/2}$ has the useful property [4] that we can analytically estimate CC*, the CC of the full dataset (i.e. after merging the random half-datasets) against the true (usually unmeasurable) intensities, using

$$CC* = \sqrt{\frac{2CC_{1/2}}{1 + CC_{1/2}}}$$

In Fig. 6.4, we show $CC_{1/2}$, CC* and <I/σ> for the CDO data. It is striking how steep the drop of the CC-based indicators is beyond a resolution of 1.6 Å. In comparison, the slope of <I/σ> is much shallower since its second derivative is positive. If a cutoff (like 0.5) at a specific value of CC* were chosen, it would be fairly accurately determined by the data. However, we prefer to use the paired refinements technique to prove that the high-resolution data improve the model; the technique can be applied to CC_{work}, CC_{free} and in this case (and other cases we investigated) indicates that $CC_{1/2}$ values somewhat lower than 0.2 still indicate the presence of data that help to improve the model when using today's refinement programs.

We can define CC_{work}, CC_{free} as CCs calculated on $F_{calc}{}^2$ of the working and free set, against the experimental data. Since they are based on the same

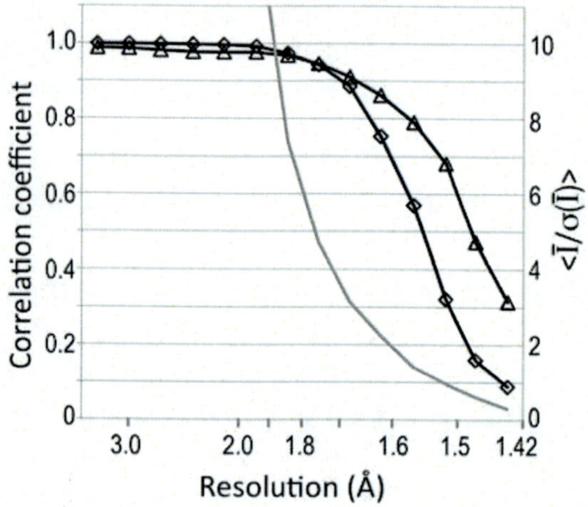

Fig. 6.4 $CC_{1/2}$ (*squares*), CC* (*triangles*) and <I/σ> (*grey*) for the CDO data (Figure from Karplus and Diederichs [4])

mathematical formulas and quantities of the same type, CC_{work} and CC_{free} can be directly compared with CC*. This quantitative relation between data and model CCs means that:

1. At the end of refinement, the CC_{work}, CC_{free} of a good model should approach CC* from lower values.
2. An inadequate model (where the term inadequate includes not only missing or wrong parts, but also e.g. cases of wrong spacegroup assignment and undetected twinning) results in CC_{work} and CC_{free} remaining less than CC*.
3. Systematic errors in data processing may (but not necessarily do) produce a higher CC* than is warranted by the data. In that case, too, CC_{work} and CC_{free} remain less than CC*.
4. If CC_{work} is higher than CC*, the model is closer to the data than the truth is to the data: this is an operational definition of overfitting; the model fits the noise.

In the case of the CDO data, there is good agreement between CC* and CC_{work}/CC_{free} at high resolution (Fig. 6.5), and it is obvious that the data quality indeed is limiting for CC_{work}/CC_{free}. At low resolution, we observe a similar gap between CC* and CC_{work}/CC_{free} as is usually observed between data and model R-values at low resolution, indicating a failure of today's crystallographic models to fully account for the experimental data.

In this particular case, the strong data originally used for the refinement of CDO can be considered a good approximation to the (unobservable) true data. If we compare the F_{calc}^2 from the model with these strong data (Fig. 6.5), we find, at first surprisingly, that their $CC_{work,free}$ is significantly higher at high resolution than the $CC_{work,free}$ against the noisy (weak) data. The model is therefore more accurate than the fit to the weak data at high resolution suggests. The reason is that the model is parsimonious, and over-determined by the data – its F_{calc}^2 are much less influenced by noise than the data.

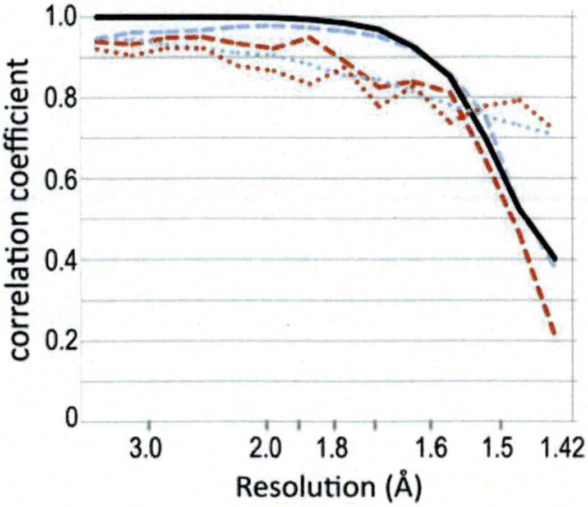

Fig. 6.5 Comparison of data and model Ccs. Shown is CC* (*black*) as well as CCwork (*dashed blue/light grey*) and CCfree (*dashed red/dark grey*). Also included (*dotted lines*) is the comparison of CC_{work}/CC_{free} (*blue/light grey, red/dark grey*) against the original (strong) CDO data (Figure from Karplus and Diederichs [4]) (Color figure online)

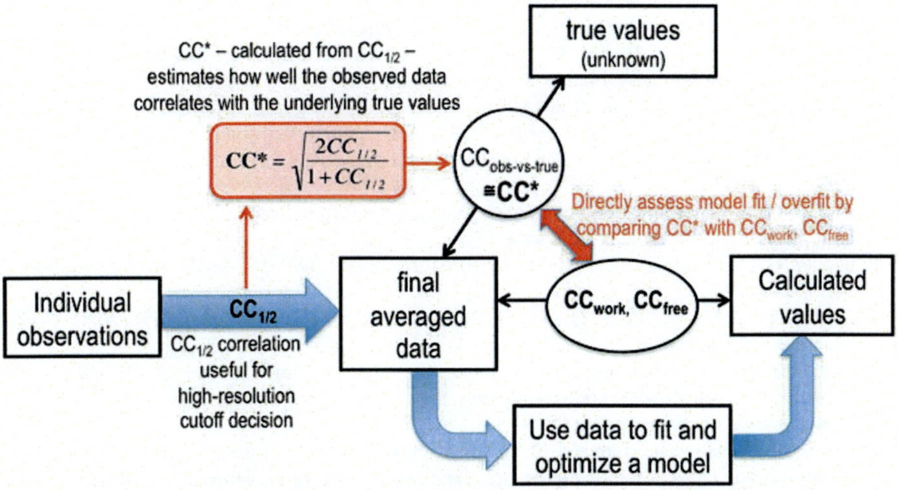

Fig. 6.6 Graphical scheme of relationships between correlation coefficients

6.5 Summary

The following summarizes our findings concerning the high-resolution cutoff of crystallographic data and the relation of their quality to that of the model. Important relationships are graphically presented in Fig. 6.6.

- Data should be used out to a higher resolution than is suggested by current conventions. It is misleading that the overall R_{work}/R_{free} from refinement may be higher, and the paired refinement technique may be used to prove that the model is actually better.

- $CC_{1/2}$ directly assesses information content of data in a statistically meaningful way
- CC^* is an upper limit for CC_{work} and CC_{free}
- Crystallographic models at high resolution may have a better correlation of their F_{calc}^2 against the (usually unknown) true data than the experimental data have, and their phases are meaningful. This is the origin of what has been called the free lunch algorithm [9].

It should also be noted that current methods for estimating the model error from high-resolution R-values yield too high estimates: the data error needs to be taken into account.

References

1. Arndt UW, Crowther RA, Mallett JFW (1968) A computer-linked cathode-ray tube microdensitometer for X-ray crystallography. J Phys E Sci Instrum 1:510
2. Diederichs K, Karplus PA (1997) Improved R-factors for diffraction data analysis in macromolecular crystallography. Nature Struct Biol 4:269–275
3. Evans P (2011) An introduction to data reduction: space-group determination, scaling and intensity statistics. Acta Crystallogr D 67:282–292
4. Karplus PA, Diederichs K (2012) Linking crystallographic model and data quality. Science 336:1030–1033
5. Rupp B (2009) Biomolecular crystallography: principles, practice, and application to structural biology. Garland Science, New York
6. Schneider TR, Sheldrick GM (2002) Substructure solution with SHELXD. Acta Crystallogr D 58:1772–1779
7. Simmons CR, Liu Q, Huang Q, Hao Q, Begley TP, Karplus PA, Stipanuk MH (2006) Crystal structure of Mammalian cysteine dioxygenase. J Biol Chem 281:18723
8. Simmons CR, Krishnamoorthy K, Granett SL, Schuller DJ, Dominy JE Jr, Begley TP, Stipanuk MH, Karplus PA (2008) A putative Fe2+-bound persulfenate intermediate in cysteine dioxygenase. Biochemistry 47:11390
9. Usón I, Stevenson CE, Lawson DM, Sheldrick GM (2007) Structure determination of the O-methyltransferase NovP using the 'free lunch algorithm' as implemented in SHELXE. Acta Crystallogr D 63:1069–1074
10. Weiss MS (2001) Global indicators of X-ray data quality. J Appl Crystallogr 34:130–135
11. Weiss MS, Metzner HJ, Hilgenfeld R (1998) Two non-proline cis peptide bonds may be important for factor XIII function. FEBS Lett 423:291–296
12. Wilson AJC (1950) Largest likely values for the reliability index. Acta Crystallogr 3:397–398
13. Wlodawer A, Minor W, Dauter Z, Jaskolski M (2008) Protein crystallography for non-crystallographers, or how to get the best (but not more) from published macromolecular structures. FEBS J 275:1–21

Chapter 7
Radiation Damage in Macromolecular Crystallography: What Is It and Why Do We Care?

Elspeth F. Garman

Abstract Radiation damage inflicted during diffraction data collection in macromolecular crystallography has re-emerged in the last decade as a major experimental and computational challenge, as even for crystals held at 100 K it can result in severe data quality degradation and the appearance in solved structures of artifacts which affect biological interpretations. Here, the observable symptoms and basic physical processes involved in radiation damage will be described and the concept of absorbed dose as the basic metric against which to monitor the experimentally observed changes outlined. Investigations into radiation damage in macromolecular crystallography are ongoing and the number of studies is rapidly increasing as the topic has now become of mainstream interest.

Keywords Radiation damage • Dose • Diffraction • Cryocrystallography

7.1 Introduction

The advent of highly intense wiggler and undulator beamlines fed from synchrotron sources has reintroduced the age-old problem of X-ray radiation damage in macromolecular crystallography (MX) even for crystals held at cryogenic temperatures (100 K). Unfortunately, such damage to macromolecular crystalline samples during the experiment is a problem that is inherent in using ionizing radiation to obtain diffraction patterns and has presented a challenge to MX since the beginning of the field. For room-temperature (RT) data collections, it often necessitates the use of many crystals to assemble a complete data set, because the crystalline order of the sample is damaged and decreases during the experiment and thus the diffracted

E.F. Garman (✉)
Department of Biochemistry, University of Oxford, Oxford OX1 3QU, UK
e-mail: elspeth.garman@bioch.ox.ac.uk

R. Read et al. (eds.), *Advancing Methods for Biomolecular Crystallography*, NATO Science for Peace and Security Series A: Chemistry and Biology, DOI 10.1007/978-94-007-6232-9_7,

intensity fades. The root cause of this damage is the energy lost by the beam in the crystal owing to either the total absorption or the inelastic scattering of a proportion of the X-rays as they pass through the crystal. The measure of this energy loss is the 'dose' measured per mass of the sample, given in SI units of grays (Gy; 1 Gy = 1 J/kg). Dose may also be quoted in terms of the non-SI unit rad (radiation absorbed dose; 1 rad = 10 mGy). In MX, dose measurements are generally of the order of a million grays (1 MGy or 100 Mrad) and the programme RADDOSE can conveniently be used to compute the absorbed dose for a protein crystal [15, 18].

The earliest investigation of radiation damage at RT in MX was carried out just over 50 years ago by Blake and Phillips [1] on a sealed-tube (copper) X-ray source. By making seven sets of successive measurements, they monitored the decay in the diffraction intensity of a particular set of reflections from crystals of sperm-whale myoglobin over a period of 300 h. They concluded that the damage was proportional to the irradiation time, which they assumed was linearly proportional to the absorbed dose. They deduced that a single 8 keV X-ray photon disrupts around 70 protein molecules and disorders a further 90 protein molecules for doses up to about 20 Mrad (0.2 MGy) absorbed after 100 h of X-ray exposure. Blake and Phillips [1] also suggested that the protein molecules suffered specific structural damage. This conclusion was reached without knowledge of either the sequence or the three-dimensional structure of the protein, and the postulate was only confirmed many years later when radiation damage to disulfide bridges was noted in electron-density difference maps calculated from data collected from des-pentapeptide insulin crystals [10].

Up until the 1990s, MX data were almost exclusively collected at RT, where at the beginning of the experiment the recommended practice was to monitor the intensity of a strong reflection at the beginning if the experiment I_0 and then periodically as the experiment proceeded and to discard the crystals once the intensity had dropped to $0.85I_0$, or at the very worst $0.70I_0$ if the particular crystals were in very short supply [2].

Now most data are collected at 100 K: starting in the 1990s cryocooling techniques for MX blossomed and were made technically more accessible for routine use in MX because of two pivotal developments: the loop-mounting method [22], in which the protein crystal is held by surface tension in a film of liquid 'cryo-buffer' across a small-diameter (1 mm down to 0.1 mm) nylon, fibre or plastic loop, and the availability of reliable open-flow unpressurized cryostats [4] with flexible stainless-steel hosing to supply a stream of cooled gaseous nitrogen at a stable temperature of around 100 K with which to surround the sample during data collection. Initially, problems with the technique included ice formation within and outside the crystal and an increase in mosaic spread, particularly when cryocooling protocols were not optimized. Methods for improving the data quality obtainable were soon developed [7, 9, 21] and there was widespread adoption of the technique. In fact it has been estimated that over 90 % of all protein structures are now determined at cryo-temperatures.

The advantages of cryocooling for MX are a very significant reduction in the rate of radiation damage; the use of a mounting technique (the loop) that is usually more gentle than the capillary method historically used for RT collection; the fact that higher resolution data can more easily be obtained because the crystal order is preserved for longer; a lower background in the diffraction experiment as it is not necessary to enclose the crystal in a glass, quartz or plastic tube to prevent dehydration; that fewer crystals (and thus a lower quantity of protein) are required for a project; that crystals can be shipped ahead of time to the synchrotron (more or less) safely; and that crystals can be flash-cooled when in peak condition for future use before they start to degrade in the crystallization drop.

These positive aspects of cryocooling commonly outweigh the disadvantages. The latter include the requirement for expensive cryostat cooling equipment, a frequent increase in crystal mosaic spread (but not necessarily if the cryoprotection concentration and crystal handling are carefully optimized), the need to invest time for optimization of cryo-buffers and cooling protocols, and the fact that there are as yet no protocols that guarantee success, although progress is being made in this direction.

The improvement in dose tolerance for a crystal held at 100 K compared with a crystal irradiated at RT has been estimated to be approximately a factor of 70 on average [16]. Thus, cryocooling is clearly a highly effective mitigation strategy. However, radiation damage is now routinely observed at synchrotrons in cryocooled crystals and the experimenter would be wise to be aware of the artifacts that can be produced. Here, the symptoms of radiation damage at cryotemperatures and the basic physical processes involved are described, the reasons why the crystallographer should care about this issue are addressed. Our current knowledge, as reflected in the published literature, is summarized in more detail in Garman [8], Ravelli and Garman [19], and an article entitled 'A beginner's guide to radiation damage' Holton [11].

7.2 What Are the Symptoms of Radiation Damage at Cryotemperatures?

Systematic studies of this phenomenon have identified two separate indicators of damage as a function of dose: global and specific damage. The former results in a loss of the measured reflection intensities (particularly at high resolution), expansion of the unit-cell volume, increasing values of the measure of the internal consistency of the data which quantifies the difference between reflection intensities that should ideally be the same (R_{meas}), an increase in both the scaling B factors for the data and the atomic B values of the refined structure, rotation of the molecule within the unit cell and often (but not always) an increase in mosaicity. Visible differences in the samples as the experiment proceeds, including colour changes, are also observed. On warming of the sample following irradiation, bubbles of gas, now proposed to be hydrogen [13] and perhaps some CO_2, are emitted and discolouration of the sample is common (see Fig. 7.1).

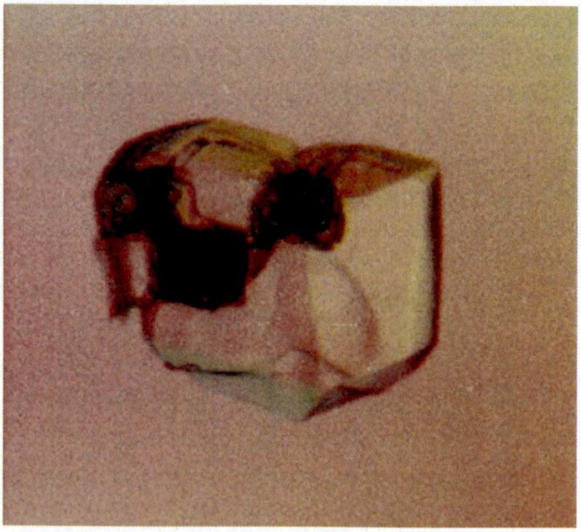

Fig. 7.1 Photograph of a 400 μm neuraminidase crystal (subtype N9 from avian influenza isolated from a Noddy Tern), space group I432, that has been irradiated on ID14-4 at the ESRF at 100 K and then allowed to warm up to RT. The three *black marks* are from the $100 \times 100\ \mu m^2$ beam; the discolouration is an indication of radiation damage (Reproduced courtesy of the IUCr from Garman [8]. http://dx.doi.org/10.1107/S0907444910008656)

Various metrics have been suggested and used for monitoring global damage, among which are the following.

(i) I_D/I_1, where I_D is the summed mean intensity (I_{mean}) of a complete data set (or equivalent sections of data) after a dose D, and I_1 is the mean intensity of the first data set. The intensity decay is weakly exponential (almost linear: see Fig. 7.2) at cryotemperatures but clearly exponential at RT. The dose which results in a drop of total intensity to half of the original value, $D_{1/2}$, has been measured to be 43 MGy for apo and holo ferritin crystals. However, the associated electron density maps exhibited significant damage of the amino acids at $0.5 \times I_1$, and an upper limit of 30 MGy has been suggested, corresponding to $0.7 \times I_1$ [17].

Note that using $I/\sigma(I)$ (where $\sigma(I)$ is the standard deviation of the signal, i.e. the 'noise') normalized to the intensity $I_1/\sigma(I_1)$ of the first data set is not a robust metric since the noise $\sigma(I)$ increases with dose and thus $I_n/\sigma(I_n)$ reduces by an amount that more than represents the true loss of diffracting power.

(ii) R_d, the pairwise R factor between identical and symmetry-related reflections occurring on different diffraction images, plotted against the difference in dose, ΔD, between the images on which the reflections were collected. The plot of R_d against ΔD is a straight line parallel to the *x* axis if there is no damage, but rises linearly in the presence of damage (see Fig. 7.3). This plot can be used to correct the intensity values of the reflections back to their 'zero-dose' values to improve the data quality [6].

(iii) The isotropic B factor (B_{rel}) has been found to be a robust measure of radiation damage at 100 K and to be linearly dependent on it [12]. An example of B_{rel} plotted against dose is given in Fig. 7.4. The relative B factors can be interpreted as proportional to the change in the mean squared atomic displacements. A coefficient of sensitivity to absorbed dose, $SAD = \Delta B_{rel}/(\Delta D 8\pi^2)$, was also

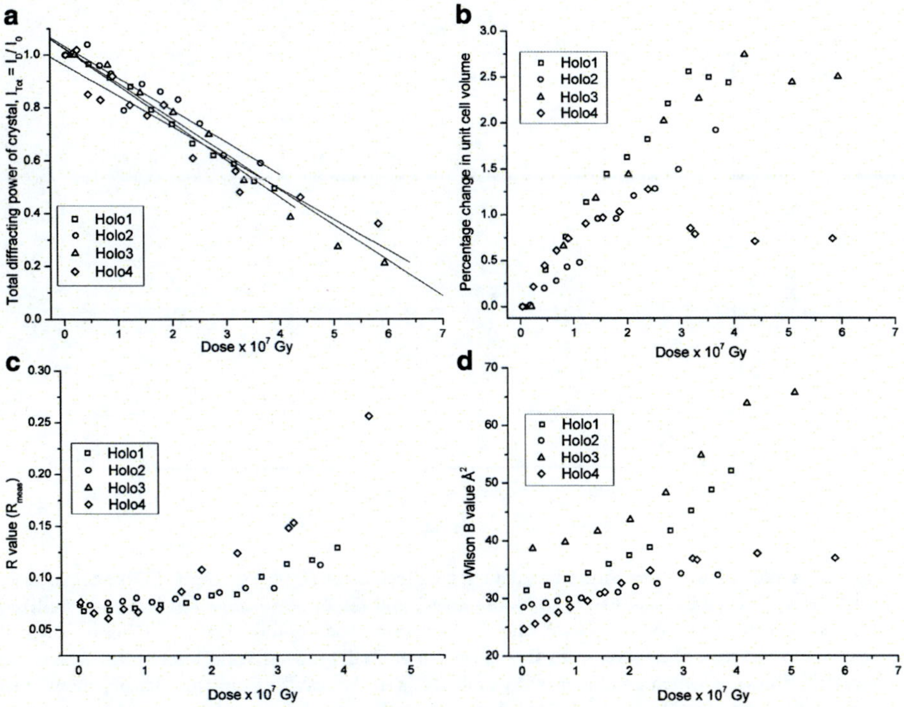

Fig. 7.2 Global radiation-damage indicators as a function of dose for four holoferritin crystals held at 100 K for data collection as detailed in Owen et al. [17]. (**a**) Mean I_n/mean I_1, (**b**) unit-cell volume, (**c**) R value and (**d**) Wilson B value. Each symbol represents the values of the respective parameter for one complete dataset (processed between 58 and 2.2 Å). A linear intensity decay against the absorbed dose is observed, with all four crystals decaying at the same rate (Reproduced courtesy of the IUCr from Garman [8]. http://dx.doi.org/10.1107/S0907444910008656)

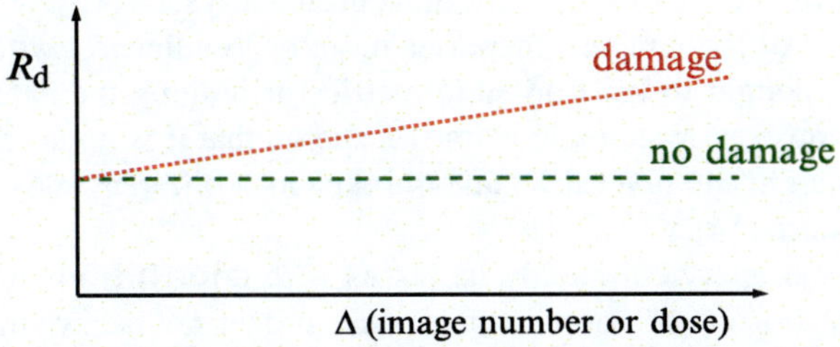

Fig. 7.3 An idealized plot of Rd, the pairwise R factor between identical and symmetry-related reflections occurring on different diffraction images, plotted against the difference in dose, ΔD, between the images on which the reflections were collected [6]. The plot is a straight line parallel to the x axis if there is no damage, but rises linearly in the presence of damage (Reproduced courtesy of the IUCr from Garman [8]. http://dx.doi.org/10.1107/S0907444910008656)

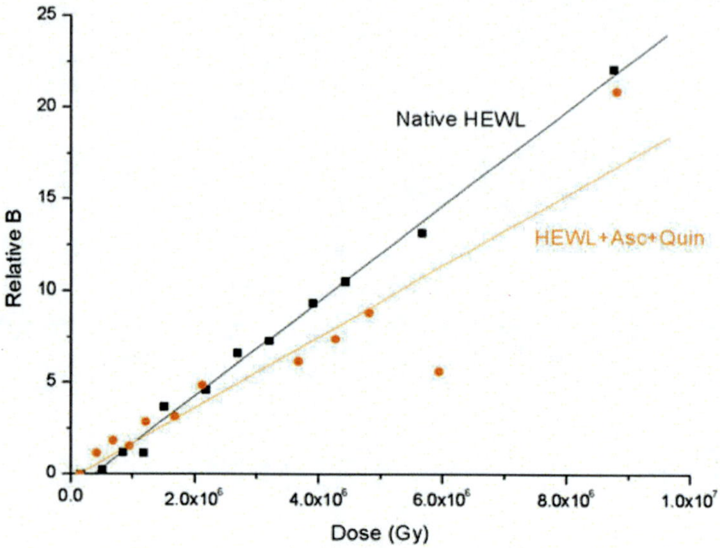

Fig. 7.4 A plot of B_{rel} (one value per data set collected on ID14-4 at the ESRF) against dose for two HEWL crystals, one native (*black squares*) and the other (*orange circles*) cocrystallized with the scavengers ascorbate (Asc) and 1,4-benzoquinone (Quin). The *solid lines* represent linear fits to the data: the increase in B_{rel} is only marginally slower with dose for the scavenger cocrystals, showing (when combined with an analysis of the resulting electron-density maps) that this particular combination is not effective in reducing the rate of damage (Reproduced courtesy of the IUCr from Garman [8] (Color figure on line). http://dx.doi.org/10.1107/S0907444910008656)

defined, where $\Delta B_{rel}/8\pi^2$ is the change in relative isotropic B factor and ΔD is the change in dose as above, i.e. SAD is the slope of the line in a graph such as that shown in Fig. 7.4. This metric relates the increase in mean-squared atomic displacements to the dose and it has been postulated that it is similar within quite a narrow range of values for most protein crystals [12].

(iv) The volume of the unit cell increases more or less linearly with dose and was originally thought to be a possible metric for judging the extent of radiation damage; however, systematic work has shown that it is not a reliable indicator since crystals of the same size and type expand at different rates with increasing absorbed dose [14].

(v) Although mosaicity commonly increases with dose, it is not a reliable metric for quantization of radiation damage, since it does not behave in a reproducible or predictable manner.

Of more direct relevance to the biological interpretation of structures than the global indicators detailed above is the fact that specific structural damage to particular covalent bonds is observed to occur in a reproducible order in many proteins [3, 20, 23]: first disulfide bridges elongate and then break, then glutamates and aspartates are decarboxylated, tyrosine residues lose their hydroxyl group and subsequently the carbon–sulfur bonds in methionines are cleaved. Covalent bonds to heavier atoms such as C—Br, C—I and S—Hg are also ruptured. Clearly, it is not feasible to monitor the specific structural damage during the experiment, since the refined structures are required. However, it is known that this damage often occurs well before there is any obvious degradation of the diffraction pattern.

The manifestations of radiation damage in the diffraction experiment can now be monitored over a range of time scales and doses. For instance, the formation of the disulfide-anion radical, $RSSR^{\bullet-}$ can be observed in real time using UV/UV–vis microspectrophotometry after a few tens of milliseconds of X-ray irradiation as a 400 nm absorption peak, and a peak corresponding to solvated electrons having a maximum absorbance at 550–600 nm also appears. This specific structural damage is often apparent in electron density maps calculated using the structure factors of a data set that took around 30 s to collect and the resulting structure represents a time and space average over the 30 s of irradiation and over all the molecules in the crystal. Metal centres are also reduced very swiftly by the X-ray beam and increasingly this can be monitored on-line during the X-ray experiment. The global intensity loss owing to radiation damage is clearly evident following the collection of several data sets in succession from the same crystal when the summed intensity for each data set is plotted normalized to the intensity of the first data set (Fig. 7.2a).

7.3 Why Should We Care?

Radiation damage in MX is an increasingly important and limiting problem for several reasons. Firstly, as the diffraction experiment proceeds, creeping non-isomorphism occurs on three simultaneous fronts: the unit-cell volume increases, there is often movement of the protein molecule within the unit cell, and structural changes are induced by the damage, so that the protein conformation is changing during the measurements. This non-isomorphism is thought to be a major cause of unsuccessful MAD (multiple-wavelength anomalous dispersion) structure determinations, since by the time the second or third wavelength is collected, the cell and atomic structure can have changed such that the reflection intensities are significantly altered. This effect can obscure the anomalous signal required for structure solution. It has been calculated that a 0.5 % change in all three dimensions of a unit cell would change the intensity of a 3 Å reflection by 15 % [5] so the MAD/SAD phasing signals (typically 4–8 % signal) would be completely overwhelmed by such a volume increase.

Secondly, the radiation-sensitivity of some crystals at 100 K means that it is not possible to collect a complete data set from a single crystal and data must be merged from several (or many) of them to measure all the unique reflection intensities. Although this was routinely the case when data were collected at RT, most crystallographers have become accustomed to being able to measure all unique reflections from just one cryocooled crystal. Use of multiple crystals to assemble a complete data set in general increases the errors arising from non-isomorphism, thereby potentially reducing the ease of structure solution as well as increasing the mounting/ dismounting time burden.

Finally, the radiation-damage-induced structural changes can affect the apparent biological properties of the macromolecule under study. Enzyme mechanisms can involve redox research papers susceptible residues, so special care is required when interpreting structures that may have been modified by X-ray damage during the data collection. For instance, irradiation can change the oxidation state of metal ions in structural/ active sites from that in their native state and cause the decarboxylation of glutamate and aspartate residues. X-ray-induced structural changes can also be misleading in studies of intermediates. In such circumstances, separating radiation damage from an enzymatic mechanism can be extremely difficult and can cast doubt on the validity of biological conclusions drawn from crystal structures [19].

In summary, radiation damage ultimately results in lower resolution structures, failed MAD structure solutions and sometimes the inaccurate interpretation of biological results if no control experiments are carried out to account for radiation-damage artifacts. It is thus an issue to be taken seriously by the structural biologist.

Acknowledgments The brief summary above represents the fruits of over 10 years of effort by a number of researchers worldwide. Radiation damage investigations are difficult and laborious and it is time consuming to obtain statistically significant results. I thank the past and present members of my research group in Oxford for their persistence and determination, and am very grateful for their inspiration, insight and energy in working towards an understanding of the various relevant phenomena.

References

1. Blake C, Phillips DC (1962) Effects of X-irradiation on single crystals of myoglobin in proceedings of the symposium on the biological effects of ionizing radiation at the molecular level. International Atomic Energy Agency, Vienna, pp 183–191
2. Blundell T, Johnson LN (1976) Protein crystallography. Academic, New York
3. Burmeister WP (2000) Structural changes in a cryo-cooled protein crystal owing to radiation damage. Acta Crystallogr D 56:328–341
4. Cosier J, Glazer AM (1986) A nitrogen-gas-stream cryostat for general X-ray diffraction studies. J Appl Crystallogr 19:105–107
5. Crick FHC, Magdoff BS (1956) The theory of the method of isomorphous replacement for protein crystals. I. Acta Crystallogr 9:901–908
6. Diederichs K (2006) Some aspects of quantitative analysis and correction of radiation damage. Acta Crystallogr D 62:96–101
7. Garman E (1999) Cool data: quantity AND quality. Acta Crystallogr D 55:1641–1653
8. Garman EF (2010) Radiation damage in macromolecular crystallography: what is it and why should we care? Acta Crystallogr D 66:339–351
9. Garman EF, Schneider TR (1997) Macromolecular cryocrystallography. J Appl Crystallogr 30:211–237
10. Helliwell JR (1988) Protein crystal perfection and the nature of radiation damage. J Crystallogr Growth 90:259–272
11. Holton JM (2009) A beginner's guide to radiation damage. J Synchrotron Rad 16:133–142

12. Kmetko J, Husseini NS, Naides M, Kalinin Y, Thorne RE (2006) Quantifying X-ray radiation damage in protein crystals at cryogenic temperatures. Acta Crystallogr D 62:1030–1038
13. Meents A, Gutmann S, Wagner A, Schulze-Briese C (2010) Origin and temperature dependence of radiation damage in biological samples at cryogenic temperatures. Proc Natl Acad Sci U S A 107:1094–1099
14. Murray JM, Garman EF (2002) Investigation of possible free-radical scavengers and metrics for radiation damage in protein cryocrystallography. J Synchrotron Rad 9(6):347–354
15. Murray JW, Garman EF, Ravelli RBG (2004) X-ray absorption by macromolecular crystals: the effects of wavelength and crystal composition on absorbed dose. J Appl Crystallogr 37: 513–522
16. Nave C, Garman EF (2005) Towards an understanding of radiation damage in cryocooled macromolecular crystals. J Synchrotron Rad 12:257–260
17. Owen RL, Rudino-Pinera E, Garman EF (2006) Experimental determination of the radiation dose limit for cryocooled protein crystals. Proc Natl Acad Sci U S A 103:4912–4917
18. Paithankar KS, Owen RL, Garman EF (2009) Absorbed dose calculations for macromolecular crystals: improvements to RADDOSE. J Synchrotron Rad 16:152–162
19. Ravelli RBG, Garman E (2006) Radiation damage in macromolecular cryocrystallography. Curr Opin Struct Biol 16:24–629
20. Ravelli RBG, McSweeney S (2000) The 'fingerprint' that X-rays can leave on structures. Structure 8:315–328
21. Rodgers DW (1997) Practical cryocrystallography. Methods Enzymol 276:183–203
22. Teng T-Y (1990) Mounting of crystals for macromolecular crystallography in a freestanding thin-film. J Appl Crystallogr 23:387–391
23. Weik M, Ravelli RBG, Kryger G, McSweeney S, Raves ML, Harel M, Gros P, Silman I, Kroon J, Sussman JL (2000) Specific chemical and structural damage to proteins produced by synchrotron radiation. Proc Natl Acad Sci U S A 97:623–628

Chapter 8
Elemental Analysis of Proteins by Proton Induced X-ray Emission (microPIXE)

Elspeth F. Garman and Oliver B. Zeldin

Abstract The identification and quantification of metals bound to proteins is a crucial problem to be solved in structural biology. This chapter will describe the technique of proton induced X-ray emission with a microfocused proton beam (microPIXE) as a tool for analysing the elemental composition of liquid and crystalline protein samples. The proton beam induces characteristic X-ray emission from all elements in the protein, which can be interpreted in terms of the metal content of the protein molecule with a relative accuracy of between 10 and 20 %. The compelling advantage of this method is that the sulphur atoms in the methionines and cysteines of the protein provide an internal calibration of the number of protein molecules present so that systematic errors are minimised and the technique is entirely internally self-consistent. This is achieved by the simultaneous measurement of the energy of backscattered protons (Rutherford backscattering), to enable the matrix composition and thickness to be determined, and so correct the PIXE data for the self-absorption of X-rays in the sample. The technical and experimental procedures of the technique will be outlined, and examples of recent measurements given which have informed a range of investigations in structural biology. The use of the technique is increasing and we are in the final stages of developing it to be a routine high-throughput method.

Keywords microPIXE • Metalloproteins • Trace element analysis • Structural genomics

E.F. Garman (✉) • O.B. Zeldin
Department of Biochemistry, University of Oxford, Oxford OX1 3QU, UK
e-mail: elspeth.garman@bioch.ox.ac.uk

R. Read et al. (eds.), *Advancing Methods for Biomolecular Crystallography*, NATO Science for Peace and Security Series A: Chemistry and Biology, DOI 10.1007/978-94-007-6232-9_8,

8.1 Introduction

The characterisation of metal atoms in proteins and other large bio-molecules is a crucial problem to be solved in structural biology. The structure–function relationship of molecules is often pivotally dependent on bound metal ions or co-factors, so that understanding function requires knowledge of their identity and concentration. In X-ray crystallography, the electron density commonly indicates the presence of a metal ion, but at present there is no optimum technique for assigning a unique atomic number to that density. In addition, there are a variety of other questions that can be answered by knowing the elemental composition of the molecule, such as whether or not DNA is bound, how much selenium–methionine has been substituted for sulphur–methionine during expression, and whether or not a protein subjected to site directed mutagenesis binds a particular metal ion. For structure determination by NMR techniques, the presence of paramagnetic atoms can make measurement impossible, and it is problematic to identify them uniquely so that they can be removed. For any of the above reasons, performing a trace element analysis on our samples may be necessary.

Metal detection and quantification in proteins imposes stringent requirements on the analytical techniques employed. Three key performance indicators need to be considered: *sensitivity* (minimum detectable limit, MDL, or limit of detection, LoD), *spatial selectivity*, and *quantitative accuracy and precision.*

The sensitivity is constrained by the mass fraction of the metal in the protein, i.e. typically one metal atom of mass around 50 Da in a molecule of around 100 kDa, or 500 $\mu g\ g^{-1}$, and so any method must be able to provide adequate analytical precision at these concentration levels.

Spatial resolution is necessary in order to distinguish small crystals or dried protein precipitate from the surrounding buffer material. The small size of typical samples requires a spatial resolution of less than 10 μm. It is also necessary to have a mapping capability in order to identify the sample or regions of contamination.

The quantitative accuracy and precision must be sufficient to give an unambiguous determination of the number of metal atoms present in a protein molecule. This determination can be facilitated by forming ratios to the sulphur present in the protein (or phosphorus in nucleic acids) and so the accuracy requirements are relaxed. Nonetheless, determining the ratio M_Z/M_S (where M_Z and M_S are the mass of the unknown metal and sulphur, respectively) to an accuracy and precision of better than 10 %, as is typically required for reliable determination of stoichiometries, can be difficult to achieve.

It is additionally highly desirable for an assay to have the ability to identify unexpected elements (i.e. with no pre-selection of analytes): multi-elemental capability is essential when trying to determine the identity of unknown metals. Speed and convenience are also very important factors in the selection of appropriate analytical methods.

The technique which forms the subject of this chapter, microbeam proton induced X-ray emission (microPIXE) [6], is the only option readily available at present which satisfies all the above constraints. PIXE relies on the interaction of high energy (2–4 MeV) protons with the electron shells of the target atoms to stimulate the emission of characteristic X-rays which can be detected to identify and quantify the atoms present [9]. This is in many ways analogous to energy dispersive X-ray analysis (EDX) using an electron microprobe, but the crucial advantage for analytical purposes is that the relatively high mass of the primary particles compared with the electrons with which they are interacting means that the continuous background of bremsstrahlung X-rays, which seriously degrades the MDL of EDX, is essentially absent. This gives PIXE the potential to achieve detection limits in the region of $\mu g\ g^{-1}$, as is required for the determination of metalloprotein stoichiometries. MeV proton beams can be focused to micrometre dimensions and this provides the basis of a high performance scanning microanalytical instrument, the nuclear microprobe [7].

The application of the microPIXE technique to protein analysis has been developed over the last 15 years. The systematics of the technique have been established [3, 4] and the method has been well proven through measurements carried out on well over 150 different proteins in crystalline or liquid form, with multiple studies being carried out on a number of them.

The compelling advantage of proton beam trace element analysis compared to other currently available techniques for proteins is the internal normalisation. Since every element heavier than fluorine can be observed simultaneously, the sulphur from the methionines and the cysteines in the protein can be used as an internal standard for the amount of protein present. Concentrations of all elements of interest can be simultaneously measured relative to the sulphur signal, and since the primary sequence of the protein is known, the number of atoms of sulphur per molecule is also determined.

The measured ratio can thus be trivially computed into a stoichiometric ratio for the number of atoms of the element of interest per protein molecule, N_X, and can then be determined in a straightforward manner using the relationship:

$$N_X = \frac{C_X}{C_S} \times \frac{M_S}{M_X} \times N_S \tag{8.1}$$

where N_x = number of atoms of X per protein molecule,
C_X = PIXE measured concentration of element X,
C_S = PIXE measured concentration of sulphur,
M_S/M_X = Mass ratio of sulphur and element X, and
N_s = Number of sulphur atoms per protein molecule, already known from the primary amino acid sequence.

Samples containing DNA offer an alternative internal standard, as the phosphorus content is proportional to the length of the DNA strand. In DNA-protein complexes, this can be used to determine the number of DNA molecules per protein molecule [1], or in place of the sulphur standard, as an internal normalisation.

The internal normalisation aspect of the method applied to proteins makes any absolute measurements unnecessary, thus minimising the systematic errors. The initial protein concentration does not have to be known, as long as it is above a minimum required concentration (see below). The internal standard provided by nature allows an accuracy of between ± 10 and ± 20 % to be achieved: this level of error would be very difficult to reach if absolute measurements were necessary.

8.2 Implementation of the Technique

Our microPIXE analyses are carried out with 2.5 MeV protons focused to a diameter of 2–3 μm using the University of Surrey Ion Beam Centre's microbeam facility (beamline described as installed at Oxford in Grime et al. [8] and subsequently moved to Surrey). Characteristic X-rays are detected using an 80 mm^2 solid state Si(Li) detector (giving high energy resolution), and the detector is fitted with a 130 μm Be filter. During exposure to the proton beam, samples are held under vacuum on a thin organic film. The film is mounted on an aluminium holder, which is held in a 'ladder' capable of holding up to four samples, within the vacuum chamber. Both liquid and crystal samples can be mounted in this way.

For a measurement, first a coarse scan (150×150 μm up to 2.0×2.0 mm^2 depending on the sample size) is collected over the protein drop or crystal by scanning the proton beam spatially in *X* and *Y*. A software window is placed round the X-ray peak in the spectrum associated with a particular element of interest, and the counts are sorted into an *XY* grid to give individual elemental maps.

Simultaneous detection of Rutherford backscattered (RBS) protons (e.g. Fig. 8.1) allows the thickness and matrix composition of the sample to be accurately determined [5]. Quantitative information is obtained by collecting spectra (e.g. Fig. 8.1) at three or four points on the sample and also on the backing film alone for 3–10 min each. These spectra are analysed using the program OMDAQ to fit the RBS spectra, and which also provides an interface to GUPIX [2] for processing the PIXE spectra. The number of atoms of each element of interest per protein molecule can then be computed from Eq. (8.1). In order to avoid the generation of spurious signals from the material of the target chamber, the beam is stopped in a Faraday cup fabricated from spectroscopically pure graphite. Beam currents are in the range 0.1–1 nA.

8.3 Experimental Considerations

Both liquid and crystalline protein samples can be investigated using microPIXE and its application to proteins is now well established [3, 4].

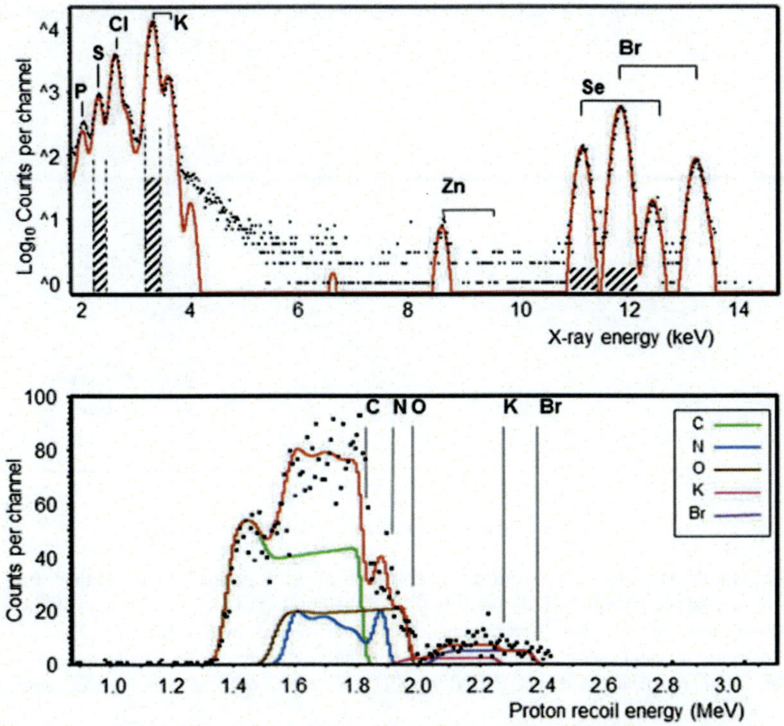

Fig. 8.1 PIXE (*top*) and RBS spectra (*bottom*) from a 2 μ diameter point on a thin liquid protein sample from which the S/Se ratio was sought, and which had been buffer exchanged from KCl into KBr to avoid overlap of the sulphur and chlorine peaks. The spectra were recorded simultaneously in 5 min using a beam of 2.5 MeV protons at a current of 100 pA. In each spectrum, the recorded data are shown as *black points* and the theoretical fit generated by the spectrum processing software is shown as a *solid red line*. In the PIXE spectrum, the identity of the major peaks is marked: note the log *y* scale of this spectrum. The *shaded regions* indicate the energy limits used to generate elemental maps such as those shown in Fig. 8.3. The RBS spectrum is modelled as a 6.15 μ thick layer of material with chemical formula $C_5N_{1.15}O_{2.28}K_{0.21}Br_{0.13}$ mounted on a 1.85 μ polymer film with composition $C_5O_{1.8}$. *Coloured lines* indicate the partial spectra corresponding to each element in the sample structure, and energy of the protons interacting with the surface layer of each element is indicated by *vertical lines* (Color figure online)

8.3.1 Required Concentration of Protein

The concentration of a liquid protein sample does not have to be known accurately for microPIXE, because the internal sulphur normalisation provides the standard. However, there is an MDL for an individual element of approximately 1–10 ppm of the dry weight, corresponding to a minimum concentration of liquid protein samples necessary for reliable quantitation by microPIXE. This minimum has been

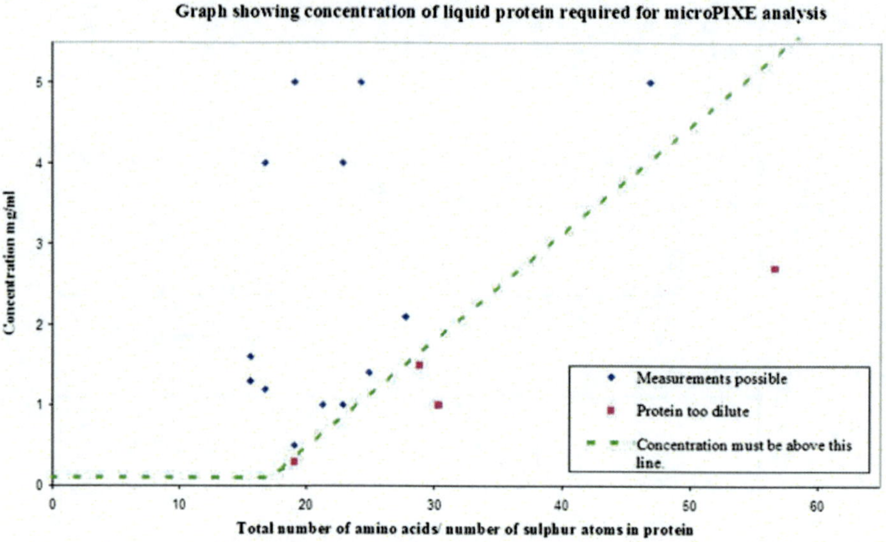

Fig. 8.2 Graph showing the minimum concentration of dried liquid protein residues (mg ml^{-1}) required for analysis by microPIXE as a function of the number of amino acids per sulphur atom in the protein molecule. The *points* represent actual experimental measurements and the *dotted line* represents an empirical guide as to whether or not analysis by microPIXE is feasible. Crystalline samples have concentrations of 100–1,000 mg ml^{-1} and are thus all of sufficient concentration for PIXE analysis

carefully investigated over the last few years, and an empirical limit has been derived. Figure 8.2 shows a graph of measurements performed on liquid samples over a range of fairly low concentrations, defined for convenience in terms of mg ml^{-1}. The concentration has been plotted against the number of amino acids per sulphur atom in the protein, known from the primary sequence information.

The sulphur signal from the samples lying below the dotted line on the plot was too weak to be measured with any confidence. All the samples lying above the line were successfully analysed and reliable quantitative results obtained. It is thus recommended that proteins are concentrated so that they definitely lie above the line before microPIXE analysis is attempted.

8.3.2 Protein Buffer and Crystal Mother Liquor

A major issue that must be addressed for every sample prior to the experimental preparations is that of the constituents of the protein buffer (for liquid samples) or the mother liquor (for crystals). Any buffers containing sulphur will affect the internal normalisation standard and thus result in an unreliable analysis. Many additives used in protein crystallisation include one or more sulphur atoms in their composition: for instance HEPES, BES and MES. This makes them undesirable buffers for PIXE measurements. This problem is exemplified in the following case study from one of our samples.

A protein was supplied for measurement at a concentration of 1.3 mg ml^{-1} in 10mM HEPES. The HEPES corresponded to a concentration of 2.38 mg ml^{-1} and thus a sulphur concentration of 0.32 mg ml^{-1}. In contrast, the 15 kDa protein contained 8 cysteines and 1 methionine, so the protein sample had a sulphur concentration of 0.025 mg ml^{-1}, 13 times lower than that of the HEPES. A reliable measurement was thus impossible, and in these circumstances buffer exchange into a non-sulphur containing solution would be essential.

Protein solutions and mother liquors also often contain salts. Most of these do not interfere with the microPIXE analysis, a notable exception being sodium chloride, which for liquid samples can give a chlorine peak stronger than the sulphur peak. Since they have consecutive atomic numbers (16 and 17), they give X-ray peaks close in energy. The low energy tail of an intense chlorine peak can affect the accuracy with which the sulphur peak can be fitted by GUPIX and quantified. In a number of cases, proteins have been successfully exchanged into potassium (or sodium) bromide. Although the chlorine peak is still large, it is no longer a major problem. Both the potassium and bromine X-ray peaks are well separated from other peaks of interest, and they can also be used to check for internal consistency, since they should be detected in a 1:1 atomic ratio.

Given a free choice, ammonium acetate is the perfect buffer, since it contains only organic elements which produce X-rays too low in energy to be detected in these experiments.

8.4 Sample Preparation

The samples are dried and then for the experiments, they are held under vacuum on thin organic polypropylene (C_3H_6) film (4 μm thick), stretched and mounted with adhesive across a 1 cm diameter hole in a 1 mm thick aluminium holder.

For liquid protein samples, a volume of between 0.1 and 0.3 μl of the necessary minimum concentration (see above) is required. This is pipetted very gently onto the film under a microscope and allowed to dry at room temperature in a covered environment to prevent dust contamination.

If heat is applied to accelerate drying, we find that the dried drop is more likely to peel off the film either before or during the measurement. Usually, three approximately 0.15 μl drops are deposited onto the film to allow for some redundancy and multiple measurements if required. The drops often dry as a ring, or the protein concentrates at the edge of the drop and the buffer crystallises out in the central part.

For crystalline samples, the crystals are deposited gently onto the film using a cryo-loop and allowed to dry. Both unwashed and washed crystals are usually investigated to check the effect of the mother liquor on the measurements. Washing is carried out by moving the crystal in a cryo-loop from its growth drop to a microbridge containing 10 μl of milliQ water (tap water contains chlorine which contaminates the samples). It is moved around the drop of water in the loop, and then transferred to a second drop of fresh water, and so on five or six times. This often results in disintegration of the crystal, but this does not matter because microPIXE does not require a crystal of diffracting quality, as it is sensitive only to the atomic species present.

8.5 Examples

During the last 15 years, we have accumulated a body of results from the analysis of more than 150 different proteins by microPIXE. Proteins have been quantitatively analysed for a range of elements, including magnesium, phosphorus from DNA, potassium, calcium, manganese, iron, cobalt, nickel, copper, zinc, selenium, iodine, barium, platinum, gold, mercury, and tungsten. We describe just one case below in order to illustrate the use of microPIXE analysis and how it has informed various aspects of structural studies on particular proteins. Many more examples can be found in Garman and Grime [4].

MicroPIXE analysis was carried out of the third KH domain of hnRNP K in complex with 'single-stranded' DNA. hnRNP K is one of the major proteins found in hnRNP particles which are ribonucleoprotein complexes containing protein and pre-messenger RNA. hnRNP K contains hnRNP K homology (KH) domains which bind both Cytosine-Thymine rich single-stranded DNA (ssDNA) and Cytosine-Uracil rich ssRNA. Co-crystallization of the third KH domain of human hnRNP K with a 15-mer ssDNA gave both rod shaped and square plate crystals in the same crystallization drop. The square plates were similar in morphology to native crystals obtained previously.

Using X-ray diffraction data from the rod shaped crystals, initial molecular-replacement trials using the structure of the native protein alone as a search model failed to allow the structure of the putative protein-DNA complex to be solved. No heavy-atom derivatives could be obtained to enable phase determination. MicroPIXE experiments showed that these crystals contained large amounts of phosphorus (Fig. 8.3a, b).

The phosphorus to sulphur atomic ratio was measured to be 4.4 and 4.7 in two different rod shaped crystals. This agreed well with a theoretical ratio of 4.67, corresponding to a stochiometry of three KH3 domains (protein molecules) per 15-mer ssDNA, taking into account the fact that the synthetic DNA has only 14 phosphates. For the square plate crystal form (Fig. 8.3c, d), the microPIXE measurements gave a maximum phosphorus to sulphur atomic ratio of below the LoD at 1.1×10^{-2}, showing that DNA was not bound.

Knowledge of the protein to DNA molecular ratio from the microPIXE measurements forced a re-examination of the X-ray data and of the self-rotation calculations. This added further evidence for a crystal asymmetric unit containing three KH3

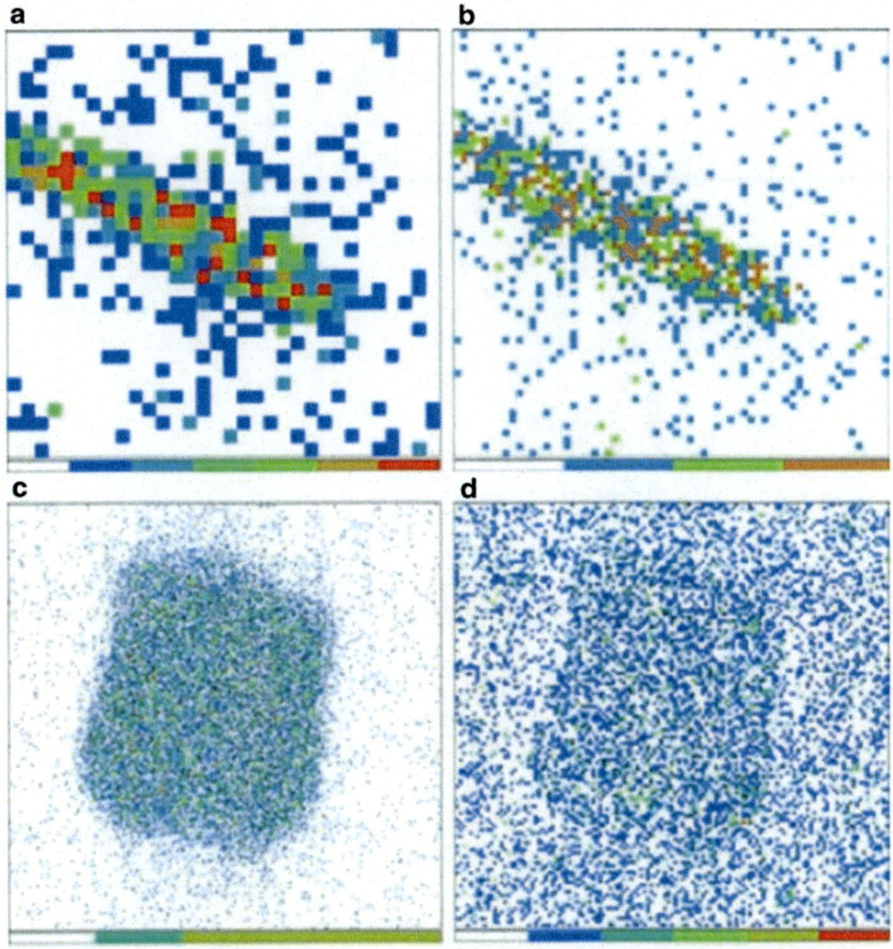

Fig. 8.3 (**a**) Sulphur and (**b**) Phosphorous elemental maps (150 × 150 μm^2) of ssDNA binding crystal (*rod shape*): DNA bound. Result: 4.4 phosphorus/sulphur atom (1 methionine per protein molecule) and thus 3 protein molecules per 15mer ssDNA. (**c**) Sulphur and (**d**) Phosphorus elemental maps (250 × 250 μm^2) of a ssDNA crystal (*plate shape*): DNA is not bound: the faint crystal outline is due to an increase in the background bremsstrahlung when scanning over the crystal (Reproduced courtesy of the IUCr from Backe et al. [1]. http://dx.doi.org/10.1107/S0907444904002628)

domains per 15-mer of DNA in the rod shaped crystals. Subsequent molecular-replacement trials using just one copy of the KH3 domain gave a cross-rotation function plot revealing three peaks, in good agreement with the self-rotation function. An electron density omit map was then calculated and gave clear difference density for the DNA, showing that there were two KH3 domains in contact with each DNA molecule and one KH3 domain positioned away from it (Fig. 8.4). Thus the structure of the KH3-DNA complex was solved with the aid of microPIXE analysis [1].

Fig. 8.4 Ribbon representation of the three KH3 domains in the asymmetric unit (coloured *red*, *yellow* and *green*) and an omit map showing the electron density around the single stranded DNA drawn at the 1σ level. Only two of the KH3 domains (*red*, *yellow*) interact with the DNA. The third KH3 domain (*green*) interacts with the second (*yellow*) via the strand β1 edges of their respective β1 sheets, an interaction also seen in crystals of the uncomplexed protein [1] (Reproduced courtesy of the IUCr from Backe et al. [1]. http://dx.doi.org/10.1107/S0907444904002628) (Color figure online)

8.6 Current Developments

The main bottlenecks presented by the current generation of microPIXE assays are in turnaround time and access to beamline facilities. These can be effectively addressed by a high-throughput approach. Current work is focusing on creating a fully integrated pipeline with the goal of making the technique available as a routine assay to laboratories worldwide. Samples are printed into micro-arrays using a non-contact printer, and data collected using automated protocols on the beamline. Data processing will be automated, allowing a large number of samples to be handled in any single 24 h run. Current proof of principle experiments indicate that it is realistic to increase the throughput from ~10 samples/day to ~200 samples/24 h. Such a large increase in throughput opens up the possibility of applying the technique to new areas of research, or as a routine assay within a structural genomics pipeline.

Acknowledgments The developments described above would not have been possible without our longstanding collaboration with Dr. Geoff Grime and the Ion Beam Centre at the University of Surrey. We are very grateful for their continuing support of this project. OBZ is funded by an

EPSRC Life Science Interface Doctoral Training Centre PhD studentship, and the high throughput methods development is supported by a grant from the John Fell Fund of the University of Oxford.

References

1. Backe PH, Ravelli RBG, Garman E, Cusack S (2004) Crystallization, microPIXE and preliminary crystallographic analysis of the complex between the third KH domain of hnRNP K and single-stranded DNA. Acta Crystallogr D 60:784–787
2. Campbell JL, Hopman TL, Maxwell JA, Nejedly Z (2000) The Guelph PIXE software package III: alternative proton database. Nucl Instrum Methods B 170:193–204
3. Garman E (1999) Leaving no element of doubt: analysis of proteins using microPIXE. Structure 7:R291–R299
4. Garman EF, Grime GW (2005) Elemental analysis of proteins by microPIXE. Prog Biophys Mol Biol 89:173–205
5. Grime GW (1996) The "Q factor" method: quantitative microPIXE analysis using RBS normalisation. Nucl Instrum Methods B 109:170–174
6. Grime GW (1999) High energy ion beam analysis methods (and background). In: Lindon JC, Tranter GE, Holmes JL (eds) Encyclopedia of spectroscopy and spectrometry. Academic, Chichester, pp 750–760
7. Grime GW (1999) Proton microprobe (method and background). In: Lindon JC, Tranter GE, Holmes JL (eds) Encyclopedia of spectroscopy and spectrometry. Academic, Chichester, pp 1901–1905
8. Grime GW, Dawson M, Marsh M, McArthur IC, Watt F (1991) The Oxford submicron nuclear facility. Nucl Instrum Methods B 54:52–53
9. Johansson SAE, Campbell JL, Malqvist KG (1995) Particle induced X-ray emission spectrometry (PIXE). In: Johansson SAE, Campbell JL, Malmqvist KG (eds) Chemical analysis: a series of monographs on analytical chemistry and its applications. Wiley, New York, p 167

Chapter 9
X-rays-Induced Cooperative Atomic Movement in a Protein Crystal

Tatiana Petrova, Vladimir Y. Lunin, Stephan Ginell, Andre Mitschler, Youngchang Kim, Grazyna Joachimiak, Alexandra Cousido-Siah, Isabelle Hazemann, Alberto Podjarny, Krzysztof Lazarski, and Andrzej Joachimiak

Abstract Protein molecules are damaged during X-ray diffraction experiments with protein crystals, which is in many cases a serious hindrance to structure solution. It is still not well understood whether radiation-induced local chemical changes lead to global structural changes in protein and what the mechanism is. We present experimental evidence at atomic resolution that irradiation causes the displacement of big parts of the protein molecule and water network. Radiation-induced structural changes in a protein molecule were studied in a series of diffraction experiments in which multiple data sets corresponding to increasing absorbed doses were collected from the same crystals of human aldose reductase (h-AR) and elastase at atomic resolution. There is a pronounced correlation between collective atomic movements and local and global damage to the crystal. Radiation-induced atomic shifts start at places with the pronounced local damage and are the largest for the damaged residues and structure fragments connected to damaged residues. An analysis of atomic displacement parameters (ADPs) revealed a distinct increase in the anisotropic character of ADP's for the atoms of some segments of the structures. This effect was pronounced for those atoms that initially had approximately isotropic ADPs and shifted over relatively large distances during irradiation. Because their displacements in different cells of the crystal occur not exactly at the same moment, this leads to an additional static disorder component.

T. Petrova (✉) • V.Y. Lunin
Institute of Mathematical Problems of Biology, Russian Academy of Sciences, Pushchino 142290, Russia
e-mail: petrova@impb.psn.ru; lunin@impb.psn.ru

S. Ginell • Y. Kim • G. Joachimiak • K. Lazarski • A. Joachimiak
Structural Biology Center, Biosciences Division, Argonne National Laboratory, Argonne, IL 60439, USA

A. Mitschler • A. Cousido-Siah • I. Hazemann • A. Podjarny
Département de Biologie Structurale et Génomique, IGBMC, CNRS, ULP, INSERM, 1 rue Laurent Fries, B.P. 163, 67404 Illkirch, France

R. Read et al. (eds.), *Advancing Methods for Biomolecular Crystallography*, NATO Science for Peace and Security Series A: Chemistry and Biology, DOI 10.1007/978-94-007-6232-9_9,

Keywords Radiation damage • Radiation-induced atomic displacements • Elastase • Aldose reductase

9.1 Introduction

The X-ray-induced damage to protein crystals during diffraction data is a serious hindrance to structure solution. A deep understanding of radiation damage to protein crystals is of great interest not only for fundamental science but also for the biological interpretation of structural data at the atomic level. This knowledge may also provide hints on how to mitigate X-ray damage.

The radiation damage to protein crystals manifests itself at different levels:

(a) as local specific chemical and structural changes in a protein molecule,
(b) as overall structural changes of a protein molecule, and
(c) as global effects related to the general crystal disorder.

On the local level, chemical and structural changes induced by radiation include the reduction of bound metal ions [1, 22] and the breakage of some covalent bonds with subsequent disordering of atoms involved in disrupted bonds. Examples of X-ray-induced covalent bond breakage are the disruption of disulfide bonds [3, 17, 21] and of covalent bonds between C and some heavy atoms, e. g., Se, Br, I, Hg [4, 13, 16], the decarboxylation of glutamate and aspartate residues [3, 5, 17], and the loss of the hydroxyl groups of tyrosines and the methylthio groups of methionines [3]. It is believed that the atoms involved in ruptured bonds become mobile and diffuse through the sample. Sometimes they can be observed in the electron density map in a new position close to the initial one [16].

The overall crystal disorder makes itself evident as an impairment of crystal diffraction properties, changes in the unit cell dimensions, and an increase in the mosaicity and Atomic Displacement Parameters (ADP, also known as atomic B factors).

All radiation-induced effects increase with the absorbed dose [18, 19], which is defined as the absorbed energy per unit cell of the crystal. It was also found that specific local structural changes occur at a very different "dose-scale", with the weakest covalent bonds being disrupted first [7]. Moreover, the residues of the same type within a given structure exhibit different susceptibility to radiation [3, 5, 6, 12], probably due to their different chemical environments. Some radiation-induced chemical changes occur long before any significant loss of diffraction [2, 17], which emphasizes the role of local damage in overall crystal decay. However, in spite of numerous studies devoted to structural damage, the link between local and global changes of the protein in the crystal and their relation to progressing crystal disorder are still poorly understood.

Based on the data on X-ray-induced changes in bond lengths in peptide crystals and radiolysis experiments, Meents and colleagues proposed that it is the hydrogen abstraction from organic molecules with subsequent formation of gaseous hydrogen bubbles that causes global changes in the crystal [8, 9].

The impact of local chemical changes on the whole protein structure and the relationship of these changes with general crystal disorder is difficult to catch experimentally because the absorbed X-ray dose causes the diffraction intensities to decay. As the intensities and resolution limit degrade, experimental errors increase and small structural changes are masked by an increase of ADP values. We studied the radiation-induced damage to a protein molecule by conducting diffraction experiments at atomic resolution using the crystals of h-AR and elastase [14, 15]. We investigated X-ray-induced overall structural changes in the protein molecule and attempted to reveal their relation with local and global damage to a protein crystal.

9.2 Experiments with an h-AR Crystal

9.2.1 Movements of Protein Domains and Water Molecules

The experiments were conducted with a crystal of the complex of h-AR with the inhibitor IDD 594 and co-factor $NADP^+$. Four complete data sets were consequently collected from one and the same crystal. The crystal was additionally exposed to an X-ray beam for 5 and 10 min before collecting data set3 and set4, correspondingly, to enhance damage and catch different stages of X-ray-induced deterioration. The total dose the crystal received by the end of the complete experiment was estimated using the program RADDOSE [11] to be 0.64×10^7 Gy. This value is approximately equal to one third of the Henderson limit. Therefore, our data sets correspond to the early stages of radiation damage to the protein crystal. The deterioration of the crystal during the experiment manifested itself mainly as a decrease of the maximum resolution limit from 1.0 Å for set1 to 1.2 Å for set4 (the number of unique observations were reduced by 41 %).

Four atomic models (M1–M4) corresponding to four stages of radiation damage were refined. A comparison of these models revealed that, along and simultaneously with the progressively increasing site-specific damage, irradiation causes displacements of big parts of the protein molecule and the water network (Fig. 9.1). These movements have a cooperative character: the displacement of individual helices, strands, and loops can be roughly treated as the movement of rigid bodies. In addition to the displacement of protein atoms, we observed the displacement of water molecules, which move in the same direction and in concert with the nearest protein atoms. It appears as if the atoms of the expanding protein molecule pull along the hydrogen-bonded network.

The collective movement of protein atoms leads to the expansion of the protein globule. A detailed analysis shows that the movement of the majority of atoms in the h-AR structure is anisotropic, and the atoms are predominantly displaced along the *a* axis. Approximately half of the protein molecule (residues of the N-terminal region) predominantly moves along the *a* axis (to the right of Fig. 9.2a), while most residues

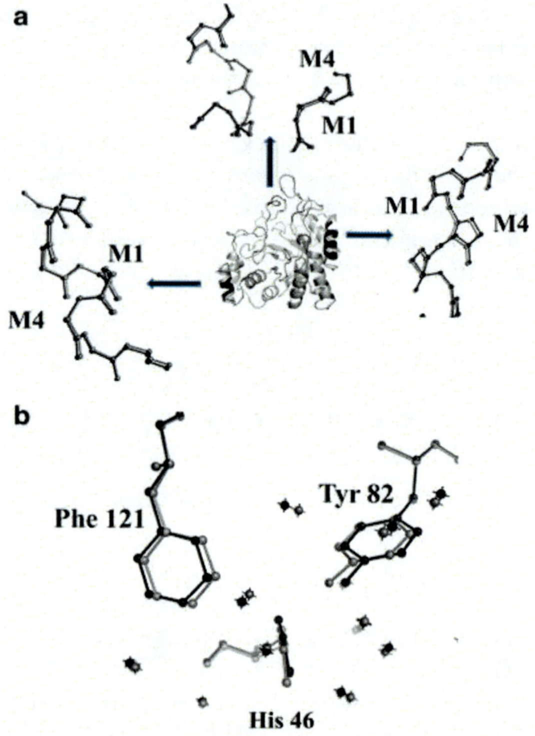

Fig. 9.1 Distinct displacements of atoms in model M4 relative to their positions in model M1. (**a**) The main chain atoms of models M1 and M4 in the enlarged regions of the loop 301–310, loop 112–115, α-helix 86–99, and α-helix 232–240. (**b**) Residues Tyr82, His46, and Phe121 of models M1 and M4 and water molecules in their vicinity. Most water molecules move approximately in the same direction as the nearest residues of the protein model

in the C-terminal region, 156–315, also move along the *a* axis but in the opposite direction. The movement of atoms within each lobe is not consistent with the rigid body movement, whereas the displacement of individual helices, strands, and loops can be approximately treated as the movement of rigid bodies. The expansion of the h-AR molecule is correlated with both the X-ray dose and the expansion of the unit cell in the *a* direction (Fig. 9.2b, c, d).

9.2.2 Local Radiation-Induced Chemical Changes and the Movement of Protein Secondary Structure Elements

The divergence of secondary structure elements is synchronized with the site specific local damage (Fig. 9.3a, b). The largest atomic displacements correspond to the atoms of helix 86–100 (Fig. 9.3b). This helix is connected by only two direct (not mediated by water molecules) hydrogen bonds with other secondary structure elements (the hydrogen bond between OG1 of Thr95 and OE1 of Glu

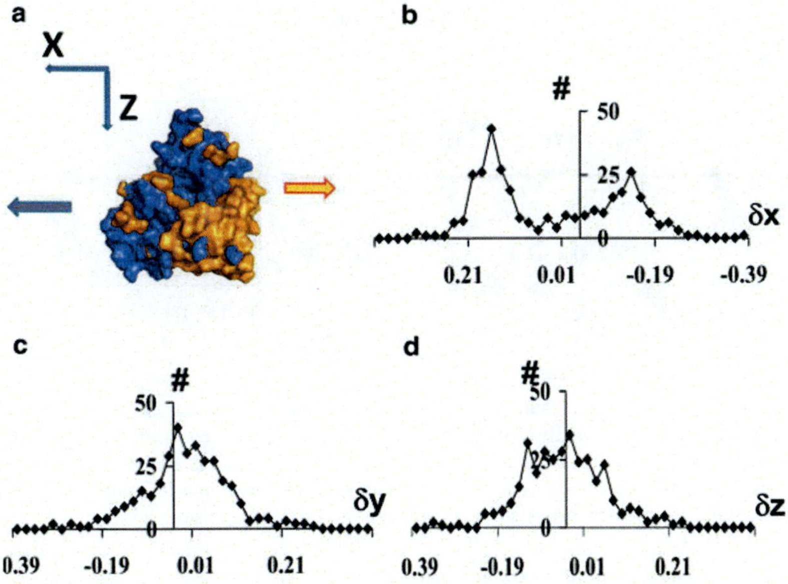

Fig. 9.2 Overall expansion of the h-AR molecule, induced by irradiation. **(a)** Surface representation of the model of h-AR. Protein residues whose Ca atoms move in the negative direction of the X axis (to the *right*) are *colored light*, and residues whose Ca atoms move in the positive direction of the X axis (to the *left*) are *colored dark*. **(b)**, **(c)**, **(d)** *Histograms* of the distribution of atoms with respect to their displacement along the 0X **(b)**, 0Y **(c)**, and 0Z **(c)** directions for the M4 model relative to the M1 model. Only atoms with the absolute displacement value exceeding 0.1 Å are included in the analysis

51 (Fig. 9.3a) and the hydrogen bond between NZ of Lys 89 and OE2 of Glu 150). Both hydrogen bonds are deteriorated during the diffraction experiment due to the damage of Glu 51 and Glu 150 by X-rays. The next largest displacements are revealed for the atoms of Asp 284, which is also damaged by X-rays during the experiment and to the atoms of the helix 231–240, which is connected to Asp 284 by hydrogen bonds (residues His 240 and Lys 239). The next largest displacements are observed for the atoms of Thr 113, which is related by halogen bond to Br of the inhibitor and the atoms of the inhibitors. This bond is deteriorated during the experiment due to the debromination of the inhibitor. Thus, there is a pronounced correlation between the site-specific damage and the radiation-induced movement of the atoms of h-AR. Radiation-induced atomic shifts start at the places with the pronounced local damage and are the largest for the damaged residues and for the structural fragments of the structure connected to the damaged residues. Based on these findings, we can conclude that there is a pronounced correlation between local damage and overall structural changes in the protein molecule.

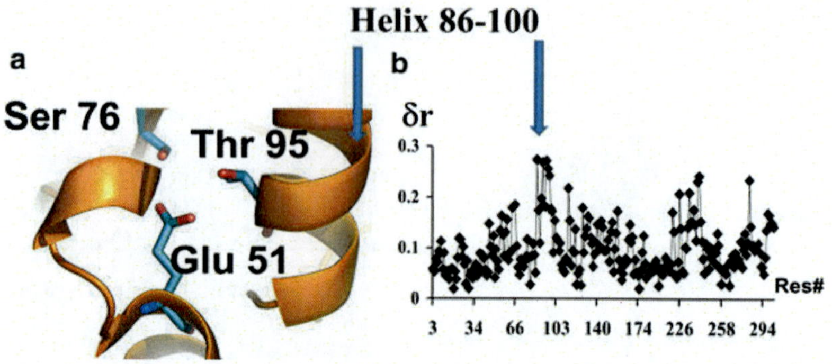

Fig. 9.3 Decarboxylation of Glu 51 involved in the contacts between different secondary structure elements of the h-AR model and concurrent displacements of atoms. (**a**) An enlarged region between α-helix 50–63, α-helix 86–99, and strand 73–78. Residues Glu51, Ser76, and Thr95 involved in hydrogen bond contacts between the secondary structure elements are shown using sticks. (**b**) The absolute values of atomic displacements of Cα atoms for model M5 relative to model M1. Only displacements of atoms in single conformations are shown

9.2.3 Analysis of Anisotropy of ADPs. Possible Link Between Local and Global Radiation Damage

Because radiation induced processes are likely to occur randomly in different unit cells of the crystal, radiation causes the overall crystal disorder, which manifests itself as an increase in atomic ADP values. For some structure segments, we found a distinct increase in the anisotropic character of ADPs. This effect is distinguished for those atoms that initially had approximately isotropic ADPs and were displaced by a relatively large distance during irradiation. The increase in the anisotropy of ADPs may be explained by the fact that atomic displacements occur not exactly at the same moment in different cells of the crystal, which leads to an additional component of crystal disorder.

For instance, an elongation of ADP's ellipsoids is clearly seen for the atoms of the inhibitor (Fig. 9.4a) and of the nearest residues. The atoms of the inhibitor (except Br atom) are displaced by a relatively large distance with dose. The occupancy of the Br atom decreased to 0.33 at the end of the experiment. The debromination causes the break of the halogen bond between Br and atom OG1 of Thr 113. The atoms of the inhibitor in debrominated copies move towards residue Thr 113. This movement in different cells of the crystal occurs with some mismatch in time. Therefore, in different crystal cells, there is either an inhibitor at its initial position or debrominated inhibitor, which is displaced relative to its initial position (Fig. 9.4b); the value of this shift is slightly different in different cells. Thus, local chemical and structural changes eventually increase the crystal disorder.

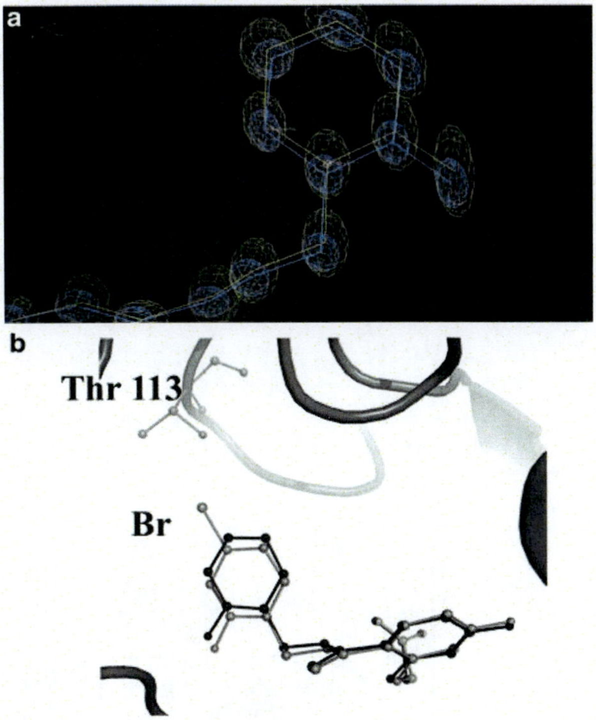

Fig. 9.4 Increase in anisotropy for the atoms of the inhibitor. (**a**) An increase in the anisotropic character of B's for inhibitor atoms of model M4 compared with the atoms of model M1. Ellipsoids of B's for the atoms of model M4 have a more elongated shape, and the direction of the long axis of the ellipsoid coincides with the direction of atomic shift. (**b**) Superposition of different copies of the inhibitor, which contain Br atom and which are debrominated. In debrominated copies, atoms of the inhibitor moved towards Thr 113

9.2.4 Model of h-AR at Different Temperatures. Comparison with X-Ray-Induced Atomic Displacements

The radiation-induced expansion of the protein globule resembles changes induced by temperature increase. In order to understand the nature of the radiation-induced structural changes in the protein molecule, we compared them with structural changes induced by temperature. A comparison of the models of AR at room temperature and 100 K shows that, in the case of temperature increase, the molecule expands mostly due to the movement of protruding loops and helices on the surface of the molecule. In the case of irradiation, a similar movement of loops and helices on the surface is observed as well. However, the largest shifts are revealed for those segments of the structure that were connected with the other segments by deteriorated during X-rays experiments bonds (Fig. 9.5a, b). The difference is due to the fact that some protein residues and crucial bonds between secondary structure elements are deteriorated during irradiation, while in the case of temperature increase these bonds remain unchanged.

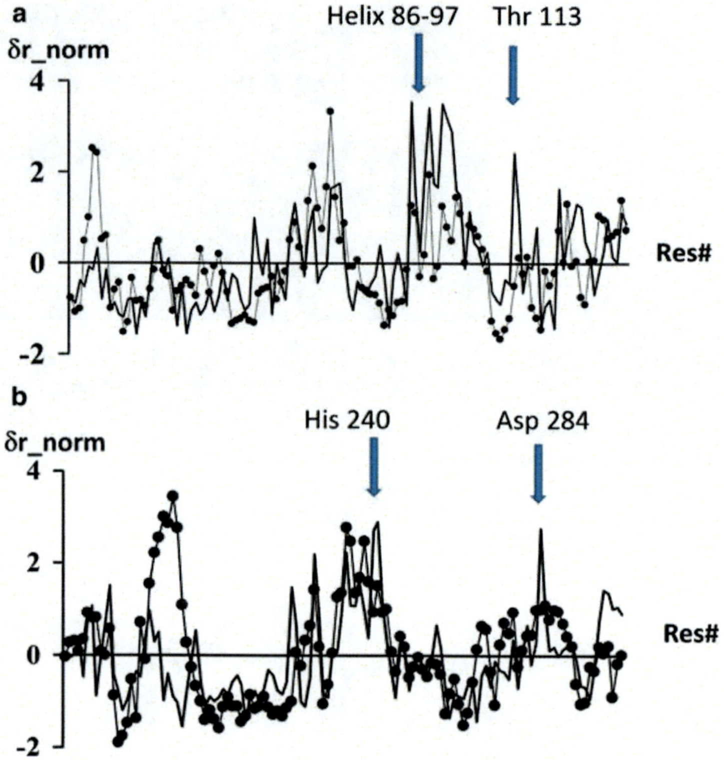

Fig. 9.5 Comparison of X-ray-induced atomic displacements (marked by *thick line*) and absolute difference of atomic coordinates for the models of h-AR at room temperature and 100 K (marked by *line with circles*). Normalized δr_norm = (r − <r>)/σ atomic displacements of Cα atoms for residues 1–144 (**a**) and 145–310 (**b**) are shown

9.3 X-Ray Diffraction Experiment with an Elastase Crystal

Another series of experiments was performed with a crystal of elastase. Elastase was chosen for the experiments because it is easily crystallized and contains four disulfide bridges. During the experiment, eight data sets (set1–set8) were consecutively collected from a single region of one and the same crystal at 100 K. Sets 1, 3, 5, 7 were collected with an attenuated beam, while sets 2, 4, 6, 8 ("killing") were collected without attenuation to enhance crystal damage. Global damage manifests itself in a decrease of the high resolution limit of the data from 1.2 to 1.82 A. A number of unique reflections decreased approximately three times.

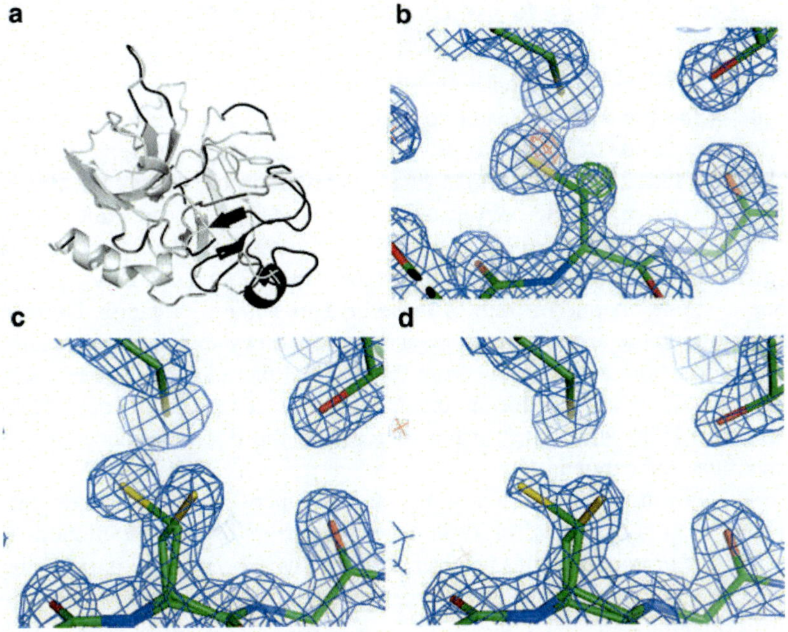

Fig. 9.6 Local and overall radiation-induced damage to the elastase molecule. (**a**) A ribbon diagram of the elastase molecule. Main chain segments that are displaced by more than 0.17 Å in model E5 relative to the model E1 are shown in *dark*. (**b**), (**c**), (**d**) Radiation-induced changes of the bridge Cys42–Cys58. A decrease in the occupancy of S^{γ} atoms and the gradual appearance of a new rotamer of Cys42. (**b**) At the beginning of irradiation, the bridge is slightly damaged. The 2Fobs-Fcalc density map for model E1 is contoured at 1σ. (**c**) A 2Fobs-Fcalc electron density map calculated for model E3 with the initial and new conformations of Cys42. (**d**) A 2Fobs-Fcalc electron density map for model E5 with the initial and new conformations of Cys42

9.3.1 *X-Ray-Induced Atomic Movements*

A comparison of the atomic coordinates for the refined models E1–E8 corresponding to different stages of radiation damage revealed X-ray-induced displacements of the atoms of elastase (Fig. 9.6a) and water molecules. The elastase molecule expands with increasing absorbed dose. The absolute values of atomic displacements are small. For example in model E7, 84 % of well ordered protein atoms with full occupancy are shifted by more than 0.1 Å and 17 % of atoms are shifted by more than 0.3 Å relative to their initial positions. The shifts increase with dose and correlate with changes in the unit cell size. Water molecules in the vicinity of the protein surface move in concert with and in the same direction as, adjacent protein atoms.

9.3.2 Local Damage to the Elastase Molecule

The most prominent radiation-induced local structural damage to the elastase molecule is the breakage of disulfide bonds. As the absorbed dose increased, we observed three kinds of effects for S^{γ}–S^{γ} bonds, namely, an increase in the S^{γ}–S^{γ} bond length (from 2.05 to 2.50 Å), a steady decrease in occupancy for all S^{γ} atoms and the appearance of new conformations for some cysteine residues (Fig. 9.6b, c, d). The second conformation of one of the cysteine residues is clearly seen in three disulfide bridges: Cys42–Cys58, Cys168–Cys182, and Cys191–Cys220. Note that one cysteine residue in each of these bridges has only a single conformation corresponding to the original position, and the second cysteine residue assumes two conformations, both are different from the original one. As the dose increases, the occupancy of the second conformation of Cys42 exceeds the occupancy of the initial conformation (Fig. 9.6d). On further increase in the dose, the occupancy values remain almost unchanged.

The displacement of secondary structure elements in the elastase molecule is synchronized with site-specific damage. It appears as if the radiation-induced movements of protein domains result in a slight divergence of the parts of the main chain that are connected by disulfide bridges, which leads to a small elongation of the disulfide bonds.

9.3.3 Comparison of X-ray-Induced Damage to Elastase at 100 and 15 K

We also studied the radiation-induced local and global damage at temperature of 15 K. In the last case, crystals were cooled by a cold helium stream. During this experiment, two series of data sets were collected from two different regions of one and the same crystal, one series from region A, and the other from region B. Each series consisted of five consecutive complete data sets (set1a–set5a from region A and set1b–set5b from region B). The experiment was designed so that the first, third, and fifth data sets were collected with an attenuated beam and at the same temperature, while the second and forth sets were collected without attenuation and at different temperatures, at 100 K from part A and at 15 K from part B of the crystal. In the last case, crystals were cooled by a cold helium stream. The dose the crystal received during data collection without attenuation was significantly higher than that during data collection with attenuated beam.

A comparison of the atomic coordinates of models A1, A2, A3, A5 and B1, B2, B3, B5 shows that the expansion of the protein molecule and the displacement of water molecules occurs at both 100 and 15 K. However, the absolute values of radiation-induced atomic displacements for models B2, B3 and B5 are smaller than for models A2, A3 and A5, correspondingly (Fig. 9.7a).

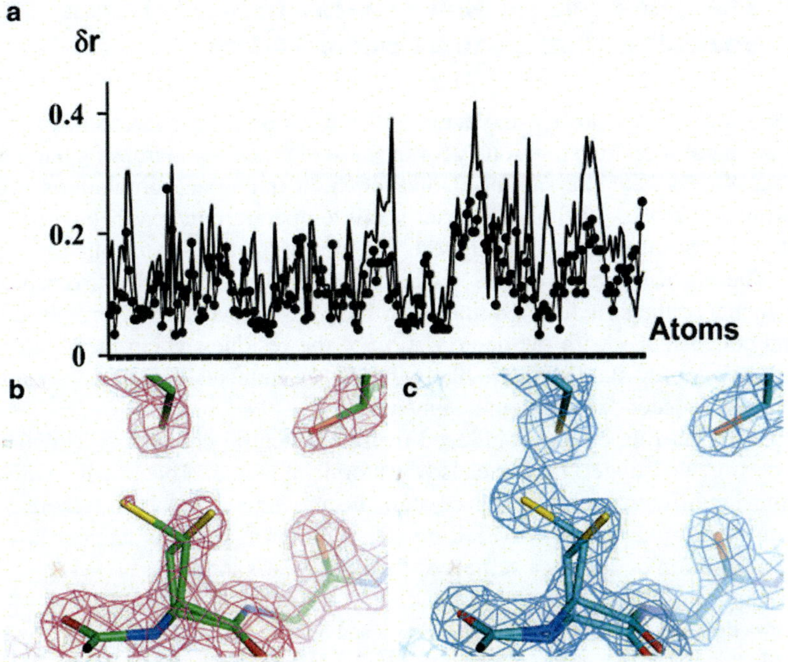

Fig. 9.7 Comparison of X-ray-induced local damage and atomic displacements at 15 and 100 K. (**a**) Atomic displacements for model A5 relative to model A1 are shown by a *continues line*; atomic displacements for model B5 relative to model B1 are shown by *line with circles*. (**b**), (**c**) Appearance of new conformations for Cys residues at 100 K (**b**) and 15 K (**c**). The 2Fobs-Fcalc maps for models A5 (**b**) and B5 (**c**) are shown in the vicinity of Cys42–Cys58 and are contoured at the same level of electron density of 0.53 e/A^3

A comparison of the models A1-A5 and B1-B5 also showed that lowering the temperature from to 100 to 15 K decreases the disulfide bond deterioration, the decrease being somewhat greater than twofold (Fig. 9.7b, c). Therefore, the radiation-induced breakage of disulfide bonds and atomic displacements occurred on approximately the same time scale (slightly ahead of local damage), which was an indirect confirmation of the interrelation of these effects.

Concerning the global damage in this experiment, lowering the temperature from 100 to 15 K, allowed us to collect data of a high resolution limit of 1.55 Å instead of 1.75 Å (in another series of experiments, 1.4 Å instead of 1.6 Å). A number of unique observations increased by about 40 %.

9.4 The Origin and the Role of *X*-ray-Induced Overall Structural Changes in a Protein Molecule

The expansion of a protein molecule and the cooperative movement of its large parts have been reported previously for the case of gradually increasing temperature. Indeed, X-rays heat the sample. However, both numerical simulations [10] and experimental studies [20] give a rather small temperature increase (less than 15 K) inside the crystal even at the highest flux density rates during our experiments [14]. Therefore, thermal effects cannot explain the observed anisotropic atom movements in the structure. On the other hand, the movement of protein domains occurs concurrently with the deterioration of the residues which mediate contacts between structural elements. It is likely that X-ray-induced local damage causes the movement of secondary structure elements.

What is the role of X-ray-induced overall structural changes in global crystal disordering? We suggest that, because the atomic movements occur with a mismatch in time in different cells of the crystal, the movement of big fragments of the molecule contributes to the overall crystal disorder. It remains unclear how great this contribution is. This is a subject of further investigations.

Acknowledgments The work of Tatiana Petrova and Vladimir Lunin was supported by Russian Foundation for Basic Research (grant number RFBR 10-04-00254-a).

References

1. Adam V, Royant A, Niviere V, Molina-Heredia FP, Bourgeois D (2004) Structure of superoxide reductase bound to ferrocyanide and active site expansion upon X-ray-induced photo-reduction. Structure 12:1729–1740
2. Banumathi S, Zwart PH, Ramagopal UA, Dauter M, Dauter Z (2004) Structural effects of radiation damage and its potential for phasing. Acta Crystallogr D 60:1085–1093
3. Burmeister WP (2000) Structural changes in cryo-cooled protein crystal owing to radiation damage. Acta Crystallogr D 56:328–341
4. Evans G, Polentarutti M, Carugo KD, Bricogne G (2003) SAD phasing with triiodide, softer X-rays and some help from radiation damage. Acta Crystallogr D 59:1429–1443
5. Fioravanti E, Vellieux FMD, Amara P, Madern D, Weik M (2007) Specific radiation damage to acidic residues and its relation to their chemical and structural environment. J Synchrotron Rad 14:84–97
6. Fuhrmann CN, Kelch BA, Ota N, Agard DA (2004) The 0.83 A resolution crystal structure of alpha-lytic protease reveals the detailed structure of the active site and identifies a source of conformational strain. J Mol Biol 338:999–1013
7. Garman EF, Owen RL (2006) Cryocooling and radiation damage in macromolecular crystallography. Acta Crystallogr D 62:32–47
8. Meents A, Dittrich B, Gutmann S (2009) A new aspect of specific radiation damage: hydrogen abstraction from organic molecules. J Synchrotron Rad 16:183–190
9. Meents A, Gutmann S, Wagner A, Schulze-Briese C (2010) Origin and temperature dependence of radiation damage in biological samples at cryogenic temperatures. Proc Natl Acad Sci USA 107:1094–1099

10. Mhaisekar A, Kazmierczak MJ, Banerjee R (2005) Three-dimensional numerical analysis of convection and conduction cooling of spherical biocrystals with localized heating from synchrotron X-ray beams. J Synchrotron Rad 12:318–328
11. Murray JW, Garman EF, Ravelli RBG (2004) X-ray absorption by macromolecular crystals: the effects of wavelength and crystal composition on absorbed dose. J Appl Crystallogr 37: 513–522
12. Nukagara M, Mayama K, Hujer AM, Bomolo RA, Knox JR (2003) Ultrahigh resolution structure of a class A beta-lactamase: on the mechanism and specificity of the extended-spectrum SHV-2 enzyme. J Mol Biol 328:289–301
13. Oliéric V, Ennifar E, Meents A, Fleurant M, Besnard C, Pattison P, Schiltz M, Schulze-Briese C, Dumas P (2007) Using X-ray absorption spectra to monitor specific radiation damage to anomalously scattering atoms in macromolecular crystallography. Acta Crystallogr D 63: 759–768
14. Petrova T, Lunin VY, Ginell S, Hazemann I, Lazarski K, Mitschler A, Podjarny A, Joachimiak A (2009) X-ray radiation induced co-operative atomic movements in protein. J Mol Biol 387(5):1092–1105
15. Petrova T, Ginell S, Mitschler A, Kim Y, Lunin VY, Joachimiak G, Cousido-Siah A, Hazemann I, Podjarny A, Lazarki K, Joachimiak A (2010) X-ray-induced deterioration of disulfide bridges at atomic resolution. Acta Crystallogr D 66:1075–1091
16. Ramagopal UA, Dauter Z, Thirumuruhan R, Fedorov E, Almo SC (2005) Radiation-induced site-specific damage of mercury derivatives: phasing and implications. Acta Crystallogr D 61:1289–1298
17. Ravelli RB, McSweeney SM (2000) The "fingerprint" that X-rays can leave on structures. Structure 8:315–328
18. Shimizu N, Hirata K, Hasegawa K, Ueno G, Yamamoto M (2007) Dose dependence of radiation damage for protein crystals studied at various X-ray energies. J Synchrotron Rad 14:4–10
19. Sliz P, Harrison SC, Rosenbaum G (2003) How does radiation damage in protein crystals depend on X-ray dose? Structure 11:13–19
20. Snell EH, Bellamy HD, Rosenbaum G, van der Woerd MJ (2006) Non-invasive measurement of X-ray beam heating on a surrogate crystal sample. J Synchrotron Rad 14:109–115
21. Weik M, Ravelli BG, Kryger G, McSeeney S, Raves ML, Harel M, Gros P, Silman I, Kroon J, Sussman JL (2000) Specific chemical and structural damage to proteins produced by synchrotron radiation. Proc Natl Acad Sci USA 97:623–628
22. Yano J, Kern J, Irrgang KD, Latimer MJ, Bergmann U, Glatzel P, Pushkar Y, Biesiadka J, Loll B, Sauer K, Messinger J, Zouni A, Yachandra VK (2005) X-ray damage to the Mn4Ca complex in single crystals of photosystem II: a case study for metalloprotein crystallography. Proc Natl Acad Sci USA 102:12047–12052

Chapter 10
Everything Happens at Once – Deconvolving Systematic Effects in X-ray Data Processing

Dominika Borek and Zbyszek Otwinowski

Abstract Diffraction intensities measurements are influenced by random errors and complex patterns of systematic effects. The systematic effects can be physically modeled if their sources are known, resulting in deconvolution of experimental data into: the signal arising from crystal structure, other signals, for instance absorption or specific radiation-induced changes, and experimental errors. The systematic effects that are not properly modeled contribute to the error estimates, effectively decreasing the, already low, phasing signal-to-noise ratio. Data processing programs, for instance Denzo and Scalepack, have built-in hierarchy that allows for optimal deconvolution of signals and errors. Their analysis relies on comparing the intensities of symmetry-equivalent reflections using multivariate statistics methods. Multicomponent modeling of variance is particularly useful for correcting the diffraction data affected by radiation damage.

Keywords Diffraction • Data Processing • Scaling • Radiation Damage • Estimation of Uncertainty

10.1 Introduction

Crystallography is a biophysical technique that provides the majority of structural models used for mechanistic interpretation of chemical actions, to design more efficient drugs, and to study the interactions of molecules with each other.

D. Borek • Z. Otwinowski (✉)
Department of Biochemistry, University of Texas Southwestern Medical Center at Dallas, 5323 Harry Hines Blvd., Dallas, TX 75390-8816, USA

Department of Biophysics, University of Texas Southwestern Medical Center at Dallas, 5323 Harry Hines Blvd., Dallas, TX 75390-8816, USA
e-mail: zbyszek.otwinowski@utsouthwestern.edu

R. Read et al. (eds.), *Advancing Methods for Biomolecular Crystallography*, NATO Science for Peace and Security Series A: Chemistry and Biology, DOI 10.1007/978-94-007-6232-9_10,

Nowadays the process of solving a macromolecular structure is highly automated, which means that for well-behaving crystals, a structural model ready for deposit is frequently obtained with limited input from the researcher, who simply applies default protocols in one of the many integrated software solutions. This is the reason why many non-crystallographers perceive structure solution of any macromolecular structure as a straightforward task, often neglecting the understanding of methods that underlie the crystallographic software.

There are several reasons why such attitude may lead to failure, two of them being especially important. First, even with methods constantly improving, there are always borderline solvable cases; it is the limits that are changing. For these difficult projects, better software may streamline calculations, but conscious intervention is essential for the successful solution of the structure. The second reason is due to inherent difficulties in solving the phase problem, which is a process of obtaining a structural model from a set or sets of X-ray diffraction intensities – an example of an ill-defined inverse problem. This means that we do not have a guarantee of solving the structure, data quality being not the only defining factor. When phase or model building problem is not solved, one must decide how to proceed. *Should I get a new crystal? Should I try a different phasing approach? Is it sufficient just to change some of the defaults in the automatic software? Should I simply give up, because this project has no future?* There are many possible choices and a good understanding of crystallographic methods is needed to make optimal decisions.

10.2 Inverse Problems and Sources of Errors

For a given structure, the physical model of the experiment defines our expectations about the collected data. Structure solution has to invert the physical model of data dependency on the crystal structure. Solving an inverse problem is achieved by building a detailed model of the process that generated the observed data, so that one can relate experimental observations to model parameters. The success of solving inverse problems greatly depends on the type of problem and the amount and quality of data. The crystallographic phase problem is a difficult case of a highly non-linear inverse problem. We could build an infinitely detailed description of all the things that happen during diffraction data collection. However, the amount of diffraction data is usually limited, so the physical model used in this inverse problem must be simplified as much as possible, without missing the most important effects. These simplifications are not straightforward and the procedures to define them are still being developed. The model must be sufficiently detailed to get a structure solution, but not so complex that the calculations lead to an ambiguous or uncertain result due to the high ratio of model's parameters to the available experimental observations. Therefore, we use hierarchical approach, in which at every stage we reassess our level of knowledge about the problem. This allows us to build a more detailed description of the problem, even with a limited amount of data. For instance, we typically assume at the beginning that our crystals are not twinned. However, if we detect the signature of twinning in the data, we should revisit the model of the phase problem to include in it the possibility of twinning. We don't do it always, because it creates severe complications, so without the signs of twinning it is not worthwhile to go through potentially less stable calculations. Problems with deciding which potential complications to consider arise when we can detect a systematic pattern of correlation in experimental data, but cannot identify a source for it. We would modify the physical model used in the process of structure solution to include a description of the detected phenomenon. However, without identifying the source of the problem, we cannot build a good physical model for it; at best, we can rely on generic correlation patterns, e.g. the ones used in local scaling. The resulting incompleteness of the description can affect the structure solution in unpredictable ways, and it is equivalent to increasing the systematic error in experimental data.

Finally, our experimental data are affected by complex patterns of errors. Errors associated with diffraction intensities have very diverse sources, and different types of errors affect the chances of obtaining the structure solution differently.

The combination of the non-linearity of the phase problem, the unavoidable incompleteness of the physical models used in crystallographic calculations, and the measurement errors that are always present in experimental data is the reason why solving macromolecular structures is still a challenging task.

10.3 Errors and Uncertainty of Diffraction Intensities

Error is the difference between the true value and the measured value of a quantity. We typically do not know the true value of the measured quantity, so the error value is not known either. Instead, we approximate error by statistical distributions and properties of these distributions define the uncertainty of measurements. Estimates of uncertainties for diffraction intensities have two main groups of contributors: those originating from random errors and those resulting from systematic errors.

Random errors are uncorrelated and described by well-defined probability distribution functions. The random error affecting diffraction intensities results from the quantum nature of X-rays. It is described by the Poisson distribution of X-ray photon counting statistics and approximated by a Gaussian function. The magnitude of the random error arising from counting statistics is equal to the square root of the expected number of photons, thus the relative error of a diffraction peak intensity measurement owing to counting statistics is equal to 1 over this number.

Systematic errors result from physical phenomena that affect the intensities of reflections in a correlated manner. A systematic error becomes a systematic effect if we discover its source and include its description in our physical model used in calculations. It means that any significant effect which is not included in the physical model used during data processing will increase estimates of uncertainty of diffraction intensities. However, including the physical description of any systematic effect in the data processing will require a good physical model for the particular effect and efficient and stable procedures to propagate its description and consequences across the hierarchical procedures used to get the structure solution.

Bayesian methods are particularly well-suited for propagation of information in multi-level procedures, where our knowledge about the problem changes between different stages of the procedure, and they are explicitly or implicitly used at every step of crystallographic calculation [1–5].

10.4 Multiplicity of Observations

Crystallographic data have a very special property – due to the symmetry of the crystal lattice, properly scaled intensities of symmetrically-equivalent reflections should be the same within measurement errors.

This property has important consequences: (1) we can average the symmetrically-equivalent reflections to get better estimates of diffraction intensities and the uncertainties associated with them and (2) we can validate whether our models describing the physics of the diffraction experiment are adequate. If our description of the experiment is inadequate, intensities of symmetrically-equivalent reflections will not agree with each other. They could be different due to higher than expected levels of measurement errors. However, the differences between these intensities may also be caused by systematic effects that were not included in the physical models applied during merging reflections. By adjusting this model, we can improve the agreement between symmetrically-equivalent reflections, which is equivalent to obtaining better estimates for merged diffraction intensities and their uncertainties.

10.5 Systematic Effects in Diffraction Data

There are three main groups of sources that contribute to the systematic effects: (1) a crystal, (2) instrumentation used in the process and (3) the diffraction process itself. For each source, we have a data model with some general initial assumptions about reasonable values of the parameters associated with it.

All data models have to be simplified. Crystals may be more mosaic than expected or may be twinned; the crystal lattice may have anisotropic order or various defects; the diffraction can be contaminated with ice. An experimental setup, which in general consists of a source of X-rays, a goniostat, cryo-stat, and a detector, may be less stable than expected. For instance, the X-ray intensity can fluctuate, a goniostat may rotate unevenly, the cryo-stream may cause vibrations of the loop holding crystal, *etc*. However, we cannot assume at the start that everything has gone wrong. We start from the simplest possible description and then add new features only if the simpler model cannot explain observed patterns in experimental data.

For instance, we know that crystals not only diffract but also absorb X-ray photons. Therefore, we can model how crystal absorption attenuates observed diffraction intensities. If the model is sufficient, intensities of symmetrically-equivalent reflections acquired at different crystal orientations will agree with each other after correction for absorption. However, if a crystal is very large or grew in unusual crystallization conditions, we may have to generate a more complex description to accommodate the more complex absorption surface.

10.6 Radiation Induced Systematic Effects in Data

During diffraction experiments, absorbed photons participate in chemical and physical reactions by interacting with atoms in the molecules of the crystal lattice and with solvent molecules in water channels. Some of the chemical reactions are temperature-dependent, so cryo-cooling can be used to slow them down. However, some processes, for instance tunneling, are temperature-independent, so they progress even in cryo-protected crystals [6].

The model of radiation-induced decay of diffraction intensities has been used in Scalepack since the very beginning [7–9] and it follows an approach established earlier in ROTOVATA/AGROVATA. In this model, collisions between the atoms building the crystal lattice and the primary and secondary photons generated during X-ray exposure results in small random movements of the atoms, and these small movements affect distances between atoms. In real space, it is described by convolution of the initial positions of the atoms with Gaussian functions describing displacements of the atoms. The scaling B-factor models these patterns in reciprocal space. We and others have shown that scaling B-factor is a good proxy for X-ray dose [10–12], which is convenient to use in X-ray experiments.

However, the scaling B-factor is insufficient to describe all effects of the X-ray radiation-induced changes acquired during data collection. Tunneling and other reactions resulting from secondary damage generate specific chemical changes in the molecules building the crystal lattice. It was shown multiple times that their patterns depend on the specific chemistry, electrostatics or even the type of crystallization solution [6, 10, 13–21]. Observed specific changes represent departures from the average dispersion of atomic positions described in data space by the scaling B-factor. In reciprocal space, specific changes induced by X-ray radiation result in changes in magnitudes of structure factors, which are specific to a particular crystal, so every protein, or sometimes even different crystal forms of the same protein, will have different patterns of specific changes induced by radiation damage. These changes significantly affect all calculations relying on intensities, for instance estimates of Bijvoet differences or dispersive differences, both used for phasing.

We asked: (1) Whether it is possible to build a physical model that would be detailed enough to correct diffraction data for the effects of specific chemical changes induced by X-rays? (2) Would the model work even though specific changes induced by X-rays are different in different crystal lattices? Finally, (3) would the model be simple enough so that calculations would work even in cases where the amount of data is very limited?

10.7 Modeling Radiation-Induced Systematic Effects

Multiple experiments indicate that the specific changes induced by radiation damage result in mostly linear changes of the intensity of scaled reflections [10–12, 20, 22, 23]. These intensities are already corrected for radiation-induced decay described with the scaling B-factor. Therefore, the observed discrepancies and their linear behavior, which depend on X-ray dose, result presumably from radiation-induced specific changes in the crystal structure. Regression-based methods could be used to extrapolate from observed patterns the corrected intensities of reflections at zero-dose [24]. However, for data sets with low multiplicity of observations, for instance crystals in triclinic or monoclinic space groups or for incomplete data sets collected from multiple crystals, such a simple method would not work. Thus, we built a framework that would solve the problem in a general case.

Specific chemical changes induced by radiation at the level of diffraction data behave like a type of non-isomorphism. A crystal structure is changing during the data collection and this change depends on the X-ray dose, which allows it to model the problem with a specific functional dependence. However, changes of the structure factors resulting in changes of intensities of symmetrically-equivalent reflections during data collection may also happen in other situations. For instance, the cryo-cooling induces a crystal lattice contraction that does not have to be uniform across the crystal. When different parts of the crystal enter the beam they will produce intensities that differ for symmetrically-equivalent reflections, but these differences will depend on rotation angle rather than on the X-ray dose. Similarly, when we collect data from multiple crystals, there will be differences between the intensities of symmetrically-equivalent reflections that originated from different crystals. In this case, it will be described by a function which includes information regarding which crystal contributed to which observed reflection. Specific types of discrepancies between intensities are also related to anomalous signal. If the level of anomalous signal is low, the reflections contributing to the Bijvoet pair have the same intensities, but when signal is increasing the intensities of the Bijvoet pair start diverging. This also can be modeled as a type of, in this case desirable, non-isomorphism. Other desirable non-isomorphisms could result from the heavy-atom or ligand-binding. The model, which includes the non-isomorphism induced by radiation, non-isomorphism induced by cryo-cooling, non-isomorphism between the crystals and non-isomorphism due to the presence of anomalous signal, requires hierarchical calculations, in which signals and uncertainties are estimated in a specific order. To estimate the significance of the non-isomorphisms involved in a particular experiment, we compare the expected level of a particular signal to its uncertainty for each merged reflection.

However, anisotropic diffraction would significantly affect estimates of the expected level of signals, so at the first step we normalize all observed intensities to include impact of this potential effect. Then, the maximum likelihood function, which includes all five types of non-isomorphisms, is optimized against differences between weighted intensities of symmetrically-equivalent reflections. The obtained estimates of signals and uncertainties are used in the next step to determine whether a particular reflection contributes to a particular type of signal, and if so how big the contribution is. Finally, the most optimal estimates for each signal can be used in the subsequent steps of the crystallographic calculations. How they are used depends on whether the software performing subsequent calculations is capable of handling continuously weighted signals. Optimal estimates of weak anomalous signals improved the structure solution in many cases. The extrapolated to zero-dose intensities are valuable when analyzing the impact of X-rays on the chemical properties of the structure, so the radiation-induced effects can be separated from desirable biological signals. The analysis of non-isomorphisms within the crystal provides a feedback when decisions have to be made about optimization of cryo-cooling conditions. Finally, the estimates of non-isomorphisms between crystals in experiments involving incomplete data from multiple crystals can be used to cluster crystals into the most isomorphous groups, which clearly benefits the subsequent structural analysis.

Acknowledgments The National Institutes of Health supported this work with grant GM053163.

References

1. Banumathi S, Zwart PH, Ramagopal UA, Dauter M, Dauter Z (2004) Structural effects of radiation damage and its potential for phasing. Acta Crystallogr D Biol Crystallogr 60:1085–1093
2. Borek D, Minor W, Otwinowski Z (2003) Measurement errors and their consequences in protein crystallography. Acta Crystallogr D 59:2031–2038
3. Borek D, Ginell SL, Cymborowski M, Minor W, Otwinowski Z (2007) The many faces of radiation-induced changes. J Synchrotron Radiat 14:24–33
4. Bricogne G (1988) A Bayesian statistical-theory of the phase problem.1. A multichannel maximum-entropy formalism for constructing generalized joint probability-distributions of structure factors. Acta Crystallogr A 44:517–545
5. Bricogne G (1997) Bayesian statistical viewpoint on structure determination: basic concepts and examples. Macromol Crystallogr A 276:361–423
6. Bricogne G, Irwin J, de la Fortelle E (1997) The Bayesian programme in X-ray crystallography: unifying experimental and mathematical sources of phasing power. FASEB J 11:A1125–A1125
7. Burmeister WP (2000) Structural changes in a cryo-cooled protein crystal owing to radiation damage. Acta Crystallogr D 56:328–341
8. Diederichs K, McSweeney S, Ravelli RBG (2003) Zero-dose extrapolation as part of macromolecular synchrotron data reduction. Acta Crystallogr D 59:903–909
9. French S (1978) Bayesian 3-stage model in crystallography. Acta Crystallogr A 34:728–738
10. French S, Wilson K (1978) Treatment of negative intensity observations. Acta Crystallogr A 34:517–525

11. Futterer K, Ravelli RB, White SA, Nicoll AJ, Allemann RK (2008) Differential specific radiation damage in the Cu II-bound and Pd II-bound forms of an alpha-helical foldamer: a case study of crystallographic phasing by RIP and SAD. Acta Crystallogr D Biol Crystallogr 64:264–272
12. Kmetko J, Husseini NS, Naides M, Kalinin Y, Thorne RE (2006) Quantifying X-ray radiation damage in protein crystals at cryogenic temperatures. Acta Crystallogr D 62:1030–1038
13. Murray J, Garman E (2002) Investigation of possible free-radical scavengers and metrics for radiation damage in protein cryocrystallography. J Synchrotron Radiat 9:347–354
14. Murray JW, Rudino-Pinera E, Owen RL, Grininger M, Ravelli RBG, Garman EF (2005) Parameters affecting the X-ray dose absorbed by macromolecular crystals. J Synchrotron Radiat 12:268–275
15. O'Neill P, Stevens DL, Garman EF (2002) Physical and chemical considerations of damage induced in protein crystals by synchrotron radiation: a radiation chemical perspective. J Synchrotron Radiat 9:329–332
16. Otwinowski Z, Minor W (1997) Processing of X-ray diffraction data collected in oscillation mode. Methods Enzymol 276:307–326
17. Otwinowski Z, Minor W (2001) Denzo & Scalepack. In: Rossmann MG, Arnold E (eds) International tables for crystallography, vol F, Crystallography of biological macromolecules. Published for The International Union of Crystallography by Kluwer Academic Publishers, Dordrecht/Boston/London, pp 226–245
18. Otwinowski Z, Borek D, Majewski W, Minor W (2003) Multiparametric scaling of diffraction intensities. Acta Crystallogr A 59:228–234
19. Ramagopal UA, Dauter Z, Thirumuruhan R, Fedorov E, Almo SC (2005) Radiation-induced site-specific damage of mercury derivatives: phasing and implications. Acta Crystallogr D 61:1289–1298
20. Ravelli RBG, McSweeney SM (2000) The 'fingerprint' that X-rays can leave on structures. Structure 8:315–328
21. Ravelli RBG, Leiros HKS, Pan BC, Caffrey M, McSweeney S (2003) Specific radiation damage can be used to solve macromolecular crystal structures. Structure 11:217–224
22. Weik M, Ravelli RBG, Kryger G, McSweeney S, Raves ML, Harel M, Gros P, Silman I, Kroon J, Sussman JL (2000) Specific chemical and structural damage to proteins produced by synchrotron radiation. Proc Natl Acad Sci U S A 97:623–628
23. Yano J, Kern J, Irrgang KD, Latimer MJ, Bergmann U, Glatzel P, Pushkar Y, Biesiadka J, Loll B, Sauer K, Messinger J, Zouni A, Yachandra VK (2005) X-ray damage to the Mn4Ca complex in single crystals of photosystem II: a case study for metalloprotein crystallography. Proc Natl Acad Sci U S A 102:12047–12052
24. Zwart PH, Banumathi S, Dauter M, Dauter Z (2004) Radiation-damage-induced phasing with anomalous scattering: substructure solution and phasing. Acta Crystallogr D 60:1958–1963

Chapter 11
Extending the Reach of Molecular Replacement

Randy J. Read, Airlie J. McCoy, Robert D. Oeffner, and Gábor Bunkóczi

Abstract Molecular replacement is already able to solve the majority of structures in the Protein Data Bank, thanks to the rapidly increasing number of template structures available and continuous improvements in the algorithms. Chances of success can be optimised by proper preparation of models, for instance by trimming poorly-conserved regions, creating an ensemble of alternative models or applying advanced homology modeling tools. The sensitivity of the molecular replacement search can be improved by using likelihood targets; these lend themselves to automation, which makes it possible to carry out extensive searches and helps to avoid user errors. The convergence radius of model completion can be extended by using methods that smoothly deform the starting model or apply advanced modeling techniques. Even more difficult structures can be solved by combining molecular replacement with other phasing methods, such as SAD phasing or multi-crystal averaging.

Keywords Molecular replacement • Likelihood • Molecular modeling • SAD phasing • Multi-crystal averaging

11.1 Introduction

When the second protein crystal structure was solved (haemoglobin; [16]), it was already seen to resemble the first protein crystal structure (myoglobin; [7]), and the seeds of the molecular replacement method were sown. In the subsequent

R.J. Read (✉) • A.J. McCoy • R.D. Oeffner • G. Bunkóczi
Department of Haematology, University of Cambridge, Cambridge, UK

Cambridge Institute for Medical Research, Wellcome Trust/MRC Building,
Hills Road, Cambridge CB2 0XY, UK
e-mail: rjr27@cam.ac.uk

R. Read et al. (eds.), *Advancing Methods for Biomolecular Crystallography*,
NATO Science for Peace and Security Series A: Chemistry and Biology,
DOI 10.1007/978-94-007-6232-9_11,

half-century it has become very clear that proteins with similar amino acid sequences have similar 3D structure, and for a long time molecular replacement has been an essential tool for the macromolecular crystallographer [22].

By now about two-thirds of protein structures are solved by molecular replacement [10] and, as the Protein Data Bank continues to expand, the method can only become more prominent. The rise in molecular replacement is also fuelled, in large part, by improvements in the algorithms, from model preparation through the molecular replacement search algorithms and on to the methods used to complete structures from poor starting models.

11.2 Model Preparation

To carry out molecular replacement, it is necessary to find a template (a related structure in the PDB) and then, possibly, to modify this template to be more similar to the target structure in the unknown crystal. Until a few years ago, the application of any molecular modelling protocol that changed the coordinates of the atoms tended to make the model worse for molecular replacement than the underlying template; in essence, there are many more ways to degrade the model than to improve it.

11.2.1 Model Trimming

One simple way to improve a template is to trim off the parts that are not expected to be conserved in the target, such as a domain or a large surface loop. At times it has been popular to trim back all the side chains to give a poly-Ala model, avoiding uncertainty about side-chain conformation; we find, in general, that this is too extreme and throws away useful signal.

Schwarzenbacher et al. [24] carried out a careful study of model trimming and drew two important conclusions. First, it is generally better to leave the side chains of conserved residues in the model, because their conformation is likely to be conserved as well, but to trim back non-identical side chains (and non-conserved surface loops). Even for non-identical residues, the first torsion angle is often conserved, so it is usually a good idea to keep the gamma atom of the residue. Second, as the sequence identity drops, it becomes essential to use the best possible sequence alignment, such as one obtained by profile-profile alignment methods, so that the right side chains and surface loops are actually modified.

Another form of model preparation is carried out in *MOLREP* [28]. Rather than simply deleting uncertain side chains, their B-factors can be inflated to reduce their influence on the calculation in a more subtle fashion [9].

The program *Sculptor* [3] combines these approaches and allows a number of different model preparation protocols to be tested. Side chains and loops can be trimmed in different ways, and B-factors can be adjusted according to surface accessibility, local sequence conservation, or a combination of both. By carrying out a series of molecular replacement calculations with a number of different variations on the starting template, the overall success rate can be increased significantly.

11.2.2 Molecular Modelling

In recent years, the sophistication of molecular modelling algorithms has finally reached the point where the starting templates can be improved for molecular replacement. Impressive results have been obtained using the *Rosetta* modelling package to improve starting models derived from NMR experiments or from the crystal structures of homologues [17].

11.2.3 Ab Initio Modelling

In fact, even an *ab initio* model created by *Rosetta* in a blind structure prediction test was shown to be sufficiently accurate to be used successfully for molecular replacement [17]. The computational resources required to fold *ab initio* models of this level of accuracy are substantial, but it has subsequently been shown that, at least in favourable cases, *ab initio* folding methods making a more modest use of CPU time can also succeed [20].

11.2.4 Ensembles

As sequence identity drops, structures become less similar and the success rate of molecular replacement also drops. However, there is also often a greater number of choices of model at a lower sequence identity level. By collecting these into an ensemble, in which the conserved features are enhanced and the variable features are downweighted, the success rate can again be boosted. The likelihood framework, discussed below, allows a statistical weighting of the contributions of members of an ensemble, which can be helpful [18].

The success rate can also be enhanced by trimming off surface loops that are not conserved among members of the ensemble, leaving a conserved core. This was essential, for instance, in solving the structure of angiotensinogen using a collection of models with about 20 % sequence identity (Fig. 11.1; [29]). An automated

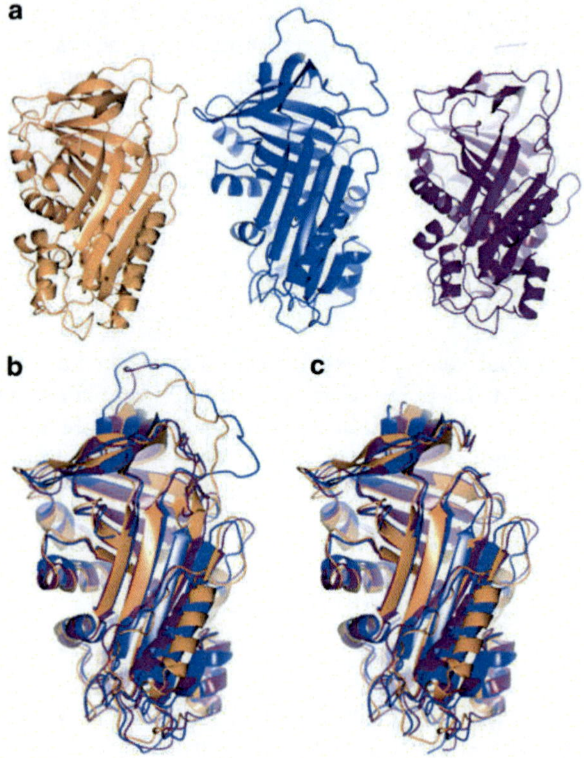

Fig. 11.1 Solving the structure of angiotensinogen [29] with a trimmed ensemble. (**a**) Individual structures of heparin cofactor II, α1-antitrypsin and thyroxine-binding globulin. (**b**) Ensemble of superimposed structures. (**c**) Trimmed ensemble. Only the molecular replacement search with the trimmed ensemble gave a clear solution

trimming option has been implemented in the *Ensembler* program (Bunkóczi and Read unpublished), along with a robust multiple-superposition method that optimises the superposition of the conserved core.

11.3 Molecular Replacement Calculations

In principle, molecular replacement is a 6*n*-dimensional search to find the orientations and positions of *n* models, but such a large space is impractical to search exhaustively. One approach is to use stochastic methods such as genetic algorithms (*EPMR*; [8]) or Monte Carlo (*QoS*; [5]) to search in the full space. However, most molecular replacement programs, such as our program *Phaser*, break the problem down into a series of 3D searches with rotation functions to find the orientation of a molecule and translation functions to find its position. For problems where the model is sufficiently accurate to yield a useful map, the signal in the individual searches is usually strong enough that the correct solution at each step is found in a relatively short list of plausible partial solutions. This enables a tree-search-with-pruning strategy [14].

11.3.1 Likelihood

Traditional molecular replacement calculations were based on the properties of the Patterson map, but the use of likelihood scores has a number of advantages [18]: the influence of data at different resolutions is weighted sensibly based on the expected quality of the model, information from partial models can be taken into account, and the likelihood score can be used robustly to rank different potential solutions, which is useful for automation strategies.

The molecular replacement likelihood functions [18] are relatively expensive to compute but, fortunately, it is possible to derive good approximations that can be computed efficiently. Likelihood-based fast rotation [25] and fast translation [13] functions can be used to generate a short list of plausible solutions, which can then be ranked using the full likelihood score.

The idea of likelihood is simple: models or hypotheses can be tested by how well they agree with the measured data. Likelihood gives a probabilistic measure of agreement with the data, *i.e.* likelihood measures the probability that the set of data would have been measured, given the model and any associated uncertainties in the model parameters or the data. A more in-depth understanding can be obtained from the review on likelihood in crystallography by McCoy [11].

11.3.2 Automation

A molecular replacement calculation can be thought of as testing a series of hypotheses about the orientation and then the position taken by molecules in the crystal. Since likelihood is an effective measure to rank hypotheses, it lends itself to decision-making in an automated molecular replacement strategy. As noted above, *Phaser* uses a tree-search-with-pruning strategy. Heuristic rules (*e.g.* the correct solution is usually above 75 % of the distance between the mean and the top in any step of the search) are used to keep a list of plausible solutions and discard the less plausible ones. Multiple alternative models for a component can be evaluated at the same time, and the best one can be chosen by its likelihood score. Even different possible choices of space group can be evaluated. If the crystal contains a complex of different components, then the search order for the different components can be evaluated by considering how well each component would explain the data.

Increasingly, molecular replacement is being implemented as part of a pipeline, such as *MrBUMP* [6], *BALBES* [10] and *AutoMR* in the *Phenix* package [1]. Ideally, such pipelines are started by supplying only the diffraction data and the sequences of the proteins in the crystal, and then they fetch the template structures, modify them, carry out molecular replacement, and even follow that with automated building and refinement.

11.3.3 Pathologies

Experience has shown that likelihood targets are more sensitive than the traditional Patterson-based methods in finding the solution. However, this sensitivity is a double-edged sword, because likelihood is also more sensitive to errors in the assumptions used to derive the likelihood targets. One such assumption is that the crystal diffracts isotropically (*i.e.* equally strongly in all directions in reciprocal space). Likelihood-based molecular replacement is severely degraded by the effects of anisotropic diffraction, unless a correction is applied. Fortunately, likelihood also provides the tools to characterise the anisotropy and correct for its effects [14], and anisotropic diffraction no longer presents a problem.

Similarly, the presence of translational non-crystallographic symmetry (tNCS) also severely violates the assumptions of the original likelihood targets. In tNCS, two or more copies of the molecule are found in the same orientation in the crystal. Depending on their relative position, and how this relates to the Bragg planes for a particular reflection, they can scatter in phase (leading to exceptionally strong reflections) or out of phase (leading to exceptionally weak reflections). Until recently, the presence of tNCS was one of the leading causes for *Phaser* to fail in cases that would otherwise be expected to succeed. Methods to characterise tNCS and account for its statistical effects on the diffraction pattern have now been implemented in *Phaser*, dramatically increasing success rates in these cases (McCoy and Read unpublished).

11.4 Model Completion

When the available models are poor (typically low sequence identity) or incomplete, or the resolution of the data is limited, it has frequently been found that the molecular replacement problem can be solved but the electron density maps are too poor to see what needs to be done to complete the structure. Fortunately, a number of recent developments have markedly improved this situation.

11.4.1 Morphing and Other Smooth Deformations

Looking at distant homologues, one often sees that the basic fold is preserved, but the relative positions and orientations of structural elements have changed slightly. Even though such movements might be difficult to see in a density map at the local level, there are weak signals that can be combined over a larger region. Tom Terwilliger (personal communication) has developed a “morphing” algorithm that takes advantage of these signals. It looks for rigid-body movements that would improve the fit to density of a window of residues along the chain, and then applies that shift to the central residue in the window. By sliding the window along the chain, a smooth transformation ("morphing") of the model is achieved. In a number of test cases, this has led to sufficient improvement in the model, and thus the phases, that further improvements to the model become clear in the density.

Refinement methods that lead to smooth deformations, such as the jelly-body method [15] or *DEN* refinement [23] are also very helpful in the initial stages of refinement from a poor molecular replacement model. This is illustrated clearly in a test case using *DEN* refinement to complete a structure that had been stuck in refinement [2].

11.4.2 Rosetta Modelling

In particularly difficult cases, the largest convergence radius in rebuilding and refining from a poor model is probably achieved by using the advanced modelling algorithms in *Rosetta* [4], combining the *Rosetta* energy functions with electron density fit scores to build into noisy density maps. The *phenix.mr_rosetta* pipeline [27] provides a convenient interface giving access to *Rosetta* modelling, molecular replacement in *Phaser*, and automated building and refinement in *AutoBuild* [26].

11.4.3 Arcimboldo

Completing the structure starting from a highly incomplete model presents similar challenges to starting from a poor but relatively complete model. The *Arcimboldo* procedure [21] is discussed elsewhere in greater detail by Isabel Usón. Briefly, this exploits the power of density modification and automated building algorithms to extend incomplete models comprising only a few helices, placed using *Phaser*.

11.5 Combined Methods

11.5.1 MR-SAD

A molecular replacement model can be used as a starting point for the computation of log-likelihood-gradient (LLG) maps to find anomalous scatterers using single-wavelength anomalous diffraction (SAD; [12, 19]). In some cases, the anomalous signal may be too weak to find the anomalous scatterers with *ab initio* substructure determination methods, but nonetheless significant phase information can be obtained once the sites have been found using SAD LLG maps, even if those are based on a poor molecular replacement model. In other cases, locating anomalous scatterers in a refined model can be a valuable tool for identifying unknown components, such as bound ions.

11.5.2 Using Density as a Model

Proteins frequently crystallise in multiple crystal forms and, at times, experimental phase information can only be obtained for one of these forms. In such cases, the electron density can be cut out of one map and used as a molecular replacement model to solve another crystal form.

Such a procedure was used in solving the structure of angiotensinogen [29]. A poor electron density map was available for crystals of the human form of this protein, combining information from molecular replacement with an ensemble of distant models at 3.3 Å resolution with SAD phases from a $GdCl_3$ derivative at 4 Å resolution. Molecular replacement with the same ensemble model did not succeed in solving the structures of crystals from rat or mouse angiotensinogen, but electron density extracted from the map of the human form did give a clear solution for two copies of angiotensinogen in one of the rat crystal forms. In turn, averaged density from this rat crystal form could be used to find two copies in the second rat crystal form, allowing 4-fold multi-crystal averaging to be initiated between the two rat crystal forms.

Molecular replacement serves two purposes for multi-crystal averaging, in such cases: it defines the rotation and translation operators that superimpose the density in one crystal on the density in the other crystal, and it provides initial phases for the second crystal form.

11.6 Future Developments

There has been rapid progress in recent years in the power and reach of molecular replacement, and there are good reasons to believe that this will continue. As density modification and model-building algorithms improve, it will become possible to solve structures from even less complete and less accurate starting points. Improvements in our understanding of the likelihood targets will feed into better automation strategies, both by allowing us to predict how good the model must be to have a chance of success, and by providing measures of confidence in partial solutions obtained along the solution path. Even if there were no improvements in the algorithms, the continued rapid growth of the PDB would ensure that there are good models for an ever-expanding set of targets.

Acknowledgments Our work on *Phaser* is supported by awards from the Wellcome Trust (082961/Z/07/Z) and the NIH (Grant No. P01GM063210). We are grateful to users who provide us with bug reports and challenging problems that push the limits.

References

1. Adams PD, Afonine PV, Bunkóczi G, Chen VB, Davis IW, Echols N, Headd JJ, Hung L-W, Kapral GJ, Grosse-Kunstleve RW, McCoy AJ, Moriarty NW, Oeffner R, Read RJ, Richardson DC, Richardson JS, Terwilliger TC, Zwart PH (2010) PHENIX: a comprehensive Python-based system for macromolecular structure solution. Acta Crystallogr D 66:213–221
2. Brunger AT, Das D, Deacon AM, Grant J, Terwilliger TC, Read RJ, Adams PD, Levitt M, Schröder GF (2012) Application of DEN-refinement and automated model-building to a difficult case of molecular replacement phasing: the structure of a putative succinyl-diaminopimelate desuccinylase from *Corynebacterium glutamicum*. Acta Crystallogr D 68:391–403
3. Bunkóczi G, Read RJ (2011) Improvement of molecular replacement models with *Sculptor*. Acta Crystallogr D 67:303–312
4. DiMaio F, Terwilliger TC, Read RJ, Wlodawer A, Oberdorfer G, Wagner U, Valkov E, Alon A, Fass D, Axelrod HL, Das D, Vorobiev SM, Iwaï H, Pokkuluri PR, Baker D (2011) Improving molecular replacement by density- and energy-guided protein structure optimization. Nature 473:540–543
5. Glykos NM, Kokkinidis M (2001) Multidimensional molecular replacement. Acta Crystallogr D 57:1462–1473
6. Keegan RM, Winn MD (2008) MrBUMP: an automated pipeline for molecular replacement. Acta Crystallogr D 64:119–124
7. Kendrew JC, Bodo G, Dintzis HM, Parrish RG, Wyckoff H, Phillips DC (1958) A three-dimensional model of the myoglobin molecule obtained by x-ray analysis. Nature 181:662–666
8. Kissinger CR, Gehlhaar DK, Fogel DB (1999) Rapid automated molecular replacement by evolutionary search. Acta Crystallogr D 55:484–491
9. Lebedev AA, Vagin AA, Murshudov G (2008) Model preparation in MOLREP and examples of model improvements using X-ray data. Acta Crystallogr D 64:33–39
10. Long F, Vagin AA, Young P, Murshudov G (2007) BALBES: a molecular-replacement pipeline. Acta Crystallogr D 64:125–132
11. McCoy AJ (2004) Liking likelihood. Acta Crystallogr D 60:2169–2183
12. McCoy AJ, Read RJ (2010) Experimental phasing: best practice and pitfalls. Acta Crystallogr D 66:458–469
13. McCoy AJ, Grosse-Kunstleve RW, Storoni LC, Read RJ (2005) Likelihood-enhanced fast translation functions. Acta Crystallogr D 61:458–464
14. McCoy AJ, Grosse-Kunstleve RW, Adams PD, Winn MD, Storoni LC, Read RJ (2007) *Phaser* crystallographic software. J Appl Crystallogr 40:658–674
15. Murshudov GN, Skubák P, Lebedev AA, Pannu NS, Steiner RA, Nicholls RA, Winn MD, Long F, Vagin AA (2011) REFMAC5 for the refinement of macromolecular crystal structures. Acta Crystallogr D 67:355–367
16. Perutz MF, Rossmann MG, Cullis AF, Muirhead H, Will G, North AC (1960) Structure of haemoglobin: a three-dimensional Fourier synthesis at 5.5-Å resolution, obtained by X-ray analysis. Nature 185:416–422
17. Qian B, Raman S, Das R, Bradley P, McCoy AJ, Read RJ, Baker D (2007) High-resolution structure prediction and the crystallographic phase problem. Nature 450:259–264
18. Read RJ (2001) Pushing the boundaries of molecular replacement with maximum likelihood. Acta Crystallogr D 57:1373–1382
19. Read RJ, McCoy AJ (2011) Using SAD data in *Phaser*. Acta Crystallogr D 67:338–344

20. Rigden DJ, Keegan RM, Winn MD (2008) Molecular replacement using ab initio polyalanine models generated with ROSETTA. Acta Crystallogr D 64:1288–1291
21. Rodríguez DD, Grosse C, Himmel S, González C, de Ilarduya IM, Becker S, Sheldrick GM, Usón I (2009) Crystallographic *ab initio* protein structure solution below atomic resolution. Nat Methods 6:651–653
22. Rossmann MG (1972) The molecular replacement method. Gordon & Breach, New York
23. Schröder GF, Levitt M, Brunger AT (2010) Super-resolution biomolecular crystallography with low-resolution data. Nature 464:1218–1222
24. Schwarzenbacher R, Godzik A, Grzechnik SK, Jaroszewski L (2004) The importance of alignment accuracy for molecular replacement. Acta Crystallogr D 60:1229–1236
25. Storoni LC, McCoy AJ, Read RJ (2004) Likelihood-enhanced fast rotation functions. Acta Crystallogr D 60:432–438
26. Terwilliger TC, Grosse-Kunstleve RW, Afonine PV, Moriarty NW, Zwart PH, Hung L-W, Read RJ, Adams PD (2008) Iterative model building, structure refinement and density modification with the Phenix AutoBuild wizard. Acta Crystallogr D 64:61–69
27. Terwilliger TC, DiMaio F, Read RJ, Baker D, Bunkóczi G, Adams PD, Grosse-Kunstleve RW, Afonine PV, Echols N (2012) phenix.mr_rosetta: molecular replacement and model rebuilding with Phenix and Rosetta. J Struct Funct Genomics 13(2):81–90. doi:10.1007/s10969-012-9129-3
28. Vagin A, Teplyakov A (1997) MOLREP: an automated program for molecular replacement. J Appl Crystallogr 30:1022–1025
29. Zhou A, Carrell RW, Murphy MP, Wei Z, Yan Y, Stanley PLD, Stein PE, Broughton Pipkin F, Read RJ (2010) A redox switch in angiotensinogen modulates angiotensin release. Nature 468:108–111

Chapter 12
Phasing Through Location of Small Fragments and Density Modification with ARCIMBOLDO

Isabel Usón, Claudia Millán, Massimo Sammito, Kathrin Meindl, Iñaki M. de Ilarduya, Ivan De Marino, and Dayté D. Rodríguez

Abstract The International School of Crystallography held a course at the Ettore Majorana Centre in Erice in 1997 on "Direct methods for solving macromolecular structures". In those days, Dual Space recycling methods, introduced by Hauptman and Weeks had allowed the breakthrough of extending atomic resolution phasing to macromolecules. The largest previously unknown macromolecule to have been phased by such methods was hirustasin at 1.2 Å resolution, with 400 independent atoms. At the time of the meeting, triclinic lysozyme at 1.0 Å, with 1,001 equal atoms was solved with SHELXD. Fifteen years later, *ab Initio* phasing has pushed the size and resolution limits of the problems it can tackle. Macromolecules with several thousands of atoms in the asymmetric unit can be solved from medium resolution data. One of the successful approaches is the combination of fragment location with the program PHASER and density modification with the program SHELXE in a supercomputing frame. The method is implemented in the program ARCIMBOLDO, described in this chapter.

Keywords *Ab initio* phasing • Fragment search • Molecular replacement • Density modification • Supercomputing

12.1 Introduction

Crystallography provides a view into the three-dimensional structure of biological macromolecules that is unsurpassed in its degree of detail and precision by any other structural technique. Nevertheless, the structural model product of the

I. Usón (✉)
ICREA at Department of Structural Biology, IBMB-CSIC, Baldiri Reixach, 13-15, Barcelona, Spain
e-mail: uson@ibmb.csic.es

C. Millán • M. Sammito • K. Meindl • I.M. de Ilarduya • I. De Marino • D.D. Rodríguez
Department of Structural Biology, IBMB-CSIC, Baldiri Reixach, 13-15, Barcelona, Spain

R. Read et al. (eds.), *Advancing Methods for Biomolecular Crystallography*, NATO Science for Peace and Security Series A: Chemistry and Biology, DOI 10.1007/978-94-007-6232-9_12,

crystallographic analysis cannot be directly calculated from the data. Even with optimal experimental data the phase problem underlies the determination as only the diffracted intensities and not the phases are amenable to the current X-ray diffraction experiment. Phases are essential for structure determination. Providing starting phases to establish an initial model is often a bottleneck in structure determination, demanding previous structural knowledge on the structure to be determined [16] or an additional experimental effort to derivatise the macromolecule or its crystals and collect data at suitable wavelengths. Phasing from the native amplitudes of one dataset, exploiting general assumption but not particular stereochemical knowledge, is what is termed *ab Initio*.

The modest resolution to which most biological crystals diffract further complicates the problem, making the construction of an objective model more difficult, as the parameters describing it are not enough overdetermined by the experimental data.

Macromolecular crystallography is computationally intensive. In the midst of the vertiginous increase in computation speed experienced in the last years, crystallography – unlike modelling – has largely turned its back on the use of large-scale parallelization. In 2009, our group launched the multisolution parallel phasing software ARCIMBOLDO. The present chapter will describe the antecedents, achievements and prospects of this method.

12.2 *Ab Initio* Solution of Macromolecular Structures and Structure Determination in a Supercomputing Frame

In the field of chemical crystallography, where most structures are composed by less than 200 independent atoms, crystal structures are generally solved by direct methods [9, 10]. Based on probabilistic relations and the possibility of evaluating many starting phase sets through reliable figures of merit, they provide an initial model that is derived exclusively from the experimental intensities measured on a native crystal, without the need of previous stereochemical knowledge or additional experimental data and are therefore termed *ab initio* methods.

In the field of macromolecular structures the situation is radically different: two barriers sever both fields. In the first place, the much larger number of independent atoms increases the complexity of the problem for biological molecules. On top of this, crystals of biological molecules tend to diffract to a much lower resolution than is required for the success of direct methods, even in the case of small molecules [13]. This shortcoming is essentially derived from the nature of macromolecular crystals: containing a large proportion of disordered solvent (typically around 50 %), their periodicity is less perfect, their size smaller, their diffraction signal to noise ratio lower, their resistance to mechanical stress and radiation damage impaired, etc.

From these two barriers, lack of resolution appeared to be the most resilient one. Indeed, thanks to the pioneering work of Hauptman and Weeks in the early 1990s the

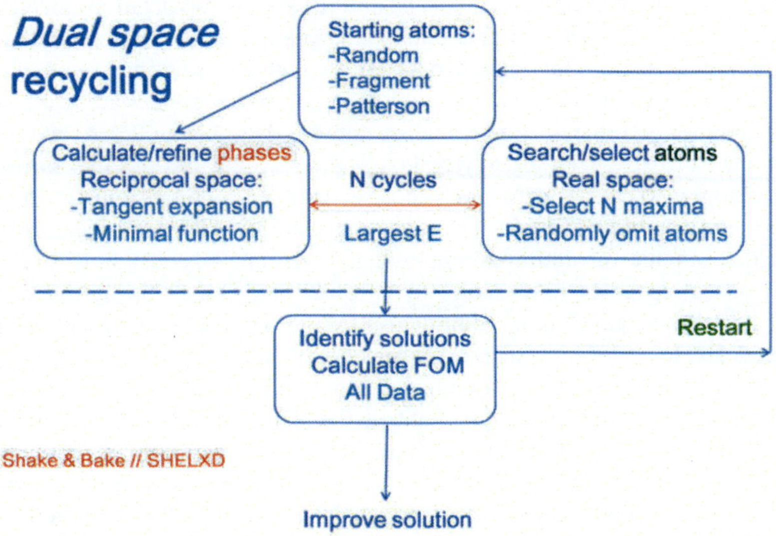

Fig. 12.1 Schematic representation of the dual-space recycling algorithm. Starting from a trial set of atoms, phases are generated. These are refined applying direct methods probabilistic formulae. From the modified phases, an electron density map can be computed, from which a new set of atoms is selected. This strategy is iterated, relying only on the strongest E values. Finally, a figure of merit is calculated using all data and the partial solution is then optimised or discarded. The process is started once again. One of the possible starts to the program is using fragments of known geometry. Some of the structures solved, like cycloamiloses were only tackled when starting from a small but accurate piece such as a diglucose fragment

use of ab initio methods could be extended to the determination of macromolecular structures diffracting to atomic resolution [12] affording a completely model bias free model. Figure 12.1 schematically illustrates the algorithm underlying dual-space recycling methods [20, 21, 24]. These have made it possible to extend the power of traditional direct methods from 200 independent atoms to over 1,000 in the case of equal atom structures (the most common case in biological molecules) while the presence of elements heavier than sulphur, such as metals in the active centre or in structural roles, constitutes a favourable circumstance, which has allowed to successfully solve structures with over 2,000 independent atoms, provided data to atomic resolution, around 1 Å are available.

In macromolecular crystallography, where atomic resolution is exceptional, these methods have found their main application in the determination of anomalous scattering or heavy atom substructures at medium resolution in SAD, MAD or SIRAS experiments [17, 27]. For the substructures, the sites are resolved even at low resolution. Substructures up to 160 independent selenium atoms have thus been determined; while conventional Patterson and direct methods run into difficulties in the location of more than 12 selenium sites (see http://www.hwi.buffalo.edu/SnB/StructureDetails.htm).

The technological advances in beamline and crystallization technologies, together with the increase in data collection time availability and beamline speed are leading to better determined datasets to higher resolution. Even if atomic resolution is likely to remain exceptional, it is foreseeable a significant increase in the number of projects where high resolution data, below 2.0 Å, will become available. Currently, they already make up almost 50 % of the PDB entries, with an additional 30 % corresponding to structures diffracting from 2.0 to 2.5 Å. Under such conditions, dual-space recycling methods are not successful, as they are tied to atomic resolution and applying the atomicity constraint in either real or reciprocal space is not useful anymore. To overcome this barrier and push the limit to lower resolution, additional information or alternative constraints are required. Exploiting the presence of heavy atoms in the structure [6], extrapolating unmeasured reflections up to atomic resolution ([5, 25, 28]); and the application of density modification techniques such as the VLD algorithm [2–4], have proven useful. Our group has pursued an alternative, based on enforcing stereochemical knowledge in the form of mainchain fragments of the predicted secondary structure elements, rather than atomicity, especially alpha-helices. This has allowed the solution of over a score previously unknown crystal structures where conventional approaches in the hands of competent crystallographers had failed. The fact that macromolecules are made up of building blocks of known geometry that can be predicted from their amino acid sequence, such as alpha helices, can be enforced as an alternative to atomicity, as a means of bringing in prior stereochemical information. One of the problems atomic resolution ab Initio methods suffer from at lower resolution is that the figures of merit are no longer reliable. Indeed, the E-based correlation coefficient [8] of partial solutions is invariably high for the expected number of atoms. In a multisolution frame, there is no use in producing correct solutions if they cannot be discriminated from among the rest, as a manual check of all solutions would not be practicable. Fragment location in combination with density modification has enabled the solution of previously unknown protein structures at resolutions up to 2 Å, and to identify the correct phases on the figures of merit characterizing the partial mainchain trace of the resulting map through its CC and number of residues traced [19]. This method has been implemented in the program ARCIMBOLDO [14], which combines multisolution location of small (10–14 amino acids), extremely accurate models such as poly-alanine alpha helices, with the program PHASER [11] and density modification and autotracing with the program SHELXE [18]. Despite this procedure being computationally intensive, it can be parallelized and run on a grid or a multiprocessor cluster. In our case, we have set up a local grid of linux computers running Condor [23] as well as a Condor grid on the supercomputer Calendula at FCSCL (http://www.fcscl.es). So the processes are distributed within a large pool of CPUs.

The program is named after the Italian painter Giuseppe Arcimboldo (1527–1593) who assembled portraits from fruits and vegetables. Analogously, the method tests many hypotheses assembled from secondary structure fragments and while most of them remain a "still life", density modification is effective in revealing and

Table 12.1 Summary of previously unsolved structures phased by ARCIMBOLDO

Protein from	Space group	Nres	Fragment	d (Å)
P. Czabotar*	$P3_121$	120	1H14	1.30
M. Graille	$P2_1$	310	1H16	1.45
K. V. Hecke	P432	165	2H14	1.60
J. Hermoso	$P6_1$	50	1H10	1.70
J. M. Pereda	C2	240	Composite frags 2H17	1.70
D. Cavalcante Hissa*, K. Gruber	$P2_1$	204	2H14	1.70
K. Zeth	$P2_1$	428	Composite frags 2H16	1.70
S. Trakhanov	$P2_12_12_1$	144	1H14	1.75
V. Arcus (4E1P) [22]	$P2_1$	112	2H12	1.80
K. Zeth	$P3_121$ twin	74	1H20	1.90
K. Zeth (4AEQ) [26]	$C222_1$	90	1H12	1.90
R. Bunker	P1	200	Model helices	1.95
S. Becker (3GHW)	$P2_1$	222	3H14	1.95
A. Thorn, G.M. Sheldrick (3SZS)	I422	327	2nmr31	1.95
J. Hermoso (2Y8P) [1]	$C222_1$	378	2hom85	2.00
X. Gomis-Rüth	$P2_12_12_1$	700	Frags + Se-MAD	2.00
N. Verdaguer	$P6_322$	50	3H14	2.10
O. Mayans	$P2_1$	240	Helices with SC	2.10
C. Artola, J. Hermoso	$P2_1$	700	Frags, modelling + BUSTER	2.70
N. Valadares, R. Garrat	Pseudo-merohedral	60–240	Coiled coils, twins	1.60–2.80

* Peter Czabotar and Denise Cavalcante Hissa are the participants whose structures were solved with ARCIMBOLDO during or as a result of the crystallographic school.
During this 45th edition of the Crystallographic School dedicated to "Present and future of biomolecular chemistry" the unsolved structure of one of the participants, Peter E. Czabotar from the WEHI institute in New Zealand, was phased with ARCIMBOLDO. Diffracting to 1.3 Å and composed of 120 amino acids, a single 14 amino acids polyalanine helix was enough to phase the whole structure. Calculations initiated at the time on the structural problem from other participant, Denise Hissa Cavalcante, from the University of Graz, culminated later on in the solution of a 204 amino acid structure at 1.7 Å

identifying the true portrait of the protein being solved. ARCIMBOLDO can be downloaded free for academics (http://chango.ibmb.csic.es/ARCIMBOLDO).

Phasing the 220 aminoacid structure of PRD2 from data to a resolution of 1.95 Å required location three helices of 14 alanines each. Three of the 1,500 generated partial solutions were correct and accurate enough to lead to the complete solution. These and other cases are summarized in Table 12.1. During this 45th edition of the Crystallographic School dedicated to "Present and future of biomolecular chemistry" the unsolved structure of a participants was phased in an analogous way and a second structure initiated during the meeting was overcome shortly after, when better data became available.

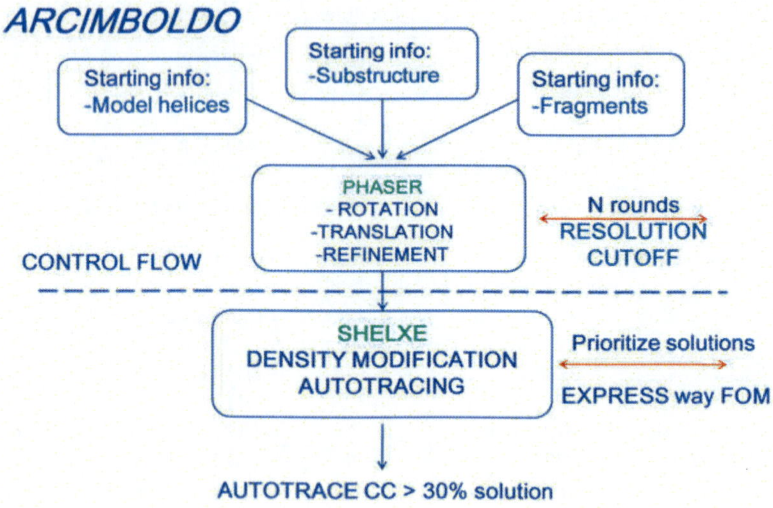

Fig. 12.2 Scheme of the multisolution parallel phasing program ARCIMBOLDO. Beyond the initial *Ab Initio* approach, complementary sources of information may be combined within the supercomputing phasing process

Moving on from its first use for the *ab initio* case, ARCIMBOLDO has expanded to other scenarios (Fig. 12.2) enabling it to tackle larger structures on poorer data [15]. User friendliness improved through the incorporation of a GUI to help the user setup and parameterize an ARCIMBOLDO run. The same frame as that used for *ab initio*, allows the exploitation of other sources of previous stereochemical knowledge, such as low homology models or experimental phases from derivatives that are too noisy to be interpretable on their own. At the same time, analysis of the figures of merit characterizing partial structures, and their geometry and phases allows to control the flow of the program by devising an "express lane" to give priority to partial solutions more likely to succeed and reduce the number of jobs to be computed. This is needed as it is possible to run 10,000 jobs in parallel on the CONDOR grid but the next order of magnitude becomes intractable.

Table 12.1 shows most of the new structures phased with ARCIMBOLDO. They are ordered according to resolution and present different complexity derived from their size and available information. The fragments used are given. These cases illustrate the various algorithms incorporated to the ARCIMBOLDO framework. Either integrated within the ARCIMBOLDO code or in separate programs, as ancillary procedures.

From a practical point of view fundamental aspects to be considered are [15]:

12.2.1 Resolution Requirement

In general, complete data of good quality to a resolution of 2 Å is required for *ab Initio* solution. Often, the resolution limits for the fragment search stages are limited to resolutions between 2.5 and 2.1 Å anyway for the rotation function and occasionally for the translation function. This speeds up the procedure and seems to be as effective. Still, the density modification expansion from a very reduced part of the total scattering mass is greatly enhanced through the availability of the higher resolution, and its success is instrumental to identifying the correct solutions. Otherwise, more previous stereochemical information or experimental phase information may help. Nevertheless, some of the phased structures summarized in the table correspond to pathological cases such as non-merohedral and merohedral or pseudo-merohedral twins.

12.2.2 Strategy Control Within ARCIMBOLDO

Running many hypotheses in parallel is prone to run out of control. Even if ARCIMBOLDO can run 10,000 jobs in parallel, 50,000 become intractable. Besides, computing time should be invested where it may succeed. Thus, an "express lane" has been .established to prioritize the most likely to succeed hypotheses. Thus, the flow control of the procedure checks average values of figures of merit to terminate part of the partial solutions or pursue their expansion. It is possible to search for different fragments, either sequentially or by evaluating different fragments, including extensive libraries, as alternatives at a given stage. It is difficult to give ideal parameters that will work for all cases but the default values provided correspond to our experience so far.

12.2.3 Combining Fragments and Experimental Phases, Strategy Control Within ARCIMBOLDO

Experimental phase information should be exploited whenever possible. In particular, phases derived from MAD/SAD or SIRAS, even if noisy or limited to the low resolution data can be very effective. It is not trivial to combine weak phase information from different sources because in most spacegroups they may relate to a different origin. There are three possibilities:

Searching for anomalous fragments against MAD or SAD data,
Searching for normal fragments once an anomalous substructure is known,
Determining the anomalous substructure once initial phases are known from a partial fragment structure finding the scatterers in the anomalous map and

combining the experimental and fragment phases within the density modification part. It requires some recycling to optimise the substructure but it is effective and much faster than autotracing.

12.2.4 Clustering Fragments

The most effective way of exploiting information already present and increasing efficiency is to fuse partial solutions at early stages, clustering phases in reciprocal space or fragments in real space. The intrinsic symmetry of the helix has to be taken into account, as well as its occurrence, several times within one structure. Thus, clustering runs the risk of merging solutions incorrectly as it depends on the origin shift.

12.3 Larger Structures: SHREDDER, Supercomputer Calendula

Low homology models tend to be unsuccessful for Molecular Replacement. Recent successes have been attained by their improvement through modelling with the software suite Rosetta [7]. Our method exploits very small fragments but requires high accuracy, thus our approach is to shred low homology models to produce manifold fragments of comparable size and let them compete against each other within the ARCIMBOLDO frame.

12.4 Composite Fragments: A Custom Builder for Libraries

The program BORGES allows to extract customized, secondary structure fragments from the PDB (e.g. all mainchain variations of the theme helix-turn-helix within given length and orientation constraints). The program further clusters the libraries, scoring and refining them against the experimental data.

12.4.1 Lower Resolution: BUSTER Refinement, SHELXE from its HL-Coefficients and Modeling

When resolution beyond 2 Å is not available, density modification of phases derived from very small fragments is not likely to succeed anymore. In this case, more sophisticated phase treatments, through Maximum-Likelihood refinement, better accounting for missing information and errors, provide better starts to the autotracing algorithm. Also, modelling side-chains onto the mainchain leads to more complete fragments and thus better starts. As it is not possible to know the sequence of the partial traces, again massive calculations consistent with a secondary structure prediction have to be generated, filtered to discard those with clearly unfavourable energies and refined in parallel.

12.4.2 Problem Cases: Twins

Coiled coil structures would appear to present ideal cases as they are completely helical. Still, they are deceptively simple as they tend to be markedly anisotropic and frequently twinned. We have encountered a number of cases where twinning involved two domains with the same mainchain structure, showing higher symmetry than the sidechains. Solution is favourable but sequence assignment and model building and refinement is complicated, especially if resolution is poor.

12.5 Future Developments

Current plans are centered in setting up a public ARCIMBOLDO web server that will accept data from external users, generate input for structure solution, run it on our grid and return the best solutions. Given the computational requirements, this is being developed in collaboration with the supercomputer FCSCL in León.

Acknowledgments Our work is supported by the Spanish MICINN, CDTI and CSIC (Grants BIO2009-10576; IDC-20101173; predoctoral grants DR, IDM, IMdI; JdC to KM), Generalitat de Catalunya (2009SGR-1036).

References

1. Artola-Recolons C, Carrasco-López C, Llarrull LI, Kumarasiri M, Lastochkin E, Martínez de Ilarduya I, Meindl K, Usón I, Mobashery S, Hermoso JA (2011) High resolution crystal structure of MltE, an outer membrane-anchored endolytic peptidoglycan lytic transglycosylase from *Escherichia coli*. Biochemistry 50:2384–2386
2. Burla MC, Caliandro R, Cammalli M, Carrozzini B, Cascarano GL, De Caro L, Giacovazzo C, Polidori G, Siliqi D, Spagna R (2007) IL MILIONE: a suite of computer programs for crystal structure solution of proteins. J Appl Crystallogr 40:609–613
3. Burla MC, Giacovazzo C, Polidori G (2010) From a random to the correct structure: the VLD algorithm. J Appl Crystallogr 43:825–836
4. Burla MC, Carrozzini B, Cascarano GL, Giacovazzo C, Polidori G (2011) Advances in the VLD algorithm. J Appl Crystallogr 44:1143–1151

5. Caliandro R, Carrozzini B, Cascarano GL, De Caro L, Giacovazzo C, Siliqi D (2005) Phasing at a resolution higher than the experimental resolution. Acta Crystallogr D 61:556–565
6. Caliandro R, Carrozzini B, Cascarano GL, De Caro L, Giacovazzo C, Mazzone A, Siliqi D (2008) Ab initio phasing of proteins with heavy atoms at non-atomic resolution: pushing the size limit of solvable structures up to 7890 non-H atoms in the asymmetric unit. J Appl Crystallogr 41:548–553
7. DiMaio F, Terwilliger TC, Read RJ, Wlodawer A, Oberdorfer G, Wagner U, Valkov E, Alon A, Fass D, Axelrod HL, Das D, Vorbiev SM, Iwaï H, Pokkuluri PR, Baker D (2011) Improved molecular replacement by density- and energy-guided protein structure optimization. Nature 473:540–543
8. Fujinaga M, Read RJ (1987) Experiences with a new translation function program. J Appl Crystallogr 20:517–521
9. Hauptman H and Karle J (1953) Solution of the phase problem I. The centrosymmetric crystal. Am Crystallogr Assoc Monograph No. 3 Dayton, Ohio: Polycrystal Book Service
10. Karle J, Hauptman H (1956) A theory of phase determination for the four types of non-centrosymmetric spacegroups 1P222, 2P22, 3P12, 3P22. Acta Crystallogr 9:635–651
11. McCoy AJ, Grosse-Kunstleve RW, Adams PD, Winn MD, Storoni LC, Read RJ (2007) *Phaser* crystallographic software. J Appl Crystallogr 40:658–674
12. Miller R, DeTitta GT, Jones R, Langs DA, Weeks CM, Hauptman HA (1993) On the application of the minimal principle to solve unknown structures. Science 259:1430–1433
13. Morris RJ, Bricogne G (2003) Sheldrick's 1.2 Å rule and beyond. Acta Crystallogr D 59: 615–617
14. Rodríguez DD, Grosse C, Himmel S, González C, de Ilarduya MI, Becker S, Sheldrick GM, Usón I (2009) Crystallographic *ab initio* protein structure solution below atomic resolution. Nat Methods 6:651–653
15. Rodríguez DD, Sammito M, Meindl K, de Ilarduya MI, Potratz M, Sheldrick GM, Usón I (2012) Practical structure solution with ARCIMBOLDO. Acta Crystallogr D 68:336–343
16. Rossmann MG (1990) The molecular replacement method. Acta Crystallogr A 46:73–82
17. Schneider TR, Sheldrick GM (2002) Substructure solution with SHELXD. Acta Crystallogr D 58:1772–1779
18. Sheldrick GM (2002) Macromolecular phasing with SHELXE. Z Kristallogr 217:644–650
19. Sheldrick GM (2010) Experimental phasing with SHELXC/D/E: combining chain tracing with density modification. Acta Crystallogr D 66:479–485
20. Sheldrick GM, Hauptman H, Weeks CM, Miller R, Usón I (2001) Ab initio phasing. In: Rossman MG, Arnold E (eds) International tables for crystallography, vol F. Crystallography of biological macromolecules. Kluwer Academic Publishers, Dordrecht, pp 333–351
21. Sheldrick GM, Gilmore C, Hauptman H, Weeks CM, Miller R, Usón I (2011) Ab initio phasing. In: Arnold E, Himmel DM, Rossman MG (eds) International tables for crystallography, vol F. Crystallography of biological macromolecules. Kluwer Academic Publishers, Dordrecht, pp 413–432
22. Summers EL, Meindl K, Usón I, Mitra AK, Radjainia M, Colangeli R, Alland D, Arcus VL (2012) The structure of the oligomerization domain of Lsr2 from *Mycobacterium tuberculosis* reveals a mechanism for chromosome organization and protection. PLoS One 7:e38542
23. Tannenbaum T, Wright D, Miller K, Livny M (2002) Condor – a distributed Job scheduler. In: Sterling T (ed) Beowulf cluster computing with Linux. The MIT Press, Cambridge. ISBN 0-262-69274-0
24. Usón I, Sheldrick GM (1999) Advances in direct methods for protein crystallography. Curr Opinion in Struc Biol 9:643–648
25. Usón I, Stevenson CE, Lawson DM, Sheldrick GM (2007) Structure determination of the O-methyltransferase NovP using the 'free lunch algorithm' as implemented in SHELXE. Acta Crystallogr D 63:1069–1074

26. Usón I, Patzer SI, Rodríguez DD, Braun V, Zeth K (2012) The crystal structure of the dimeric colicin M immunity protein displays a 3D domain swap. J Struct Biol 178:45–53
27. Weeks CM, Adams PD, Berendzen J, Brünger AT, Dodson EJ, Grosse-Kunstleve RW, Schneider TR, Sheldrick GM, Terwilliger TC, Turkenburg MG, Usón I (2003) Automatic solution of heavy-atom substructures. Meth in Enzymol 374:37–83
28. Yao JX, Dodson EJ, Wilson KS, Woolfson MM (2006) ACORN: a review. Acta Crystallogr D 62:901–908

Chapter 13
SAD/MAD Phasing

Zbigniew Dauter

13.1 Introduction

Currently, the most common ways of solving novel macromolecular crystal structures are based on the anomalous signal provided by some atoms present in the investigated structures. They can be implemented as the Single- or Multi-wavelength Anomalous Diffraction (SAD or MAD) method. Instead of collecting diffraction data from the native crystal and a number of derivatives, as in the classic Multiple Isomorphous Replacement (MIR) approach, these techniques utilize one or more data sets, recorded from only one crystal containing suitable anomalous scatterers. Whereas with MIR the protein phases are estimated from the additional scattering of the heavy atoms present in the derivative crystals, in SAD and MAD they are calculated from the wavelength-dependent quantitative differences in the anomalous scattering contribution of certain atoms contained in the crystal.

The potential usefulness of the anomalous signal for phasing novel structures has been known since the early days of protein crystallography [1]. However, the anomalous signal is usually much smaller than the isomorphous signal provided by heavy atoms such as Hg, Pt, or Au, so initially it was only used as auxiliary information in phasing by the Isomorphous Replacement approach; the photographic methods of data collection were simply not accurate enough to measure the minute anomalous (a.k.a. Bijvoet or Friedel) differences for use as the sole source of reflection phases.

Historically, the first protein structure solved exclusively from the anomalous signal was crambin phased by SAD signal of sulfur using data measured on a four-circle, single-counter diffractometer [5]. Also, in the early 1980s B.C. Wang

Z. Dauter (✉)
Synchrotron Radiation Research Section, MCL, National Cancer Institute,
Argonne National Laboratory, Argonne, IL 60439, USA
e-mail: dauter@anl.gov

R. Read et al. (eds.), *Advancing Methods for Biomolecular Crystallography*,
NATO Science for Peace and Security Series A: Chemistry and Biology,
DOI 10.1007/978-94-007-6232-9_13, © Springer Science+Business Media Dordrecht 2013

proposed that SAD phasing could be effectively combined with the process of iterative modification of the solvent region in the protein crystals (solvent flattening; [16]).

The theory of MAD was also formulated early by Karle [6], but its practical implementation is attributed to Hendrickson [3], and coincided with the introduction of automatic and accurate Imaging Plate and then CCD detectors. The usage of anomalous signal was quickly accepted as a method of choice for the solution of novel structures, first as the MAD version, and in the last decade as SAD. Currently, the majority of novel macromolecular crystal structures are currently solved by the SAD technique.

13.2 SAD Phasing

The structure factor equation $F_h = \sum_j f_j \exp(2\pi i\, h \bullet r_j) = \sum_j f_j \left[\cos(2\pi i\, h \bullet r_j) + i \sin(2\pi\, h \bullet r_j)\right] = A_h + i\, B_h = |F_h| \exp(i\varphi)$ can be represented as a summation of vectors in the Argand diagram, as in Fig. 13.1, where $f_j \exp(2\pi i\, h \bullet r_j)$ is the contribution of an individual atom j, located at the position r_j in the unit cell, to the scattering of a reflection with index h (or, fully, h,k,l in three-dimensional space). The scalar product $(h \bullet r_j)$ stands for $(hx_i + ky_i i + lz_j)$. f_j is the scattering power contributed by one atom (the length of vector f_j in Fig. 13.1) and the term $\exp(2\pi i\, h \bullet r_j)$ defines the phase φ_j of this contribution (the direction of this vector). Here and in the following the terms related to the atomic displacements or vibrations (the B factors) are neglected for simplicity.

If the structure contains some "special" atoms, for example heavier (H) than the common protein (P) atoms, their contribution F_H can be grouped separately in the structure factor equation and in the Argand diagram (Fig. 13.2).

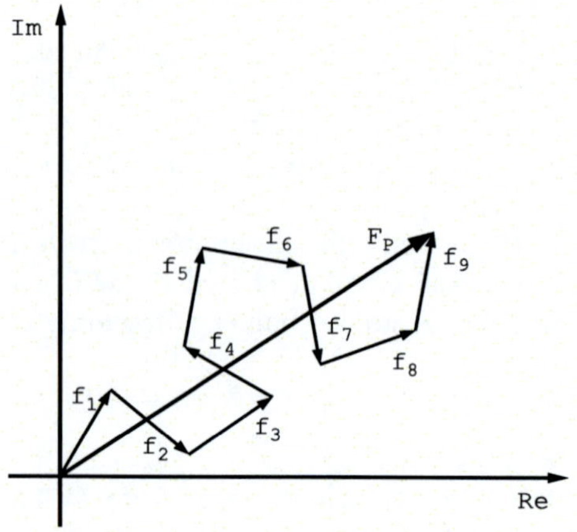

Fig. 13.1 Atomic scattering factors f_j of individual atoms in the structure add-up as vectors, resulting in the total protein structure factor F_P

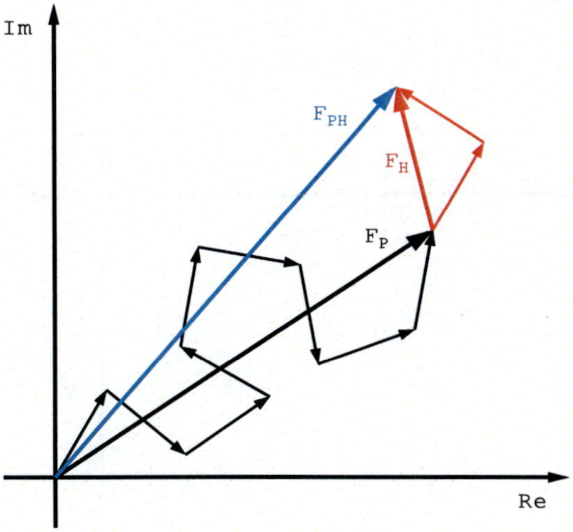

Fig. 13.2 The total scattering factor F_{PH} can be divided into contribution from common protein atoms F_P and from heavy or otherwise special atoms F_H

$$F_{PH} = \sum_j^P f_j \exp\left(2\pi i\, h \bullet r_j\right) + \sum_j^H f_j \exp(2\pi i\, h \bullet r_j) = F_P + F_H$$
$$= |F_P| \exp(i\varphi_P) + |F_H| \exp(i\varphi_H)$$

Under certain circumstances, when the energy of the incident X-rays is higher than the excitation energy of the inner electrons of some atoms, those atoms become the source of additional, resonant or "anomalous" scattering with its phase shifted forward with respect to the "normal" scattering f_j. The anomalous scattering can therefore be represented as additional terms contributing to the atomic scattering factor:

$$f = f^o + f' + if''$$

where f' is the real part of the anomalous correction in phase with the normal scattering factor f^o, and if'' is the imaginary part representing the forward phase shift by 90° with respect to f^o, as illustrated in Figs. 13.3 and 13.4

$$f = \left[|f^o| + |f'| + i\,|f''|\right] \exp(i\varphi) = \left[|f^o| + |f'|\right] \exp(i\varphi) + |f''| \exp[i(\varphi + 90°)]$$

which is evident after a simple trigonometric rearrangement, since i exp(iφ)

$$= i\,[\cos(\varphi) + i\ \sin(\varphi)] = i\ \cos(\varphi) - \sin(\varphi)$$
$$= i\ \sin(\varphi + 90°) + \cos(\varphi + 90°) = \exp[i(\varphi + 90°)]$$

The normal atomic scattering factor f^o depends on the scattering angle θ, but not on the X-ray wavelength. In contrast, the anomalous corrections f' and f'' are

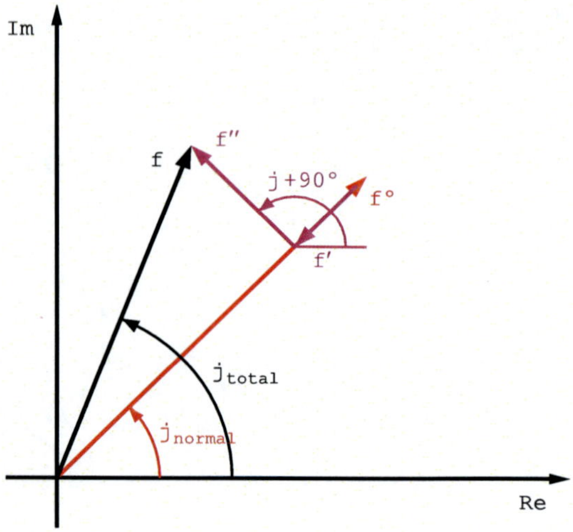

Fig. 13.3 Diffraction contributions of an anomalously scattering atom. The real correction f′ is usually negative and its direction is therefore antiparallel to f°, but the imaginary correction f″ is rotated positively by 90° with respect to f°

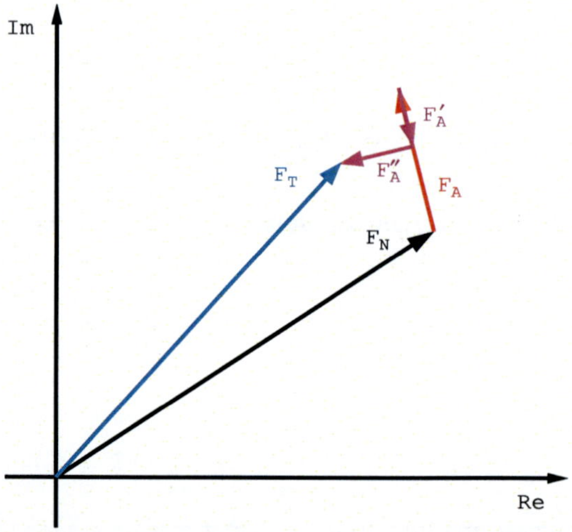

Fig. 13.4 The total structure factor F_T consisting of the contributions of non-anomalously scattering atoms F_N and some anomalous scatterers F_A

independent of θ, but vary with changing X-ray energy (i.e. wavelength). Figure 13.5 shows schematically that normal scattering originates from all electrons, scattering elastically as electron cloud, diffused around an atom, so that in consequence interference of rays scattered from individual electrons diminishes with increasing scattering angle. The anomalous contribution is provided by a single electron, which is small in comparison with the wavelength of the X-ray, so that there is no angular dependence of its scattering; instead, its excitation critically depends on the energy of the X-ray quanta.

It is possible to estimate the anomalous effect by recording the fluorescence emitted by the investigated sample while scanning the wavelength (and energy) of the X-rays. The f″ component changes in pace with the absorption and fluorescence, whereas the f′ component is proportional to the first derivative of f″. A typical

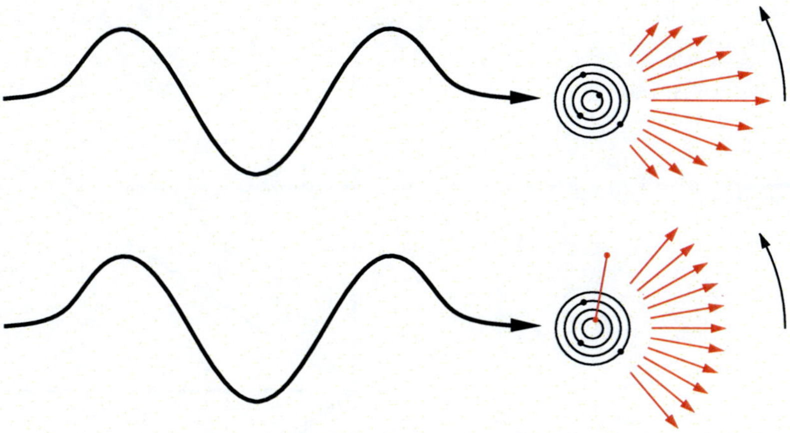

Fig. 13.5 Normal atomic scattering (*top*) results from interference of all electrons and diminishes with increasing diffraction angle. The anomalous effect (*bottom*) does not change with the angle, but depends on the energy of the X-ray quanta

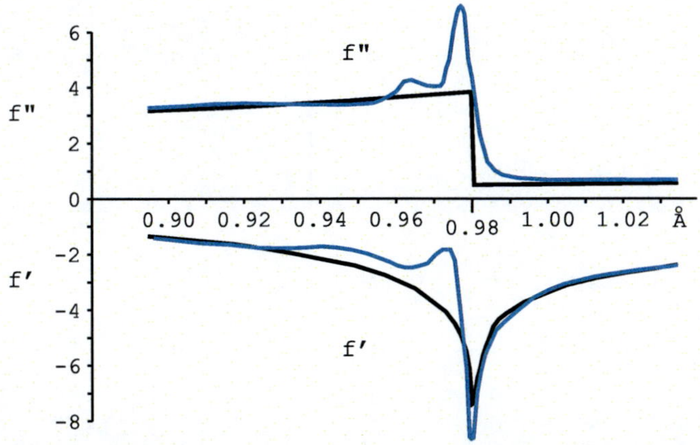

Fig. 13.6 The excitation spectrum (sometimes called the "fluorescence" spectrum) of a sample containing selenium. The *black line* is calculated theoretically for a single atom in vacuum, the *blue line* is from experimental measurements (Color figure online)

spectrum obtained from a selenomethionine-containing protein crystal is shown in Fig. 13.6, together with a theoretical curve calculated for an isolated Se atom. Interactions with atoms surrounding the anomalous scatterer in the real chemical compound modulate the energy levels of Se resulting in the additional fine features in the spectrum, notably the "white line" with f″ increased to the "peak" value above the absorption edge.

Normally the structure factor amplitudes of centrosymmetrically related reflections constituting a Friedel pair, F(h) and F(-h), are equal. However, the presence of the imaginary anomalous contribution if″ breaks of this Friedel's law, as shown in Figs. 13.7 and 13.8.

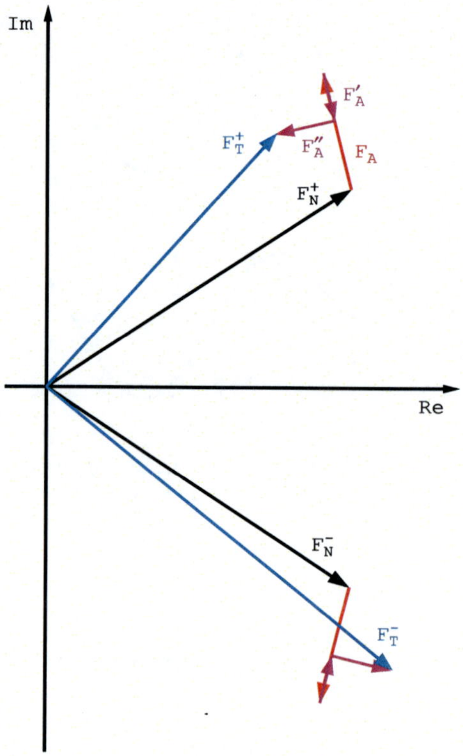

Fig. 13.7 Since the anomalous vector F''_A is rotated forward by 90° for F(h) with its phase φ(h) and for F(-h) with its phase –φ(h), the amplitudes of the Friedel-related pair are different, |F(h)| ≠ |F(-h)|

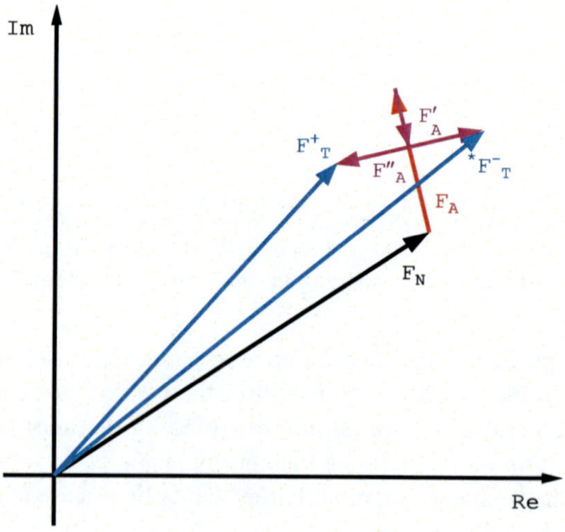

Fig. 13.8 It is customary to represent the F(-h) vector as its complex conjugate, on the other side of the Argand diagram, illustrating clearly that Friedel's law is broken in the presence of anomalous effect

In the diffraction experiment, the total reflection amplitudes $|F_T|$ are measured. As shown in Fig. 13.8, in general the amplitudes of Friedel-related reflections are different. In fact, this depends on the phase difference between the contributions of

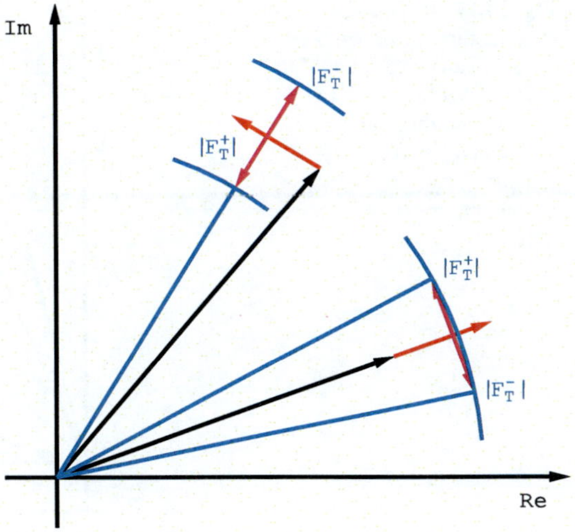

Fig. 13.9 The measurable anomalous effect is maximal when the vectors of the normal and anomalous scatterers are orthogonal and is minimal when they are (anti)parallel

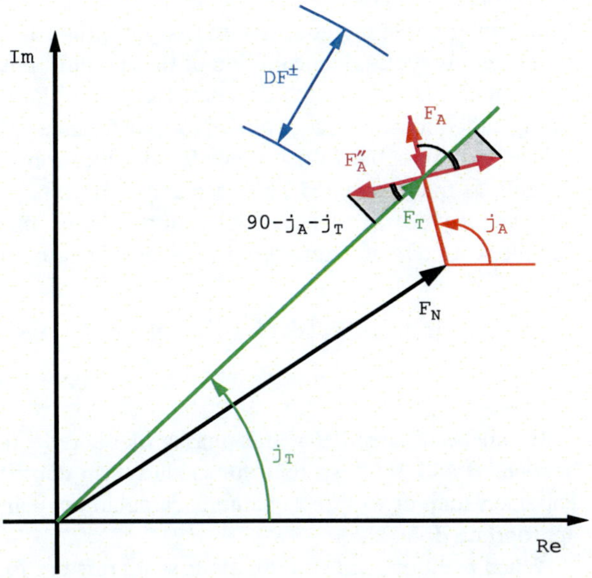

Fig. 13.10 The dependence between the measurable Bijvoet difference $\Delta F^{\pm}$ and the anomalous scatterers contribution F_A is sinusoidal

the normal F_N and anomalous F_A atoms. Figure 13.9 shows that when the phase difference between F_N and F_A is ±90°, the measured anomalous difference $|F^+{}_T| - |F^-{}_T|$ (customarily called Bijvoet difference) is maximal and equal to 2 F''_A. If the phase difference between F_N and F_A is close to 0° or 180° (their vectors are parallel or antiparallel), the Bijvoet difference is minimal and not measurable. There is a sinusoidal relation between the apparent Bijvoet difference $\Delta F^{\pm}$ and the phase difference $(\varphi_T - \varphi_A)$:$\Delta F^{\pm} = 2F''_A \sin(\varphi_T - \varphi_A) = 2\,(f''/f^o)\,F_A \sin(\varphi_T - \varphi_A)$ illustrated in Fig. 13.10.

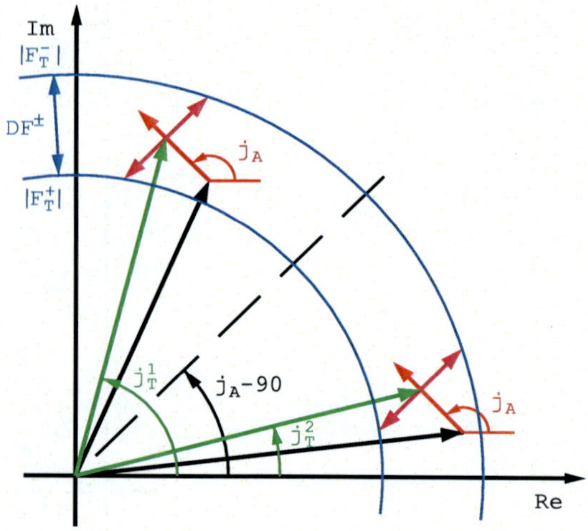

Fig. 13.11 If the anomalous substructure is known, then the size and orientation of the *red/pink* vectors are also known, and they can be positioned in two places indicating two alternative protein phases (Color figure online)

To locate the positions of the anomalous scatterers in a crystal, by the use of Patterson or direct methods, the most appropriate are the values of F_A, representing the normal scattering contribution of these atoms, but these values are not available within a single wavelength data set. Direct and Patterson methods rely mostly on strong reflections. Since large Bijvoet differences always correspond to large F_A values, it is possible to use them for the location of the anomalous scatterers in the cell, as first suggested by Rossmann [11]. After appropriate rearrangement, the observed Bijvoet difference consists of two terms, one directly proportional to $|F_A|^2$ and the second related through the cosine of phase difference.

$$|\Delta F^{\pm}|^2 = 4(f''/f^o)^2|F_A|^2\sin^2(\varphi_T - \varphi_A) = 2(f''/f^o)^2|F_A|^2 + 2(f''/f^o)^2|F_A|^2\cos^2(\varphi_T - \varphi_A)$$

It can be assumed that the phases of all reflections are distributed randomly between 0 and 360°, so that the second term contributes only random noise to a Patterson map or to direct methods calculations using $|\Delta F^{\pm}|$ instead of the more appropriate $|F_A|$ values.

When positions of the anomalous scatterers are found, it is possible to calculate their contribution to the total diffraction. However, even in the ideal case of error-free measurements, this does not provide a unique solution to the phase problem since, in general, there are two possible arrangements satisfying the vector relations. They lead, for each reflection, to two possible total protein phase angles, φ_T, symmetrically placed around the φ_A−90° direction in Fig. 13.11. The alternative values of φ_T are [9]

$$\varphi_T = 90° + \varphi_A \pm \delta \text{ where } \delta = \cos^{-1}\left(\Delta F^{\pm}/2F''_A\right)$$

Fig. 13.12 Since it is not known which of the two alternative phases is correct, it possible to compute the map using the averaged structure factor

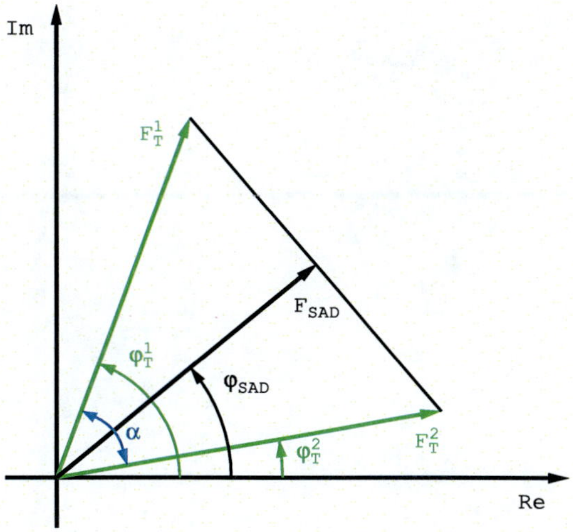

Only for the largest Bijvoet differences this SAD ambiguity degenerates to a single solution with $\varphi_T - \varphi_A = \pm 90°$.

Since it is not known which of the alternatives is correct, one can calculate a Fourier map using structure factors that are vectorial averages of the two possibilities, as indicated in Fig. 13.12, with their amplitudes weighted by the corresponding figure of merit, FOM

$$F_{SAD} = |F_{SAD}| \exp(2\pi i\ \varphi_{SAD}) = |F_T|\,\text{FOM}\ \exp(2\pi i\ \varphi_{SAD})$$
$$\text{where FOM} = \cos\left[(\varphi_{T^1} - \varphi_{T^2})/2\right] \text{ and } \varphi_{SAD} = (\varphi_{T^1} + \varphi_{T^2})/2$$

The figure of merit FOM is the highest (and approaches unity) when the two alternative phases are close to each other, and it approaches zero when they differ by close to 180°.

The average F_{SAD} vectors approximate the sum of two alternative solutions and the corresponding Fourier map is a superposition of two maps, one originating from the correct phases, representing the true structure, and the other one from the wrong phases representing featureless noise, as illustrated in Fig. 13.13. It may be therefore possible to distinguish in such a map the regions belonging to either the protein or to solvent. As proposed by Wang [16], modifying the solvent region and making it flat (with a constant value of the electron density) and subsequently calculating the inverse Fourier transform, one should obtain a new set of phases, closer to the true ones, thus likely indicating the correct solution. An iterative application of such solvent flattening (and other types of density modification) should eventually produce an interpretable map, revealing the whole structure, or at least a large part of it. This procedure also discriminates between the two possible enantiomeric solutions, producing meaningful results only for one of the two enantiomeric constellations of the anomalous substructure.

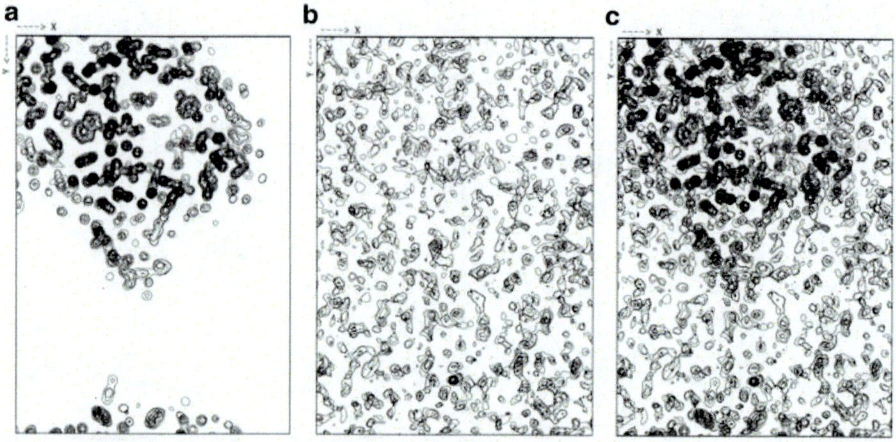

Fig. 13.13 The correct structure factors Fgood produce an interpretable electron density map of the structure, while the wrong structure factors Fwrong lead to featureless noise. However, structure factors that are averages of the two give a map in which the regions corresponding to the protein and to solvent are discernible

Fig. 13.14 In the presence of measurement errors the phases are not indicated sharply, but spread with certain probability, marked by the outer *red* curve (Color figure online)

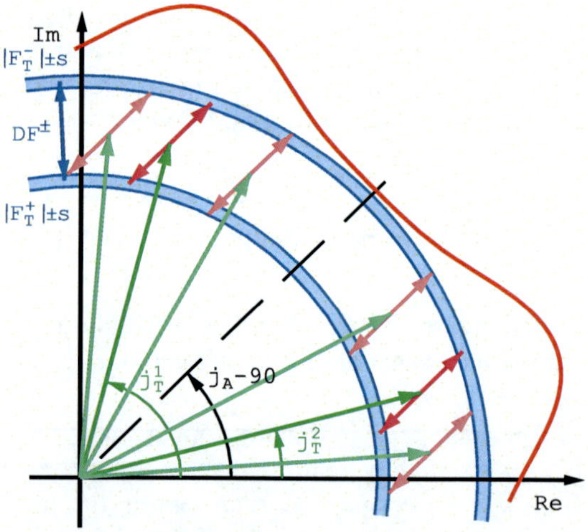

In practice, the estimation of SAD phases is not so straightforward and unequivocal, since the reflection intensities measured in a diffraction experiment inevitably contain errors. The anomalous scattering signal is usually small, at most at the level of a few percent of the total crystal diffraction intensity. The anomalous diffraction data should be therefore collected very carefully, with proper estimation of the errors (standard uncertainties) associated with all measured intensities. If measurement errors are taken into account, the modified SAD vector diagram (Fig. 13.14) shows that instead of two sharp solutions, there are two regions where the total protein phase, φ_T, may lie.

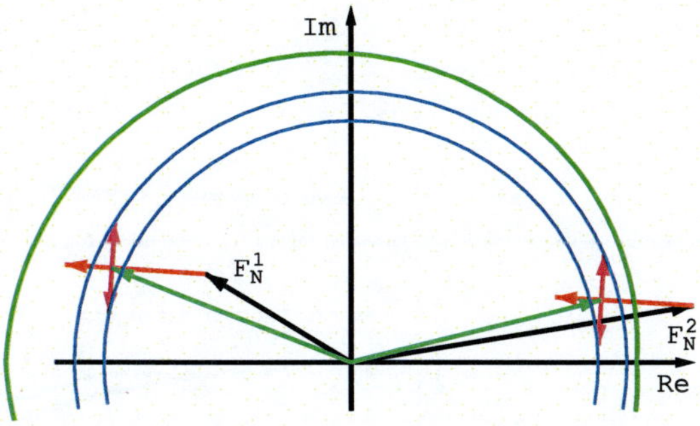

Fig. 13.15 The two SAD solutions are not symmetric, the one with F_N and F_A vectors parallel is more likely than the other, as marked by the outer *green* probability curve (Color figure online)

The probability P that the protein phase has a particular value φ_T for the known Bijvoet difference, $\Delta F^{\pm}$, the calculated anomalous scatterers phase, φ_A, and the imaginary contribution, F''_A, is according to Hendrickson & Teeter [5]

$$P_{anom}(\varphi_T) = N\ \exp\left\{-\left[\Delta F^{\pm} + 2F''_A \sin(\varphi_T - \varphi_A)\right]^2 / 2E^2\right\}$$

where E is the standard error and N a normalization factor. This can also be expressed using the Hendrickson-Lattman [4] coefficients, A, B, C, D:

$$P_{anom}(\varphi_T) = N\ \exp(A\ \cos\varphi + B\ \sin\varphi + C\ \cos\ 2\varphi + D\ \sin\ 2\varphi)$$

As illustrated in Fig. 13.15, the two alternative SAD phase indications are not fully symmetric. The case where $|\varphi_T - \varphi_A| < 90°$ and the F_N and F_A vectors are parallel, is more probable than when $|\varphi_T - \varphi_A| > 90°$ and the F_N and F_A vectors are antiparallel, since the amplitude $|F_N|$ of the normal scatterers in the former case is smaller than in the latter case. This results from the fact that the part of the structure, namely the anomalous substructure, is known. The theory of protein phase probability resulting from a known partial structure has been worked out by Sim [13, 14]

$$P_{part}(\varphi_T) = N\ \exp\{2\left[(F_T F_A)/F_{N^2}\right]\cos(\varphi_T - \varphi_A)\}$$

where F_N represents the scattering amplitude of the unknown part of the structure and N is the normalizing factor. The probability resulting from the known partial structure depends on the cosine function of the phase difference $\varphi_T - \varphi_A$ and, as a result, the combined probability is different for the two alternative solutions for φ_T (Fig. 13.16). Ramachandran and Raman [9] first postulated that the total phase closer to the anomalous phase should be chosen for map calculation, which agrees with the Sim probability indication based on the partial structure.

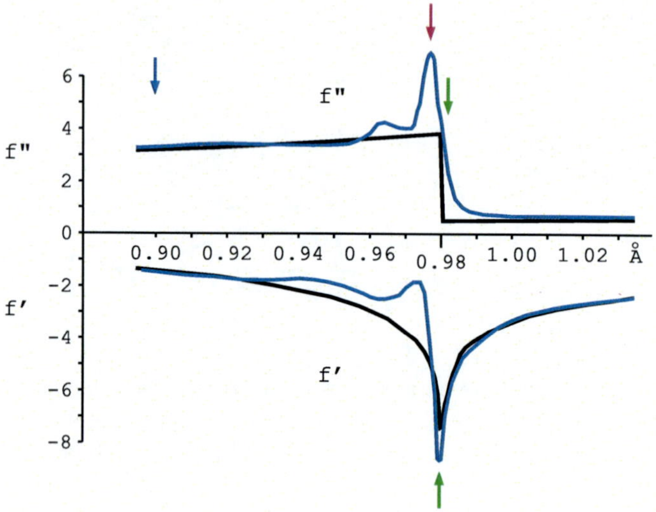

Fig. 13.16 The fluorescence spectrum of Se with indicated three wavelengths, corresponding to the peak (*pink top arrow*), inflection (*green/light grey arrows*) and high energy remote (*blue/dark grey arrow*) points of the spectrum (Color figure online)

13.3 MAD Phasing

The anomalous substructure provides different contributions to reflection structure factors at different wavelengths, since the f′ and f″ values depend on the wavelength, as for example shown in Fig. 13.16 for selenium.

The largest differences between f′ and f″ can be obtained if data sets are measured at the wavelengths corresponding to the peak, edge (inflection point), and a high energy (shorter wavelength) point on the X-ray absorption curve of the anomalous scatterer present in the crystal, as indicated in Figs. 13.16 and 13.17.

The original MAD technique, developed by Hendrickson [2], was based on the analytical approach of Karle [6], where the unknown amplitudes and phases were calculated algebraically.

Various vectors representing a single structure factor with a significant anomalous diffraction contribution are shown in Fig. 13.18. The blue vectors F_T^+ and F_T^- are the total measured amplitudes for both Friedel mates, F_T^o is the (unknown) amplitude corresponding to the total diffraction without any anomalous contribution and F_A, F_A' and F_A'' are the normal and anomalous contributions of the anomalous substructure. To solve the structure it is necessary to estimate the amplitude $|F_T^o|$ and its phase φ_T as well as $|F_A|$ and φ_A. From the law of cosines

$$|F_{T+}|^2 = |F_{T^o}|^2 + y^2 - 2\,|F_{T^o}|\,y\,\cos\alpha$$

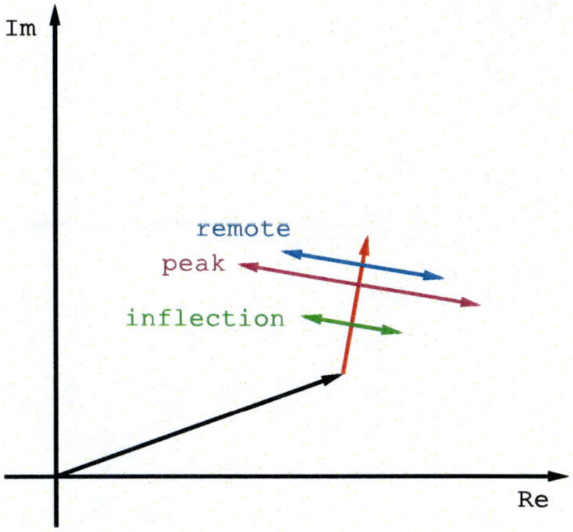

Fig. 13.17 Vector representation of the three points in the spectrum: peak with largest f″, inflection with largest f′ and remote with substantial f″ and small f′

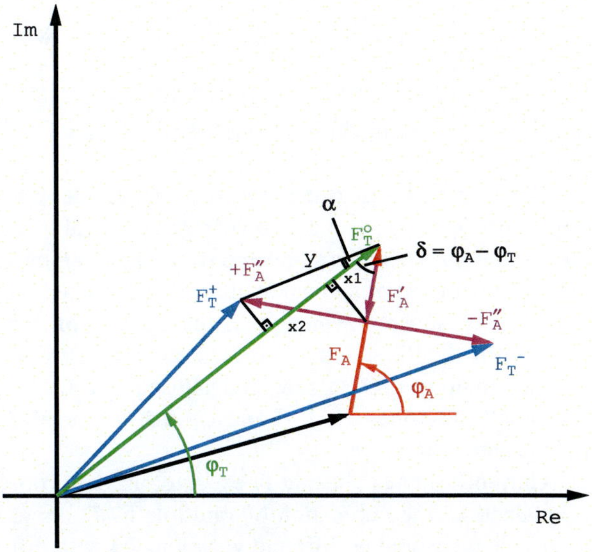

Fig. 13.18 Structure factors and phases used in the algebraic approach to MAD

where

$$y^2 = |F'_A|^2 + |F''_A|^2 = |F_A|^2\left[\left(f'^2 + f''^2\right)/f^{o2}\right] \quad \text{since} \quad F'_A = F_A(f'/f^o)$$

$$\text{and} \quad F''_A = F_A(f''/f^o)$$

$$y \cos\alpha = x_1 + x_2 = F'_A \cos\delta + F''_A \sin\delta = F_A(f'/f^o)\cos\delta + F_A(f''/f^o)\sin\delta$$

After substituting $\delta = \varphi_A - \varphi_T$ and taking into account that F_A' is negative in Fig. 13.18, the resulting equation [3] takes the form (for both Friedel mates)

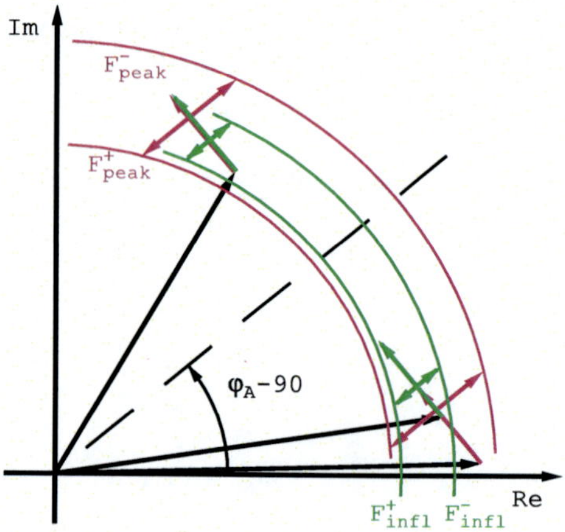

Fig. 13.19 If two measurements recorded at two different wavelengths with different combinations of f′ and f″ are available, then there is only one common case between the two solutions, indicating the correct phase

$$\begin{aligned} |F_T{}^{\pm}|^2 &= |F_{T^o}|^2 + a(\lambda)\,|F_A|^2 &\quad \text{where } a(\lambda) &= \left(f'^2 + f''^2\right)/f^{o2} \\ &+ b(\lambda)\,|F_{T^o}|\,|F_A| \cos(\varphi_T - \varphi_A) & b(\lambda) &= 2\,f'/f^o \\ &\pm c(\lambda)\,|F_{T^o}|\,|F_A| \sin(\varphi_T - \varphi_A) & c(\lambda) &= 2\,f''/f^o \end{aligned}$$

For each such equation there are three wavelength-independent unknowns, $|F_T{}^o|$, $|F_A|$ and $(\varphi_T - \varphi_A)$, one known measured value of $|F_T{}^+|$ (or $|F_T{}^-|$), and three parameters $a(\lambda)$,$b(\lambda)$,$c(\lambda)$ related to the anomalous scattering factors and common for all reflections at a given λ. If a Friedel-related pairs of reflections are measured for two or more wavelengths, this system of four or more equations can be solved for the three unknowns. The estimated amplitude F_A is then used to locate the anomalous substructure and subsequently to calculate its phase φ_A, which eventually provides the $F_T{}^o$ and φ_T that are needed to calculate the Fourier map of the entire structure.

Currently MAD phasing is executed in the probabilistic version, which can be explained as a combination of multiple SAD approaches with the use of several data sets measured at different wavelengths with different values of f′ and f″. Each wavelength will indicate a different combination of the alternative phases, of which only one will agree amongst all the wavelengths, providing therefore a solution to the SAD ambiguity, as illustrated in Fig. 13.19.

A number of computer programs use sophisticated probabilistic approaches for estimation of the MAD phases, taking into account various effects, such as uncertainties of the measured amplitudes, potential non-isomorphism between data sets, etc. (SHARP [7], SOLVE [15], PHASER [10], SHELXE [12], MLPHARE [8]).

References

1. Blundell TL, Johnson LN (1976) Protein crystallography. Academic, London
2. Hendrickson WA (1985) Analysis of protein structure from diffraction measurement at multiple wavelengths. Trans Am Cryst Assoc 21:11–21
3. Hendrickson WA (1991) Determination of macromolecular structures from anomalous diffraction of synchrotron radiation. Science 254:51–58
4. Hendrickson WA, Lattman EE (1970) Representation of phase probability distributions for simplified combination of independent phase information. Acta Cryst B 26:135–143
5. Hendrickson WA, Teeter MM (1981) Structure of the hydrophobic protein crambin determined directly from the anomalous scattering of sulfur. Nature 290:107–113
6. Karle J (1980) Some developments in anomalous dispersion for the structural investigation of macromolecular systems in biology. Int J Quantum Chem Symp 7:357–367
7. La Fortelle E, de Bricogne G (1997) Maximum-likelihood heavy-atom parameter refinement for multiple isomorphous replacement and multiwavelength anomalous diffraction methods. Methods Enzymol 276:472–494
8. Otwinowski Z (1993) Proceedings of the CCP4 study weekend. In: Wolf W, Evans PR, Leslie AGW (eds) Isomorphous replacement and anomalous scattering. Daresbury Laboratory, Warrington, pp 80–86
9. Ramachandran GN, Raman S (1956) A new method for the structure analysis of non-centrosymmetric crystals. Curr Sci 25:348–351
10. Read RJ, McCoy AJ (2011) Using SAD data in phaser. Acta Cryst D 67:338–344
11. Rossmann MG (1961) Position of anomalous scatterers in protein crystals. Acta Cryst 14:383–388
12. Sheldrick GM (2008) A short history of SHELX. Acta Cryst A64:112–122
13. Sim GA (1959) The distribution of phase angles for structures containing heavy atoms. II. A modification of the normal heavy-atom method for non-centrosymmetric structures. Acta Cryst 12:813–815
14. Sim GA (1964) A note on the determination of phases by anomalous dispersion. Acta Cryst 17:1072–1073
15. Terwilliger TC, Berendzen J (1999) Automated MAD and MIR structure solution. Acta Cryst D 55:849–861
16. Wang BC (1985) Resolution of phase ambiguity in macromolecular crystallography. Methods Enzymol 115:90–112

Chapter 14
Macromolecular Phasing: Solving the Substructure

Tim Grüne

Abstract The phase problem is one crucial step in order to create an atomic model from crystalline diffraction data. There are mainly three methods to solve the phase problem for macromolecular diffraction data: molecular replacement, isomorphous replacement, and anomalous dispersion. The latter two methods are so-called experimental phasing methods, both of which require firstly to solve a substructure and secondly to expand the phases from the substructure to the whole structure. This article explains the term 'substructure', how the substructure data are extracted from the total data, and how the substructure is used to solve the phase problem. While these steps may only take a few minutes in real-time, understanding the underlying basics is important in order to properly carry out the data collection.

Keywords Direct methods • Harker construction • Phase problem • Substructure • Tangent formula

14.1 Substructure

Any subset of the atoms present in the asymmetric unit of a crystal structure is called a "substructure". In the context of macromolecular phasing one usually refers to the *substructure* as those atoms which contribute to experimental phasing (Fig. 14.1):

heavy atoms for isomorphous replacement (MIR, SIR, SIRAS, ...)
anomalous scatterers for anomalous dispersion methods (MAD, SAD, SIRAS, ...)

In a *very* simplified view, crystal structures can be solved by direct methods from one single (native) data set, if the atoms are separated further than the resolution of

T. Grüne (✉)
Institut fuer anorganische Chemie, Tammannstr. 4, D-37077 Goettingen, Germany
e-mail: tg@shelx.uni-ac.gwdg.de

R. Read et al. (eds.), *Advancing Methods for Biomolecular Crystallography*, NATO Science for Peace and Security Series A: Chemistry and Biology,
DOI 10.1007/978-94-007-6232-9_14,

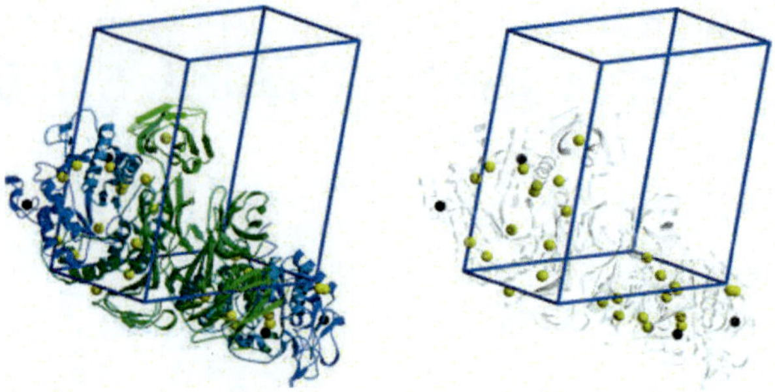

Fig. 14.1 *Left*: A protein structure and its unit cell with the S-atoms marked by CPK presentation. *Right*: The S-atoms can be used to acquire phases for this crystal. They form the *substructure* for this protein in the context of phasing

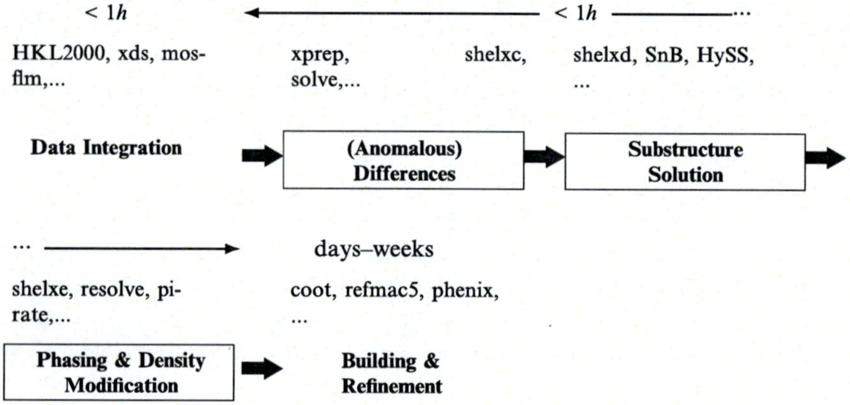

Fig. 14.2 Simplified view of macromolecular structure solution with experimental phasing. While the phasing step takes only a relatively short time, it is both a crucial step and requires as good data as possible

the data set[1]. For real crystals this restricts direct methods to about 1.2 Å [3], the order of magnitude of chemical bond lengths.

Experimental phasing methods, however, are based on the creation of an **artificial** data set which would have been produced by the substructure in within the unit cell of the real crystal;

1. Collect data, *e.g.* native and derivative, or peak and inflection point
2. Extract the substructure contribution to the data
3. Solve the substructure by direct methods
4. Use the substructure coordinates to expand the phases to the full data-set.

as depicted in Fig. 14.2. The extraction of the artificial data and the subsequent extension of the phases make use of rather crude approximations that require the input data set to be of as good quality as possibly for the respective phasing scenario.

[1] This statement is an **untested** extrapolation of "Sheldrick's Rule" [3]

14.2 The Phase Problem

$$\rho(x, y, z) = \frac{1}{V_{\text{cell}}} \sum_{h,k,l} |F(h, k, l)| e^{i\phi(h,k,l)} e^{-2\pi i(hx+ky+lz)} \tag{14.1}$$

Equation 14.1 mathematically expresses the relationship between an data of a crystallographic X-ray experiment and the chemical model which is finally submitted to the PDB: The "independent atom model" (IAM) calculates the electron density $\rho(xyz)$, which is represented by the chemical model, from

1. the reflection intensities $I(hkl) = \sqrt{|F(h, k, l)|}$
2. a phase angle $\phi(h, k, l)$ associated with each reflection
3. the unit cell parameters (hidden in the fact that (x, y, z) are *fractional coordinates*)

The intensities are the direct result of the X-ray experiment. The phase angle $\phi(h, k, l)$, however, cannot be measured directly. It is essential for the calculation of the electron density map – hence the origin of the term **phase problem** in crystallography.

14.3 Macromolecular Solutions to the Phase Problem

Typically for small molecule crystallography are small unit cell dimensions (few atoms in the asymmetric unit) and high resolution data. This allows for solving the phase problem with *direct* methods, nowadays usually based on the tangent formula published by [6].

$$\tan(\phi_{\mathbf{h}}) \approx \frac{\sum_{\mathbf{h}'} |E_{\mathbf{h}'} E_{\mathbf{h}-\mathbf{h}'}| \sin(\phi_{\mathbf{h}'} + \phi_{\mathbf{h}-\mathbf{h}'})}{\sum_{\mathbf{h}'} |E_{\mathbf{h}'} E_{\mathbf{h}-\mathbf{h}'}| \cos(\phi_{\mathbf{h}'} + \phi_{\mathbf{h}-\mathbf{h}'})} \tag{14.2}$$

The tangent formula is a relation between the phases reflection triplets which allows to refine a starting set of phases. The starting set of phases is picked at random or semi-random in many attempts/trials. Despite this random picking they often refine close enough to the real phases.

The success rate of direct methods depends on the resolution and the unit cell dimensions, and for macromolecular crystals they are often not an option. The most common methods to solve the phase problem in macromolecular crystallography are

- Molecular Replacement (MR): placement of a similar structure in the unit cell and calculation of the phases from its coordinates

- Isomorphous replacement (MIR, SIR, SIRAS, RIP, ...): comparison of data sets in the presence and absence of heavy atoms
- Anomalous Dispersion (MAD, SAD, SIRAS, ...): exploitation of the breakdown of Friedel's law at element specific wavelengths

The latter two methods are called "experimental phasing methods". They are based on the extraction of an artificial substructure data set from one or more data sets as it would have been created if the substructure atoms where the only content of the unit cell. Because the substructure atoms are usually further away than the (anomalous) resolution, direct methods can be applied to determine their positions in the asymmetric unit.

14.4 Solving the Phase Problem with a Substructure

Both the structure factor amplitude $|F(h,k,l)|$ and the phases $\phi(h,k,l)$ for the substructure can be calculated for each reflection once the substructure coordinates are known. It remains to determine the phase angle Φ_T for the actual data set.

Historically isomorphous replacement problems were then solved using the **Harker construction**, which is explained in most text books about macromolecular crystallography [1, 2, 5].

In the case of anomalous dispersion, the determination of the artificial substructure data set and of Φ_T is based on a relationship published by [7] and [8]:

$$|F^+|^2 = |F_T|^2 + a|F_A|^2 + b|F_A||F_T| + c|F_A||F_T|\sin\alpha \tag{14.3}$$

$$|F^-|^2 = |F_T|^2 + a|F_A|^2 + b|F_A||F_T| - c|F_A||F_T|\sin\alpha \tag{14.4}$$

$$|F^+|^2 - |F^-|^2 = 2c|F_A||F_T|\sin\alpha \tag{14.5}$$

The angle α used in Eq. 14.5 is defined in Fig. 14.3. It fulfills

$$\Phi_T = \Phi_A + \alpha \tag{14.6}$$

The quantities $|F^+|^2$ and $|F^-|^2$ form the Bijvoet pairs. They are the actually measured quantities.

$|F_T|$ is the non-anomalous contribution of **all atoms**, *i.e.* without the contributions of f' and f''.

$|F_A|$ is the non-anomalous contribution of the substructure atoms. It represents the artificial substructure data set, from which the substructure can be solved by direct methods.

The constants $a = \frac{f''^2 + f'^2}{f^2}$, $b = \frac{2f'}{f}$, and $c = \frac{2f''}{f}$ depend only on the atomic scattering factor components $f_{\text{total}}(\theta,\lambda) = f(\theta) + f'(\lambda) + if''(\lambda)$ and are determined by the fluorescence scan or from tabulated values [4].

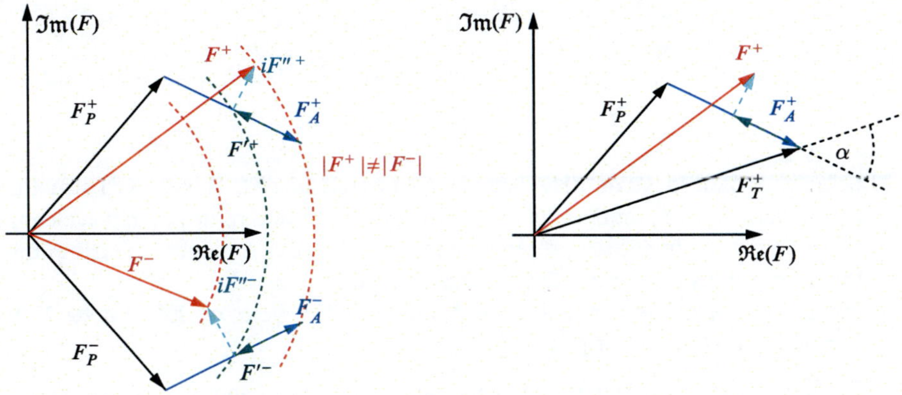

Fig. 14.3 *Left*: Friedel's Law ($|F(hkl)| = |F(\bar{h}\bar{k}\bar{l}), \Phi(hkl) = -\Phi(\bar{h}\bar{k}\bar{l})$) holds for all contributions except for the anomalous part if_v''. *Right*: Definition of the phase angle α as the angle between the total and the substructure **non-anomalous** structure factor

14.4.1 MAD/ MIR/ SIRAS Phasing

A MAD data set consists of at least two measurements at different wavelengths. This results in four equations (Eqs. 14.3 and 14.4 for each wavelength) for three unknown variables ($|F_A|$, $|F_T|$, and α)[2].

Therefore, $|F_A|$ and α can be determined **exactly** up to experimental error, *i.e.* even with an anomalous signal up to only 7 Å MAD provides an unbiased, correct electron density map provides the heavy atom positions are determined **as exact as possible**.

Density modification can reduce the effect of experimental errors but also extend the phases to the resolution range of the whole data set.

A similar argumentation holds for MIR and SIRAS.

14.4.2 SAD/ SIR Phasing

N.B.: The following SAD phasing. For SIR phasing there are minor differences, but the idea is similar.

In the case of SAD, even though there are less equations than unknowns, one can still get to an initial electron density map.

shelxd applies a couple of very sophisticated tricks to reduce Eq. 14.5 to one single unknown ($|F_A|$) and the difference of the Bijvoet pairs.

[2] All three unknowns are based on **non-anomalous** contributions to the total structure factor, hence all three a **wavelength independent!**

1. small anomalous difference:

$$|F_T| \approx 1/2(|F_+| + |F_-|)$$
$$\Rightarrow |F_+| - |F_-| \approx c|F_A| \sin \alpha$$

2. using **normalised structure factors** $|E|$ instead of $|F|$. Normalised structure factors have proven more efficient in direct methods and remove the wavelength dependency and hence the presence of the factor c (this statement oversimplifies the impact of using normalised structure factors)
3. the largest normalised anomalous differences will have $\alpha \approx 90°$ or $\alpha \approx 270°$, depending on the sign of $|F_+| - |F_-|$

   Putting everything together and using only the strongest anomalous differences,

$$||E_+| - |E_-|| \approx \text{sign}(|E_+| - |E_-|)|E_A| \tag{14.7}$$

Equation 14.7 provides the desired artificial substructure data set which can be solved by "conventional" direct methods.

14.5 "Phase Extension": Density Modification

The previous section describes how to extra a "virtual substructure data set", represented in the above equations by $|F_A|$, from the Bijvoet pairs in the presence of an anomalous scatterer. The substructure coordinates can be solved from this data set by direct methods. The phase angle Φ_A can be calculated from the substructure coordinates.

Since the phase angle Φ_T is the sum of α and Φ_A, the electron density map can then be calculated.

MAD/ MIR/ SIRAS provide exact and close estimate of α, and the quality electron density is limited by experimental errors and the accuracy of the substructure coordinates. In this case, density modification serves to reduce the impact of these errors to get an accurate as possible electron density map to start model building.

SAD/ SIR are based on crude assumptions and approximations, so the substructure coordinates are likely to have large errors. Furthermore the angle α can only be estimated to be either 90° or 270° (SAD; 0° or 180° for SIR). An electron density map based on these values will not be interpretable to the human eye. Density modification is **essential** in order to improve these estimates.

References

1. Blundell TL, Johnson LN (1976) Protein crystallography. Academic Press, London
2. Drenth J (2007) Principles of protein X-ray crystallography. Springer, New York
3. Morris RJ, Bricogne G (2003) Sheldrick's 1.2Å rule and beyond. Acta Crystallogr D Biol Crystallogr 59:615–617
4. Prince E (ed) (2002) International tables for crystallography, volume C. Union of crystallography, 3rd edn. Springer, Heidelberg
5. Rupp B (2009) Biomolecular crystallography: principles, practice, and application to structural biology. Garland Science, New York
6. Karle J, Hauptman H (1956) A theory of phase determination for the four types of non-centrosymmetric space groups 1P222, 2P22, 3P12, 3P22. Acta Crystallogr 9(8):635–651
7. Karle J (1980) Triplet phase invariants: formula for acentric case from fourth-order determinantal joint probability distributions. Proc Natl Acad Sci U S A 77(1):5–9
8. Hendrickson WA, Smith JL, Sheri S (1985) Direct phase determination based on anomalous scattering. Methods Enzymol 115:41–55

Chapter 15
Advanced Applications of Shelxd and Shelxe

Tim Grüne

Abstract The programs shelxc, shelxd, and shelxe by G. Sheldrick provide a powerful pipeline for experimental phasing of various phasing scenarios, like MAD, SIRAS, or MR-SAD. They are command line programs, which is the reason why many people are discourage from using them and exploiting their power. This tutorial shows that shelx c/d/e are not very difficult to use and only little knowledge is required to solve the phase problem and create and initial poly-Alanine model and an interpretable map with very little effort.

Keywords Shelxc • Shelxd • Shelxe • Anode • Auto-tracing • Free-lunch algorithm • Experimental phasing • Shelx c/d/e tutorial

15.1 Tutorial Data

A tutorial for phasing with instructions and data is available through my web-site http://shelx.uni-ac.gwdg.de/~tg/teaching/anl-ccp4/. This tutorial is much more detailed than this short introduction.

15.2 Introduction: Phasing with shelx c/d/e

The triad shelx c/d/e [18] by G.M. Sheldrick form a set of programs to carry out experimental phasing for macromolecular crystallographic data. They can be used for

- MAD / SAD (multi-/single-wavelength anomalous dispersion)
- SIRAS (Single Isomorphous Replacement with Anomalous Signal)

T. Grüne (✉)
Institut fuer anorganische Chemie, Tammannstr. 4, D-37077 Goettingen, Germany
e-mail: tg@shelx.uni-ac.gwdg.de

R. Read et al. (eds.), *Advancing Methods for Biomolecular Crystallography*, NATO Science for Peace and Security Series A: Chemistry and Biology, DOI 10.1007/978-94-007-6232-9_15,

- SIR (Single Isomorphous Replacement)
- MR-SAD (partial MR-solution to improve SAD phasing [12])
- RIP (Radiation Induced Phasing [14])

They perform the tasks

shelxc : data preparation, statistics for evaluation, *e.g.* resolution cut-off
shelxd : application of direct methods to find the substructure coordinates
shelxe : density modification to extend and improve the substructure phases to all of the data set.

All three programs are *command line programs*, *i.e.*, they are run from a terminal. They do not produce graphical output. They are designed to run *fast* and *robust* but maintain an enormous flexibility (the program shelxd was originally written and is still being used to solve small molecule structures and could be extended to protein phasing with virtually no modification).

When called without any arguments, both shelxc and shelxe print short usage instructions to the terminal.

15.2.1 shelxc

Shelxc reads the data formats

`.hkl` native shelx [17]-format
`.sca` HKL2000 [3] scalepack output
`XDS_ASCII.HKL` XDS [7] output file

shelxc extracts the anomalous difference from the Add "Bijvoet" to dictionary pairs, pre-calculates or approximates the angle α, and sets up the input file required to run shelxd.

An example input file `my_shelxc.input` for a MAD experiment:

```
NAT jia_nat.hkl
HREM jia_hrem.sca
PEAK jia_peak.sca
INFL jia_infl.sca
CELL 96.00 120.00 166.13 90 90 90
SPAG C2221
FIND 8
SFAC SE
```

would be called as

```
shelxc mymad < my_shelxc.input | tee shelxc.log
```

to create the files

`mymad.hkl` Native data used by shelxe

mymad_fa.hkl Anomalous differences and an estimate of the phase angle α. shelxe calculates the phase angleϕ_P from ϕ_A (calculated from the substructure coordinates found by shelxd) and α:

$$\phi_P(hkl) = \phi_A(hkl) + \alpha(hkl) \tag{15.1}$$

Ignore "mymad" during the current session_fa.ins Instructions for shelxd.

The log-file shelxc.log contains important information, *e.g.* the pairwise anomalous correlation coefficients by which one should decide about the resolution cut-off for shelxd (SHEL keyword where CC_{ano} >30 %) and one can detect outliers that ought to be rejected, *e.g.* due to radiation damage.

15.2.2 shelxd

The command

```
shelxd mymad_fa
```

asks shelxd to read the data from mymad_fa.hkl and the instructions from mymad_fa.ins.

shelxd applies (small molecule) *direct methods* to find the substructure coordinates.

shelxd automatically writes a logfile mymad_fa.lst, so no redirection (as in the case of shelxc) necessary.

While shelxd is running, it writes the currently best substructure solution to mymad_fa.res, based on the correlation coefficient between the coordinates and the data. mymad_fa.res contains the substructure coordinates in fractional coordinates that are later read by shelxe.

```
REM Best SHELXD solution: CC 60.74 CC(weak) 49.22   CFOM 109.96
TITL mymad_fa.ins MAD in C2
CELL  0.98000  109.02   61.75 71.74 90.00   97.08   90.00
LATT  -7
SYMM -X, Y, -Z
SFAC SE
UNIT   192
SE01   1  0.758774  0.508636 0.246391 1.0000  0.2
SE02   1  0.792908  0.398262 0.138903 0.8845  0.2
                 [...]
SE10   1  0.925819  0.231575 0.191291 0.5569  0.2
SE11   1  0.495239  0.183609 0.416278 0.5352  0.2
SE12   1  0.643097  0.029221 0.210653 0.4897  0.2 <---
SE13   1  0.811539  0.048553 0.227752 0.1453  0.2 <---
SE14   1  0.600281  0.156860 0.149628 0.0764  0.2
HKLF 3
END
```

The sixth column contains the occupancy of the corresponding atom. A sharp drop (here between SE12 and SE13) is a promising sign of a correct solution. Because weak atoms are weighted down by the occupancy shelxd can even find a substructure composed of different atom types, *e.g.* Se- and S-atoms.

The correlation coefficient (CC and CCweak) in the first line measures the reliability of the solution. For MAD phasing a CC>40 % is almost certainly a solution; for SAD phasing the line is at CC>30 %.

15.2.3 shelxe

shelxe does not, as opposed to shelxd, require an instructions file but takes all parameters from the command line option. For the above example it would read

```
shelxe mymad mymad_fa -s0.65 -h12 -a -q
```

The native data are read from `mymad.hkl`, `mymad_fa.res` provides the heavy atom positions (and thus Φ_A), and `mymad_fa.hkl` provides the angle α.

`-a` run 5 (default) cycles of poly-ALA autotracing. This feature is most useful and currently available in the β-version of shelxe (send an email to gsheldr@shelx.uni-ac.gwdg.de)

`-q` by default, shelxe searches for tri-peptides during the auto-tracing cycles. –q lets shelxe explicitly search for α-helices. Unless you know there are no helices (*e.g.* no protein at all), this option should always be used since it significantly improves the result.

shelxd cannot distinguish between the original and the inverted hand of the substructure. Therefore shelxe must be run a second time with the option '–i' to invert the substructure coordinates.

The output file `mymad.pdb` contains the poly-ALA trace from shelxe, the file `mymad.phs` the improved phases which can be displayed by coot [5] as electron density map.

15.3 Substructure Recycling

The anomalous signal usually does not reach as far as the actual data set. It may well be that it drops at, say, 3 Å below CC_{ano} 30 %, even though the crystal itself diffracts to 2 Å. Since shelxe uses the full resolution range for density modification, it can improve the position of the substructure atoms. The improved coordinates are written to the `.hat`-file (`mymad.hat`). It has the same format as the `.res`-file. By renaming the `.hat`-file to the `.res`-file and rerunning shelxe with the same options, the result can sometimes improve.

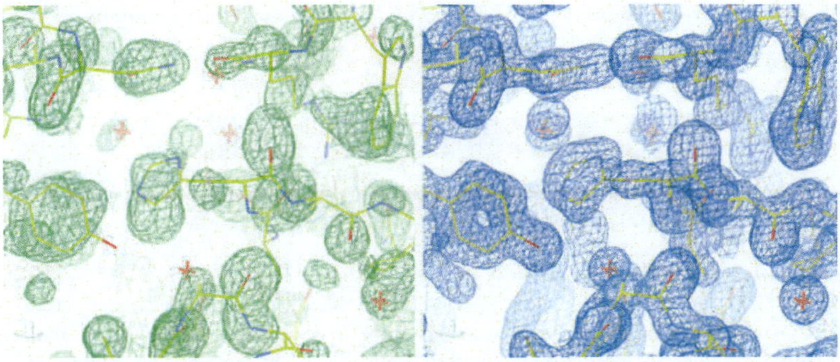

Fig. 15.1 The free-lunch algorithm. *Left*: "conventional" experimental phases after density modification with a `MapCC` of 0.57. *Right*: Expansion of the data to 1.0 Å with calculated data (`MapCC` = 0.94) [20]

Sometimes only a large number of iterations leads to a positive result [1] [9].

Sharp/autoSHARP [1, 21] provide more versatile possibility to improve the substructure coordinates.

15.4 Free-Lunch-Algorithm

The free-lunch algorithm was published independently in [2] and [6]. The modified electron density is used to calculate structure factors (both amplitudes and phases) to higher resolution than originally measured. The methods works well with native data measured to 2 Å and better. The map resulting from this type of density modification is sometimes easier to interpret and can be used for automated model building (Fig. 15.1).

NB: These calculated data must not be used for refinement!

15.5 Autotracing

The β-version of shelxe[2] is capable of producing a poly-ALA trace into the modified electron density map. To my knowledge shelxe was the first program to iteratively combine model building and density modification which has proven to be extremely

[1]NB: this was before the auto-tracing capabilities of shelxe by which the effect of recycling the substructure has become less important.

[2]β-versions of the shelx-Programs are available to registered shelx-users upon email request to either George Sheldrick or me. Registration is free of charge for academics and only requires the sending of a fax to the institute. See http://shelx.uni-ac.gwdg.de for details.

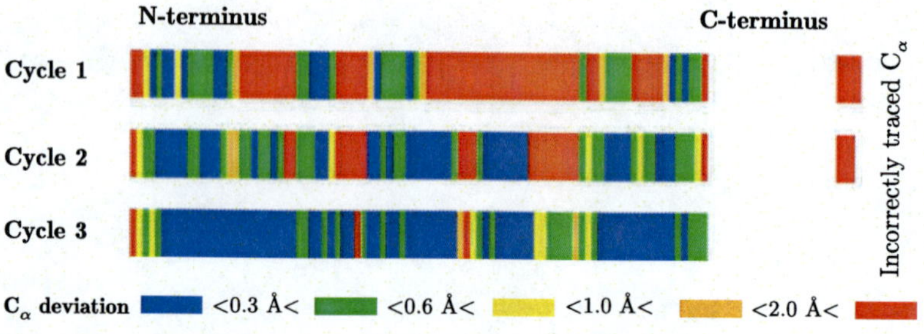

Fig. 15.2 Autotracing of Fibronectin: starting from a very noisy S-SAD map shelxe traced more than 94 % the mainchain correctly within 1.0 Å after only three cycles (Figure courtesy G. Sheldrick)

powerful, and even very poor phase information can lead to a valid poly-ALA trace also at moderate resolution (3′ish–4′ish Å).

The autotracing of shelxe can improve phases from very poor starting phases.

1. Find potential α-helices in the density and try to extend them at both ends. Then find other potential tripeptides and try to extend them at both ends in the same way.
2. Tidy up and splice the traces as required, applying any necessary symmetry operations.
3. Use the traced residues to estimate phases and combine these with the initial phase information using sigma-A weights, then restart density modification. The refinement of one B-value per residue provides a further opportunity to suppress wrongly traced residues.

There is not much configuration possible: simply starting shelxe with the option '–a' starts the auto-tracing option and shelxe will build as much as possible.

The option '–q' makes shelxe look for helices (as secondary structure element) and not just for tri-peptides.

The option '–tX' makes shelxe search helices and peptides X-times longer than the default.

The resulting trace is output to `mymad.pdb`. The header lists the correlation coefficient of the trace against the data and the number of peptides and fragments.

A CC>25 % and/or an average fragment length >10 is a good sign for a correct solution. At moderate solution (better than 2.5 AA) one should also start seeing side-chains.

The power of autotracing is demonstrated by the Fibronectin test structure [16] (2 Å resolution at λ=1.77 Å, short wavelength dataset to 1.5 Å resolution). The progress and success of autotracing is show in Fig. 15.2. In the presence of NCS, the results are sometimes further improved by the '–nN' switch. It makes shelxe search for up to N-fold NCS in the substructure and apply the NCS operators to the traces.

15.6 MR-SAD

MR-SAD was first described in [12] and has been made widely available *via* the autorickshaw server [11]. It can be used when an MR solution provides too weak information in order to start model building but still contains valid information (of course one cannot tell *a priori*, but it is usually worth a try).

The solution produced by the MR-program of choice must be renamed to 'mymad.pda' (appropriate for the 'mymad' example used throughout here) and shelxe started as *e.g.*

```
shelxe mymad.pda mymad_fa -s0.65  -a10 -q
```

The MR-solution will be used to locate the substructure which are in turn used to produce experimental phases. Thus the *model bias from the poor MR-solution is removed*, but the phase information it contains is used to produce a better substructure than would be possible with SAD alone.

15.7 Anode – Post-Analysis

Anode [19] was written by G. Sheldrick as post-analysis program of the anomalous signal in the data. The command line

```
anode name
```

expects the files `name.ent` or `name.pdb` as the current model, and `name_fa.hkl` as source of the angle α in order to calculate the substructure phases $\Phi_A(hkl)$ by inverting Eq. 15.1. The anomalous difference map is written to `name.pha` and a list of main peaks as putative substructure to `name_fa.res`. The latter can be used together with `name.hkl` to rerun shelxe and e.g. test in the case of MR-SAD whether SAD would have been sufficient as phase information provided accurate enough substructure coordinates.

15.8 Collaboration with Other Programs

15.8.1 Add "Arcimboldo" to Dictionary

Add "Arcimboldo" to dictionary [15] is one of the fancier recent development in macromolecular crystallography because it allows *ab initio* solution of protein structures at 2 Å and better, *i.e.* the phase problem is overcome and a single native dataset can be used without the danger of model bias as in molecular replacement.

Arcimboldo uses phaser to place short model α-helices in the asymmetric unit. While these provide usually too little phase information for an interpretable map, they can be used by the autotracing capabilities of shelxe to be extended into a poly-ALA trace.

This methods appears to work to resolution of about 2 Å and better.

15.8.2 *hkl2map*

All shelx-programs are command line programs which sometimes deters people from using them. Some GUIs exist to facilitate learning shelx c/d/e. Most recommended is hkl2map [13]. It sets up the necessary scripts and plots useful graphics from the shelx output that help making decisions like finding the correct resolution cut-off for shelxd, *etc.* Its usage is fairly intuitive.

15.8.3 *Sharp/ AutoSharp*

Sharp [1] and autoSharp [21] both use shelxd to locate the substructure which sharp further refines. The refinement and combination with various density modification programs like dm [4] and auto-building programs like arp/warp [8] can be crucial in borderline cases and are particular useful for low-resolution structures.

15.8.4 *Autorickshaw*

The web-service autorickshaw [11] offers a "brute-force" approach to phasing and was one of the first pipelines to offer MR-SAD. The pipeline applies a vast number of combinations of various programs in order to get to a solution. It is probably the best demonstration that there is not *the best program* – it always depends on your particular data.

15.8.5 *Crank*

Like autorickshaw, Crank [10] combines and tests several combinations of programs and is suitable for SAD, MAD, and SIRAS phasing. It is available through CCP4 [4] and as such can be run standalone.

Acknowledgements Discussions with and presentations from G. Sheldrick are the main source of information for this document. Some parts of this document were copied verbatim from his presentations with permission.

References

1. Bricogne G, Vonrhein C, Flensburg C, Schiltz M, Paciorek W (2003) Generation, representation and flow of phase information in structure determination: recent developments in and around SHARP 2.0. Acta Crystallogr D Biol Crystallogr 59:2023–2030
2. Caliandro R, Carrozzini B, Cascarano GL, De Caro L, Giacovazzo C, Siliqi D (2005) *Ab initio* phasing at resolution higher than experimental resolution. Acta Crystallogr D Biol Crystallogr 61:1080–1087
3. Carter CW Jr, Sweet RM (eds) (1997) Macromolecular crystallography, vol 276(A), Methods in enzymology. Academic Press, New York, pp 307–326
4. Collaborative Computational Project Number 4 (1994) The CCP4 suite: programs for protein crystallography. Acta Crystallogr D Biol Crystallogr 50:760–763
5. Emsley P, Lohkamp B, Scott WG, Cowtan K (2010) Features and development of *Coot*. Acta Crystallogr D D Biol Crystallogr 66:486–501
6. Jia-xing Y, Woolfson MM, Wilson KS, Dodson EJ (2005) A modified *ACORN* to solve protein structures at resolutions of 1. 7 Å or better. Acta Crystallogr D Biol Crystallogr 61:1465–1475
7. Kabsch W (2010) XDS. Acta Crystallogr Biol Crystallogr 66:125–132
8. Morris RJ, Perrakis A, Lamzin V (2002) ARP/wARP's model – building algorithms. I. The main chain. Acta Crystallogr D Biol Crystallogr 58:968–975
9. Nanao MH, Sheldrick GM, Ravelli RBG (2005) Improving radiation-damage substructures for RIP. Acta Crystallogr D Biol Crystallogr 61:1227–1237
10. Ness SR, de Graff RAG, Abrahams JP, Pannu NS (2004) CRANK: new methods for automated macromolecular crystal structure solution. Structure 12:1753–1761
11. Panjikar S et al (2005) *Auto-Rickshaw*: an automated crystal structure determination platform as an efficient tool for the validation of an X-ray diffraction experiment. Acta Crystallogr D Biol Crystallogr 61:449–457
12. Panjikar S et al (2009) On the combination of molecular replacement and single-wavelength anomalous diffraction phasing for automated structure determination. Acta Crystallogr D Biol Crystallogr 65:1089–1097
13. Pape T, Schneider TR (2004) HKL2MAP: a graphical user interface for phasing with SHELX programs. J Appl Cryst 37:843–844
14. Ravelli RBG, Nanao MH, Lovering A, White S, McSweeney S (2005) Phasing in the presence of radiation damage. J Synchrotron Radiat 12:276–284
15. Rodríguez DD et al (2009) Crystallographic *ab initio* protein structure solution below atomic resolution. Nat Methods 6(9):651–653
16. Rudiño-Piñera E et al (2007) The solution and crystal structures of a module pair from the staphylococcus aureus-binding site of human fibronectin-a tale with a twist. J Mol Biol 368:833–844
17. Sheldrick GM (2008) A short history of *SHELX*. Acta Crystallogr A 64:112–122 G.M. Sheldrick,
18. Sheldrick GM (2010) Experimental phasing with SHELXC/D/E: combining chain tracing with density modification. Acta Crystallogr D Biol Crystallogr 66:479–485
19. Thorn A, Sheldrick GM (2011) *ANODE*: anomalous and heavy-atom density calculation. J Appl Cryst 44:1285–1287
20. Usón I, Stevenson CEM, Lawson DM, Sheldrick GM (2007) Structure determination of the *O*-methyltransferase NovP using the 'free lunch algorithm' as implemented in *SHELXE*. Acta Crystallogr D Biol Crystallogr 63:1069–1074
21. Vonrhein C, Blanc E, Roversi P, Bricogne G (2007) Automated structure solution with autoSHARP. Methods Mol Biol 364:215–230

Chapter 16
Use of a Weak Anomalous Signal for Phasing in Protein Crystallography: Reflection from Personal Experience

Felix Frolow

Abstract Determination in the past and the contemporary re-determination of several protein structures are considered in the context of our personal experience with the use of single-wavelength anomalous dispersion (SAD) techniques. A unifying feature of these cases is that in each event, the anomalous signal generated by metal ions was not optimized to the peak of the absorption edge, but was, nevertheless sufficient for structure re-determination. Examples are described with the emphasis on educational aspects. Determinants of success are discussed in each case. In addition, briefly discussed are some concepts related to the quality of the diffraction data used in experimental phasing, with special attention to SAD phasing.

Keywords Protein crystallography • Diffraction experiment • Non-optimized anomalous signal • Phasing • SAD • HKL-2000

16.1 Introduction

In the present and past Erice courses many concepts related to the determination of structures of protein molecules and complexes by experimental phasing have been discussed. Single-wavelength anomalous dispersion (SAD) and multi-wavelength anomalous dispersion (MAD) methods, which characterize contemporary crystal solution practice, have been extensively elaborated. Here we discuss several examples of structure re-determination of macromolecules and their complexes of various sizes using SAD method. Previously, these structures were determined with much effort, by the use of MIRAS and SIRAS techniques. These cases are

F. Frolow (✉)
Department of Structural Biology and Biotechnology, The George Wise Faculty of Life Sciences, Tel Aviv University, Tel Aviv 69978, Israel
e-mail: mbfrolow@post.tau.ac.il

R. Read et al. (eds.), *Advancing Methods for Biomolecular Crystallography*, NATO Science for Peace and Security Series A: Chemistry and Biology, DOI 10.1007/978-94-007-6232-9_16,

characterized by the anomalous signal generated by metal ions, but non-optimized to the peak of the absorption edge. Reasons for success with both past and present modes of structure determination will be analyzed for each case, and some concepts important for the successful experimental phasing of macromolecules will be discussed. We will focus on few topics related to data quality that are essential for a proper conduct of the experimental phasing but are often overlooked in the adrenaline-induced urgency and enthusiasm when working on a synchrotron experimental floor. Graphical results are from HKL-2000 [7], SHELXC/D/E [9], and COOT [2] programs. Some graphical material are provided here and the rest in the Power Point presentation that can be found in the site of the course.

16.2 Ferredoxin from *Haloarcula marismortui*

16.2.1 Background

Haloarcula marismortui is an archaebacterium that flourishes in the world's saltiest body of water, the Dead Sea. The cytosol of this organism is a supersaturated salt solution in which proteins are soluble and active. We determined the crystal structure of a 2Fe-2S ferredoxin from *H. marismortui* and found that the structure is analogous to those of plant-type 2Fe-2S ferredoxins of known structures, but with two important differences. First, the entire surface of the protein, except the vicinity of the iron-sulphur cluster, is coated with acidic residues. Secondly, two amphipathic helices are inserted near the N-terminus, forming a separate hyperacidic domain whose postulated function is to provide additional surface carboxylates for solvation. These data, together with the fact that bound surface water molecules have on the average 40 % more hydrogen bonds than a typical non-halophilic protein crystal structure, support the notion that haloadaptation is dependent on better water-binding capacity [4].

16.2.2 Methodology, Past and Present

Previously (in the early 1990s) we solved the structure of ferredoxin at 3 Å with difficulty, using a weak Pt derivative and an anomalous signal of the [2Fe-2S] cluster of the native crystal. We collected data on a CAD-4 diffractometer equipped with a Cu sealed tube, at a wavelength of 1.5418 Å. We used the PHASES package [5] because of its unique ability utilize the anomalous signal of the native crystal in the phasing procedure. We used the FRODO [6] program to trace the structure, albeit with difficulty, on an Evans & Sutherland PS300 graphics workstation. After acquiring the data to 1.9 Å resolution on a Xentronix multiwire area detector, we were able to extend the phases by the refinement procedure using XPLOR (A.T. Brunger, 1990, New Haven, Yale University). Subsequent improvement in the crystallization of ferredoxin enabled us to collect the high-resolution data to 0.87 Å at CHESS in 2001, and the structure was swiftly re-determined using the SHELXC/D/E pipeline.

16.2.3 Conclusions/Significance

The data set was collected on an ADSC area detector from a distance of 50 mm, in 400 diffraction images, covering 80° with 0.2° per frame, at a wave length of 0.949 Å. Collinear with the rotation axis, a stream of cold nitrogen was used to cool the crystal. We carved a segment of exceptional quality from the data, comprising 300 diffraction frames, 80,596 unique reflections with the redundancy of 2.49, 88 % completeness, $R_{sym} = 0.024$, and exhibiting an excellent anomalous signal (Fig. 16.1a). Segmentation of the data and quality assessment were assisted by the information from HKL-2000 in the graphical form, which was very easy to understand (Fig. 16.1b).

16.3 Bacterioferritin from *Escherichia coli*

16.3.1 Background

Bacterioferritin (also known as cytochrome b1) of *Escherichia coli* is a hollow, nearly spherical shell made up of 24 identical protein subunits and 12 haem groups. We solved this structure in a tetragonal crystal form at 2.9 Å resolution. Each haem is bound in a pocket formed by the interface between a pair of symmetry-related subunits. The quasi-twofold axis of the haem group is closely aligned with the local twofold axis relating these subunits. The axial ligands of the haem group are sulfur atoms of two equivalent methionyl residues (Met 52) from the symmetry-related subunits. A cluster of four water molecules is trapped in the gap between the upper edge of the haem and two extended protein loops that close off the haem from the outer aqueous environment. This was the first structure of a bis-methionine ligated haem-binding site and the first case of a twofold symmetric haem-binding site.

16.3.2 Methodology, Past and Present

In 1994, we determined the structure of bacterioferritin using the SIRAS method based on a UO_2-acetate heavy-atom derivative with four distinct sites in each monomer [3]. The main difficulty in this project was to determine the subset structure of the heavy atoms. Owing to complications related to the relatively high symmetry ($P4_22_12$), as well as to the relatively large number of heavy atoms in the

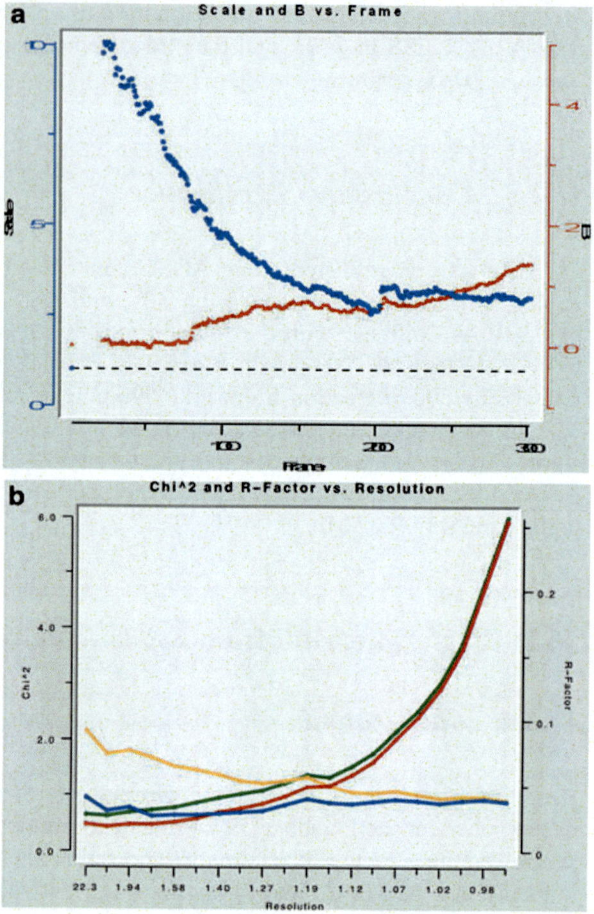

Fig. 16.1 Demonstration of selected results for the processing of diffraction data of ferredoxin that was measured at CHESS. (**a**) Linear (strongly falling due to the short beam lifetime of CHESS) and exponential (representing radiation decay, almost constant) components of the scale factor. (**b**) Presence of the excellent anomalous signal as exhibited by results of HKL2000

asymmetric unit (44 atoms) and the pseudo-symmetry (close to cubic), it was impossible to start de-convolution of the heavy-atom structure from the Patterson function, which exhibited very crowded Harker sections. By experimenting with the pseudo-cubic platinum derivative, we were able to determine partial structure, which after transformation into the tetragonal space group of bacterioferritin produced phases sufficient to locate heavy atoms. By applying methods of electron density modification, including solvent flattening and non-crystallographic symmetry (NCS) averaging, we produced an electron map of sufficient quality to trace the molecule.

In 1995, we collected data from the UO_2-acetate derivative to extend the resolution using imaging plates mounted on a Weissenberg camera at the KEK Photon Factory in Tsukuba, Japan. We used a wavelength of 1 Å and a parallel but broad beam of 0.15 mm. The beamline we used was not equipped with a cryo-device but was kept at 1 °C and in very dry atmosphere with slight overpressure. Despite the large size of the imaging plate, we were able to collect data only to 2.5 Å.

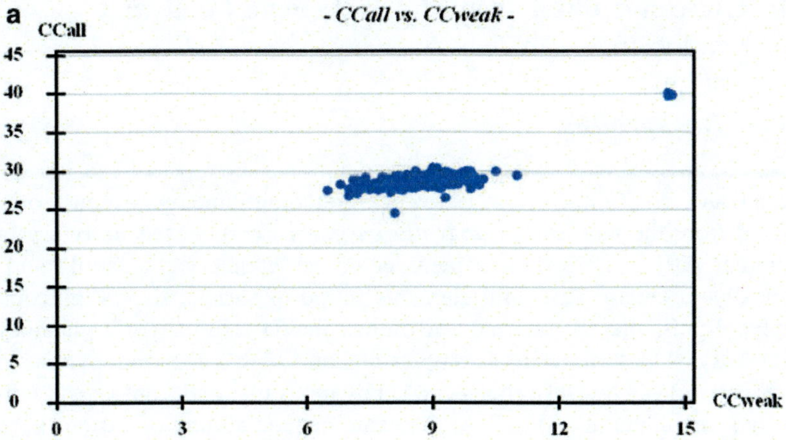

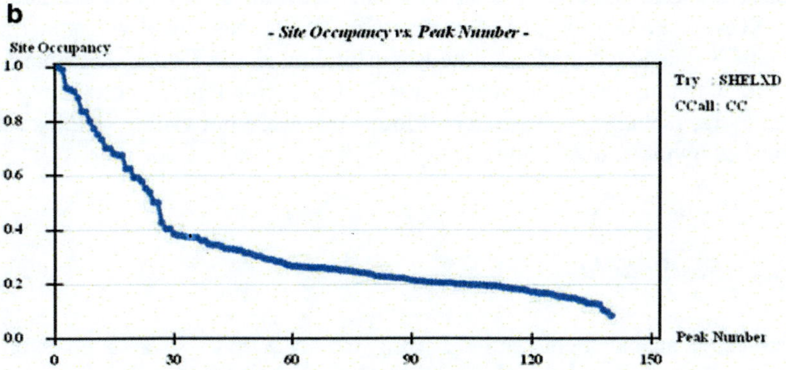

Fig. 16.2 Demonstration of selected results for the determination of heavy atom substructure of bacterioferritin molecule. (**a**) Just few trials in 1,000 attempts using SHELXF produce the correct substructure determination. (**b**) Around 100 sites are visible, and can be divided into two groups: group 1, of 28 atoms with occupancies changing rapidly between 1.0 and 0.4, and group 2, of 62 atoms with occupancies changing slowly between 0.4 and 0.1

We recently used this data set to re-determine the structure using the SHELXC/D/E pipeline in SAD mode (Fig. 16.2a, b). Although the heavy-atom substructure was determined very rapidly, an electron-density map at 2.5 Å resolution is not easy to trace automatically. Substantial manual intervention was required.

16.3.3 Conclusions/Significance

Structure determination of complicated systems such as bacterioferritin is made less work-intensive because of development of the SAD method and its efficient implementation in SHELXC/D/E.

16.4 Concanavalin A from the Jack-Bean *Canavalia ensiformis*

16.4.1 Background

Concanavalin A (ConA) is a lectin (carbohydrate-binding protein) originally extracted from the jack-bean *Canavalia ensiformis* and consequently crystallized by Sumner [10] in 1919. It is a member of the legume lectin family and binds specifically to certain structural elements found in various sugars, glycoproteins and glycolipids, mainly internal and non-reducing terminal α-D-mannosyl and α-D-glycosyl. The structure of a complex of ConA with α-D-mannopyranoside was determined in 1989 by Derwenda et al. [1] by molecular replacement and refined in 1994 by Naismith et al. [8] to 2 Å. We examined the affinity of various sugars for the ConA binding site and re-determined the structure of the ConA complex using SHELXC/D/E SAD approach. A measurable anomalous signal was generated by Mn and Ca ions associated with the sugar-binding site in ConA. The added value obtained with this re-determination was in the corrections of certain regions of the structure, including some external loops that were not visible in the structures determined by molecular replacement.

16.4.2 Methodology

We collected an excellent data set using the RAXISIV++ area detector. Data were complete to 1.6 Å resolution with $R_{sym} = 0.027$ and redundancy of 4.29. An anomalous signal of good quality exhibited itself on the graphical display of HKL-2000 GUI. The SHELXD program was remarkably successful in enabling us to determine correct heavy-atom subset in almost every trial (Fig. 16.3a, b). Experimental phases after electron-density modification in SHELXE were used for auto-tracing, revealing the complete structure by the end of the process.

16.4.3 Conclusions/Significance

Structure determination by the molecular replacement method may, especially if the model used is not entirely correct, introduce significant model bias that will be almost impossible to eliminate. The ultimate solution to this problem is to attempt to use experimental phases derived from anomalous signal.

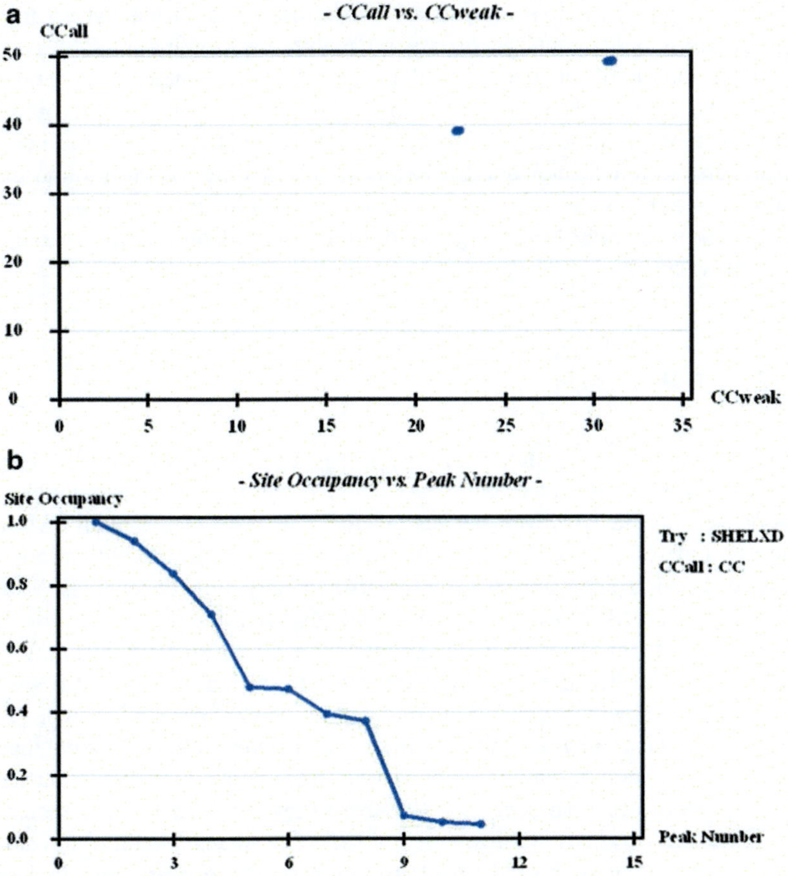

Fig. 16.3 ConA: selected results of the heavy atom substructure determination. (**a**). Correct substructure is determined in 87 % of trials. The remaining 13 % are similar and easily lead to the correct protein structure. (**b**) Clearly visible four Mn and Ca atoms with occupancies of 1–0.75 and 0.43–0.39 respectively

16.5 Photosystem I from Higher Plants

16.5.1 Background

Oxygenic photosynthesis is the principal producer of both oxygen and organic matter on earth. The conversion of sunlight into chemical energy is driven by two multi-subunit membrane protein complexes termed photosystem I and II. We determined the crystal structure of the complete photosystem I (PSI) from a higher plant (*Pisum sativum* var. Alaska) to 4.4 Å resolution. Its intricate structure shows 12 core subunits, 4 different light-harvesting membrane proteins (LHCIs) assembled in a half-moon shape on one side of the core, 45 transmembrane helices, 167 chlorophylls, 3 Fe-S clusters, and 2 phylloquinones. About 20 chlorophylls are positioned in strategic locations, in the cleft between LHCIs and the core. This structure provides a framework for exploration not only of energy and electron transfer but also of the evolutionary forces that shaped the photosynthetic apparatus of terrestrial plants after the divergence of chloroplasts from marine cyanobacteria one billion years ago.

16.5.2 Methodology

The core structure of the plant and bacterial PSI are relatively similar. Applying the molecular replacement method we found a solution with two molecules in the asymmetric unit. However, because of the large size of the plant PSI and the unknown mode of association of LHCIs and accompanying modifications in the core, we decided to determine the structure using heavy-atom derivatives. Molecular replacement was instrumental in generating phases for detection of the heavy-atom superstructures of each of 20 different heavy-atom derivatives. When we calculated anomalous difference map for each derivative, in each case we found 6 equivalent prominent peaks with a large gap to the next peaks. These peaks were present also in the anomalous map of the native structure. We understood that these peaks were generated by an anomalous signal from Fe4-S4 inorganic clusters that are present in the stromal part of PSI, where they serve to transfer electrons to shuttling ferredoxins (part of the activity of PSI). The slow and tedious development of the structure was supported by contribution of the phases generated by this anomalous signal. This contribution was of the paramount importance (unpublished results), because heavy-atom derivatives were found to be not very isomorphous to the native crystal and of partial occupancy with almost undetectable anomalous contribution.

16.5.3 Conclusions/Significance

Even for the structure determination of very large systems such as plant PSI, a well-defined anomalous signal may have ultimate influence on structure determination.

16.6 Phasing Using an Anomalous Signal from Metal Ions of Low Occupancy – The Case of Tetragonal Hen Egg-White Lysozyme

16.6.1 Background

We postulate that cocktails of ions of the various transition metals, such as Co, Ni, Cu and Zn, can be used during the crystallization process of protein molecules to increase the probability of their specific incorporation into the protein structure, thereby enabling determination of diffraction phases by the use of an anomalous diffraction signal associated with the ion's presence. We performed a feasibility study on the phasing of the protein diffraction data using an anomalous signal from the Zn^{+2} ion. The research was aimed at determining the lower limit of Zn^{+2} concentrations in the crystallization solution. Using the model protein molecule hen egg-white lysozyme (HEWL), we obtained groundbreaking results showing that even with only a fraction of the occupancy ($\sim$0.2) of Zn^{+2} binding sites it was possible to allow us to determine the protein's structure by SAD techniques using SHELXC/SHELXD/SHELXE. This finding implied that the concentration of transition-metal ions introduced into the crystallization solution could be much lower (approximately 0.2 M) than was previously thought.

16.6.2 Methodology

Diffraction data from HEWL crystals were measured in two separate sessions of synchrotron radiation at ID23-1 and ID29 beamlines at the European Synchroton Radiation Facility in Grenoble, France. An ADSC Q315 detector was used in both cases. The X-radiation wavelength was determined for each diffraction experiment by analysis of the X-ray fluorescence scattering from each individual crystal before the diffraction data were measured. For data collection, crystals were mounted on a MiTeGen stiff micromount made of polyimide. Consequently, they were flash-cooled in a nitrogen stream produced by an Oxford Cryostream low-temperature generator at a temperature of 100 K.

16.6.3 Conclusions/Significance

A surprising result was that relatively low concentrations of Zn^{+2} acetate (0.2–0.3 mM) in the protein crystallization solution were sufficient to produce crystals with anomalous signals that allowed us to determine the protein structure. This low concentration was associated with a low molar ratio of Zn^{2+} to protein and a relatively low occupancy of Zn^{2+}-binding sites. Our results enabled us to define the concentration limits of transition metals in crystallization solution formulations. We have developed a limited kit of such solutions to be used in crystallization experiments.

16.7 Utilization of Unexpected Anomalous Scatters: CBM3b-ScaA from *Acetivibrio cellulolyticus*

16.7.1 Background

The carbohydrate-binding module (CBM) from the major scaffoldin subunit ScaA of the cellulosome of *Acetivibrio cellulolyticus* binds cellulose and is classified as the family 3b CBM module. The CBM3b was overexpressed, purified and crystallized. The crystals belonged to hexagonal space group $P6_122$ with unit-cell parameters $a = b = 52.50$, $c = 193.97$ Å, and diffracted to a resolution of 1.08 Å with synchrotron radiation.

16.7.2 Methodology

The structure was determined by molecular replacement and refined to a final R and R_{free} of 0.103 and 0.145 respectively [11]. In addition to a "classical" CBM3 calcium-binding site, the structure contained a nickel-binding site located in the vicinity of the N terminus (remains of the His-tag sequence designed for cleavage by thrombin). The structure could also be determined independently by the SAD method using data collected at the Ni absorption peak wavelength of 1.48395 Å and even, in a favorable case, at wavelength 0.97625 Å (Fig. 16.4a, b).

16.7.3 Conclusions/Significance

The excellent quality of the diffraction data enabled us to determine the structure using SAD. This could be accomplished without experimental optimization of the anomalous signals of Ni2+ and Ca2+ ions (i.e., collection of data at the Ni or Ca absorption edge).

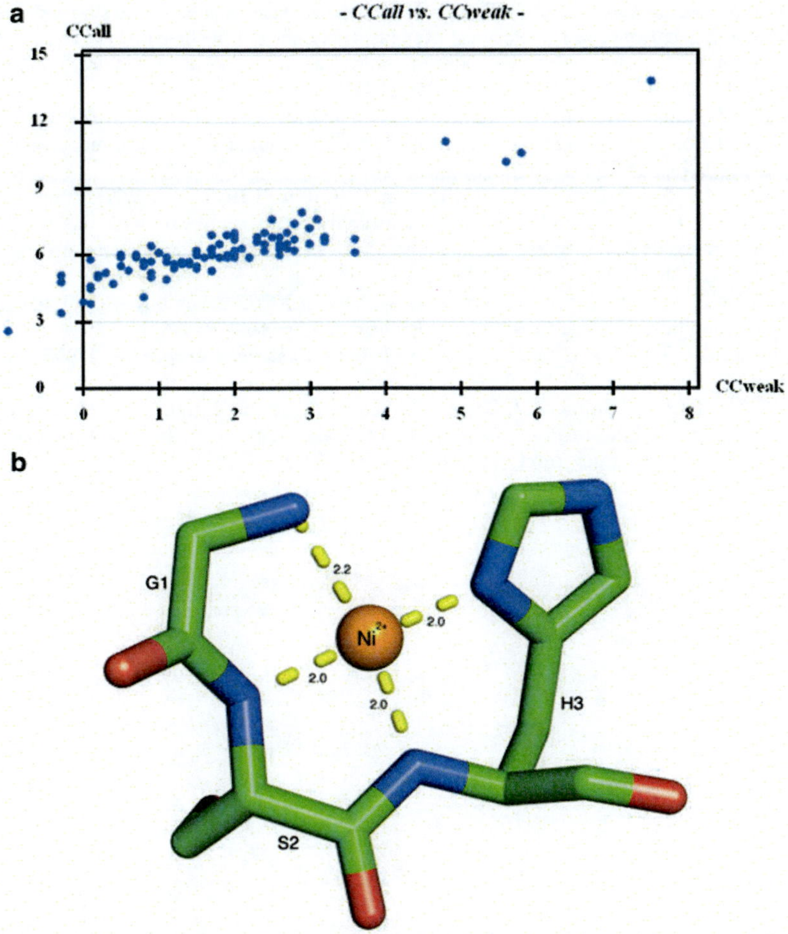

Fig. 16.4 CBM: selected results of the structure determination. **(a)** Trials of heavy-atom substructure determination for CBM from *Acetivibrio cellulolyticus* produce outstanding but marginal substructure solution. **(b)** N-terminal part of the structure exhibiting the incorporated Ni^{2+} ion

References

1. Derewenda Z, Yariv J, Helliwell JR, Kalb AJ, Dodson EJ, Papiz MZ, Wan T, Campbell J (1989) The structure of the saccharide-binding site of concanavalin A. EMBO J 8:2189–2193
2. Emsley P, Lohkamp B, Scott WG, Cowtan K (2010) Features and development of Coot. Acta Crystallogr D Biol Crystallogr 66:486–501
3. Frolow F, Kalb AJ, Yariv J (1994) Structure of a unique twofold symmetric haem-binding site. Nat Struct Biol 1:453–460
4. Frolow F, Harel M, Sussman JL, Mevarech M, Shoham M (1996) Insights into protein adaptation to a saturated salt environment from the crystal structure of a halophilic 2Fe-2S ferredoxin. Nat Struct Biol 3:452–458

5. Furey W, Swaminathan S (1997) PHASES-95: a program package for processing and analyzing diffraction data from macromolecules. Methods Enzymol 277:590–620
6. Jones TA (1985) Diffraction methods for biological macromolecules. Interactive computer graphics: FRODO. Methods Enzymol 115:157–171
7. Minor W, Cymborowski M, Otwinowski Z, Chruszcz M (2006) HKL-3000: the integration of data reduction and structure solution–from diffraction images to an initial model in minutes. Acta Crystallogr D Biol Crystallogr 62:859–866
8. Naismith JH, Emmerich C, Habash J, Harrop SJ, Helliwell JR, Hunter WN, Raftery J, Kalab AJ, Yariv J (1994) Refined structure of concanavalin A complexed with methyl alpha-D-mannopyranoside at 2.0 A resolution and comparison with the saccharide-free structure. Acta Crystallogr D Biol Crystallogr 50:847–858
9. Sheldrick GM (2010) Experimental phasing with SHELXC/D/E: combining chain tracing with density modification. Acta Crystallogr D Biol Crystallogr 66:479–485
10. Sumner JB (1919) The Globulins of the Jack Bean, *Canavalia ensiformis*. J Biol Chem 37: 137–142
11. Yaniv O, Halfon Y, Shimon LJ, Bayer EA, Lamed R, Frolow F (2012) Structure of CBM3b of the major cellulosomal scaffoldin subunit ScaA from *Acetivibrio cellulolyticus*. Acta Crystallogr Sect F Struct Biol Cryst Commun 68:8–13

Chapter 17
Ab Initio Low Resolution Phasing

Vladimir Y. Lunin, Natalia L. Lunina, and Alexandre G. Urzhumtsev

Abstract Low resolution *ab initio* phasing technique may give the first crystallographic images of macromolecules or their complexes when other crystallographic approaches fail to do it due to poor diffraction quality of crystals, in particular those for membrane proteins. This phasing technique uses a set of observed structure factor magnitudes and some general (mathematical) properties of correct low-resolution Fourier syntheses to restore phase values. The paper gives a review of such general properties and discusses the features of their application.

Keywords *Ab initio* phasing • Low resolution • Connectivity • Likelihood • Density histograms • Few Atoms Model • Cluster analysis

17.1 Introduction

Phase determination is a necessary step to transform a set of diffraction magnitudes into images of the electron density distribution. In macromolecular crystallography conventional ways to solve this problem involve either additional diffraction experiments (using modified wavelengths or crystal content) or knowledge of a model of a homologous object. There exist also methods capable to solve the structure using a single set of structure factor magnitudes and some general

V.Y. Lunin (✉) • N.L. Lunina
Institute of Mathematical Problems of Biology, Russian Academy of Sciences, Pushchino 142290, Russia
e-mail: lunin@impb.psn.ru

A.G. Urzhumtsev
IGBMC, CNRS-INSERM-UdS, 1 rue Laurent Fries, B.P.10142, 67404 Illkirch, France

Physics Department, Faculté des Sciences et des Technologies, Université de Lorraine, B.P. 239, 54506 Vandoeuvre-lès-Nancy, France

R. Read et al. (eds.), *Advancing Methods for Biomolecular Crystallography*, NATO Science for Peace and Security Series A: Chemistry and Biology, DOI 10.1007/978-94-007-6232-9_17, © Springer Science+Business Media Dordrecht 2013

properties of the electron density distribution ('atomicity' as a rule). In this paper we call these phasing methods '*ab initio*' or 'direct' although sometimes these terms are reserved for a more broad meaning. Such methods are routine in 'small molecules' crystallography. Last decades they came into the macromolecular field [18], however their application requires a high-resolution data set (about 1 Å, usually). In this paper we consider an opposite case, namely low resolution phasing, when the low-resolution edge reflections only are used [11, 13, 14]. The phasing of such reflections cannot provide one with a detailed structure of the studied object but the information obtained may play a significant role for further success of the structure determination. In its current state, the low-resolution *ab initio* phasing procedure is laborious and requires a complete set of very low resolution data which are often missed using standard data collection setups. This procedure is not a routine tool of a structural analysis but can be used when standard methods fail.

In this paper we use the term *low resolution* to note up to a few hundred reflections of the lowest resolution for the given crystal. Depending on the size of the unit cell a Fourier synthesis calculated with these structure factors may present different information on the object studied. If the structure factors correspond to the resolution approximately $d_{min} > 16$ Å, this information concerns mostly the macromolecular envelope and its position in the unit cell simplifying the translation and eventually rotation search in the molecular replacement and facilitating the use of complementary sources of information like electron microscopy reconstructed images. If the resolution reaches approximately 8 Å, Fourier syntheses may show α-helixes while β-sheets may be identified at the resolution of about 4–5 Å. Syntheses of an intermediate resolution 16 Å $> d_{min} >$ 8 Å are the most difficult for interpretation and overcoming this resolution interval presents the largest difficulties in *ab initio* phasing. Some examples of low-resolution phasing may be found in [12, 13, 15].

The basic idea of the discussed approach is exploration of the configuration space of all possible phase sets by Monte Carlo type procedures. Expected general-type properties of electron density distribution in macromolecular complexes are used as *selection criteria* to filter randomly generated phase sets and to form a collection of admissible ones. Cluster analysis and averaging allow getting an approximate solution and preparing advanced steps of the phasing procedure.

17.2 Basic Definitions

We suppose that the input of the phasing procedure is a set of low resolution structure factors magnitudes $\{F^{obs}(\mathbf{s})\}$, $\mathbf{s} \in S$, and we aim to find structure factor phases $\{\varphi(\mathbf{s})\}$, $\mathbf{s} \in S$ that allow calculating Fourier synthesis

$$\rho(\mathbf{r}) = \frac{1}{V_{cell}} \sum_{s \in S} F^{obs}(\mathbf{s}) \exp[i\,\varphi(\mathbf{s})] \exp[-2\pi i\,(\mathbf{s}, \mathbf{r})],$$

an approximation to the true electron density distribution. Below, we say *magnitudes* and *phases* implying the structure factor magnitudes and phases.

A *phase variant* or simply a *variant* (of the solution of the phase problem) stands for any set of phases $\{\varphi(\mathbf{s})\}$, $\mathbf{s} \in S$. The final goal is to find a variant that is as close as possible to the true solution of the phase problem. Sometimes it is convenient to consider variants as points in M-dimensional configuration space of all possible phase sets. (Here M is the number of reflections in the set S).

A phase variant is considered as a *good one* if it is close enough to the true solution, a variant is *bad* if it differs significantly from the true solution. This definition is not strict and is used to describe phasing tendencies qualitatively. It is obvious that the comparison of a variant with the true phases may be done only when testing the method and not in the process of a structure determination.

A *population* is any set of phase variants. We start each cycle of the phasing procedure by generating randomly a large *start population* and then apply some *selection criteria* to reduce the start population. The leading idea is to get a *population enriched by good variants* in comparison with the start one.

Random variants may be generated in different ways. In the first one all phases are considered as independent random variables and, in absence of any phase information, as equally probable. If a phase probability distribution for a reflection $\mathbf{s}$ is available, it may be used to generate random phase values distributed no more uniformly. Usually, a unimodal von Mises distribution (known also as 'circular normal', 'Sim' *etc.* distribution) is used.

Alternatively, the phases are calculated from some randomly generated models. For example, a phase set may be calculated from a model containing a small number of huge 'globs' with randomly generated coordinates [8]. The glob coordinates may be distributed uniformly in the whole unit cell at the beginning of phasing and inside a molecular mask later when some phase information becomes available.

17.3 Selection Criteria

By a *selection criterion* we mean a rule that is applied to reduce the number of variants in a population. Very often this criterion estimates numerically a quality of the phase set and the selection rule consists in the rejecting a variant if the value of this criterion is below (or above) some cut-off level. In classical direct methods such criteria are called 'figures of merit', but in this paper we reserve the term *figure of merit* for the reliability of the phase value of a particular reflection rather than for the quality of the whole phase set.

Two ways to generate phase variants lead to two different types of selection criteria. First, for each trial phase set, a Fourier map can be calculated using experimental magnitudes. A selection rule may check whether the map looks like a macromolecular one. If phase sets are calculated from a model, then corresponding calculated structure factor magnitudes are also available, and their similarity to the observed values can be used as well to estimate the corresponding phase set.

Main ideas on possible low resolution criteria are discussed below.

17.3.1 Fourier Syntheses Histograms

The histogram of a Fourier synthesis indicates how frequently different values appear in the synthesis (see *e.g.* Lunin [3] and references therein). The histogram corresponding to a macromolecular crystal has a specific asymmetric shape. It depends on the synthesis resolution and solvent content and may be predicted with a reasonable accuracy. When the true (or predicted) histogram is known, it can be used to select appropriate Fourier syntheses. For example, a selection criterion may be defined as the correlation of the true histogram with that for the Fourier synthesis calculated with the observed magnitudes and trial phases.

17.3.2 Connectivity

Another approach to judge the quality of a phase set is based on topological properties of regions of high electron density, *e.g.* their connectivity. A visual inspection of continuous regions in the Fourier maps and the absence of noisy peaks are used routinely to estimate the map quality. Baker et al. [1] suggested a formal scheme how to use the connectivity for phase improvement. These ideas can be incorporated also into low-resolution phasing [10]. For a Fourier synthesis calculated with the observed magnitudes and trial phases, each trial phase set becomes associated with a mask region composed by the grid points with the synthesis values above some cut-off level. The simplest topological characteristics of the region are the number of connected components in the region and their size. If the synthesis resolution is low and the cut-off level is high enough, it is expected that the mask consists of a small number of 'globs' corresponding to individual molecules. The coincidence of the number of connected components in the mask with the desired number (for example, a known number of molecules in the unit cell) may be used as a connectivity-based selection criterion. More sophisticate criteria can be introduced as well.

17.3.3 Likelihood

Statistical likelihood estimates the probability to reproduce experimental values in a framework of a suggested statistical hypothesis. The likelihood of a trial mask region may be defined as the probability to get the values of calculated magnitudes equal to the observed ones when placing atoms randomly and uniformly inside the mask and calculating structure factors from such a random model [9]. The higher the probability, the more likely the mask. Such an approach may be used to select a mask among several alternatives. In an advanced approach this is done considering more general prior probability distributions for atomic coordinates [2]. The likelihood turns to a phase-set selection criterion if every phase set is associated with a mask. Such mask may be constructed *e.g.*, as a region of highest values in the Fourier synthesis calculated with observed magnitudes and trial phases [16]. The likelihood may be calculated by a straightforward Monte-Carlo type procedure [16] or using an analytical [5] or more complicated saddle-point approximations [2] of the likelihood function. The set of reflection used to calculate the likelihood may be different from the set of reflection for trial phases.

17.3.4 Few Atoms Method

At a low resolution the content of a macromolecular crystal may be represented reasonably by a small number of large isotropic Gaussian scatterers [8], by one sphere as an extreme case. The number of scatterers necessary for an appropriate modeling depends on the molecular shape and the resolution. Their size, represented by the *ADP* parameter (*B* factor), can be estimated at the first step of the procedure. At the beginning the coordinates of the spheres can be generated randomly and uniformly in the unit cell. For an advanced search when the molecular envelope is already known, the spheres can be generated inside this envelope. More complicate generation rules can be applied as well. For a generated model, a set of structure factors is calculated and their magnitudes are compared with the experimental data. If these values are close enough, the corresponding phase set is selected for further analysis. In the simplest case when the molecule is approximated by a single sphere, a systematic search of the position of this sphere in the unit cell becomes possible instead of a random one. However the result of such a search is rather unpredictable: the chosen center of the sphere may be in the middle of the molecule, may belong to the solvent region, or even belong to an interface of the molecules.

17.4 General Properties of Low Resolution Selection Criteria and Multi-filtering Phasing Method

All the selection criteria mentioned above (as well as some others tried in order to select phase sets) show similar features when applied at low resolution [11, 13]:

- The best value of a selection criterion may correspond to a bad phase set.
- A selection criterion value for a good phase set may be rather bad.
- Local refinement of model parameters or phases may lead to significant improvement of the selection criterion value without any improvement in the phases.

Owing to these features, searching for the solution of the low-resolution phase problem by minimizing (or maximizing) a selection criterion is unreliable. At the same time, we observed that

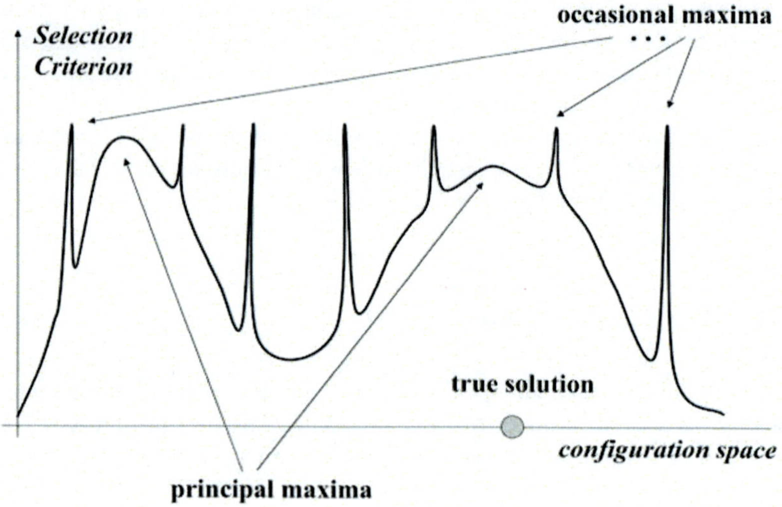

Fig. 17.1 Schematic illustration of a low-resolution selection criterion in the configuration space of phase sets. The true solution is in a vicinity of one of broad 'principal' maxima. In a favorable case there is a single 'principal' maximum, however a few of them can coexist in the general situation. There are also a large number of sharp 'occasional' maxima. Usually a local phase refinement is trapped by one of them

- the population selected on the base of an appropriate criterion is enriched, *i.e.* it has a higher percent of good variants in comparison with a random start population.

The features of the selection criteria and phasing procedures can be summarized in the following conclusion concerning the 'profile' of such criteria illustrated by Fig. 17.1. A selection criterion has a lot of local very narrow 'occasional' maxima. A search procedure can generate a phase variant in a neighborhood of such a maximum; this explains why local refinement can increase the criterion value without real phase improvement. These occasional maxima look like modulations of a basic 'surface' of the criterion. This surface contains a few large 'principal' maxima, one of them corresponding to the correct phase solution; usually, it is the largest one. If the selection cut-off value of the criterion has been correctly chosen, the selected population of phase variants contains a lot of phase variants centered on the correct solution. Therefore, this solution can be obtained by averaging the selected variants. In the general case, several principal maxima can be present, the selection reveals several groups of phase variants; they should be averaged separately giving several possible solutions to be taken and verified at further steps.

Figure 17.2 illustrates a general scheme of the phasing procedure. The cycle may be repeated several times varying the random-phase generating mode, selection rules, the set of reflections to be phased, *etc.*

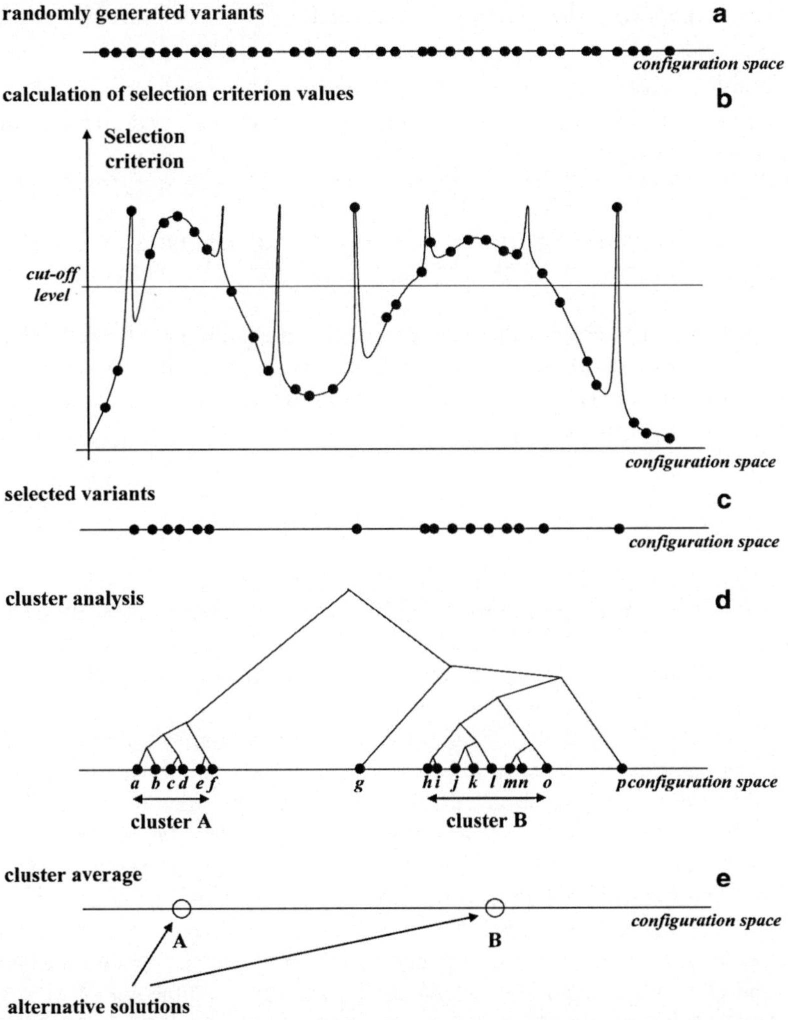

Fig. 17.2 Determination of the principal maxima of low-resolution selection criteria. **(a)** A starting population of points in the configuration space (of all phase sets) is chosen covering this space uniformly. Probability to generate such a point in an occasional maximum is small but non-zero. **(b)** The criterion value is calculated for every point of the start population. The points beyond a threshold are chosen for further analysis. **(c)** The selected points may form several compact clusters corresponding to principal maxima. A small number of points corresponding to occasional maxima may be present too. **(d)** Cluster analysis procedures allow removing occasional points as isolated ones and revealing one or a few compact clusters in the configuration space. **(e)** Averaging procedure performed separately for every cluster provides one with a few alternative solutions corresponding to the principal maxima of the selection criterion

17.5 Processing the Selected Variants

The simplest procedure consists in averaging all selected variants. For every reflection **s** from the set S the best (centroid) phase $\varphi^{best}(\mathbf{s})$ and its figure of merit $m(\mathbf{s})$ are calculated as

$$m(\mathbf{s}) \exp\left[i\,\varphi^{best}(\mathbf{s})\right] = \frac{1}{K}\sum_{j=1}^{K} \exp\left[i\,\varphi_j(\mathbf{s})\right],$$

where $\varphi_j(\mathbf{s})$ is the phase in j-th selected variant corresponding to the reflection **s** and K is the number of variants in the selected population. The figure of merit reflects the divergence of the phases corresponding to the same reflection in different selected variants. It is worthy of noting that

$$m(\mathbf{s}) = \frac{1}{K}\sum_{j=1}^{K} \cos\left(\varphi_j(\mathbf{s}) - \varphi^{best}(\mathbf{s})\right),$$

so that $m(\mathbf{s}) = 1$ if the phase for this reflection is the same in all selected phase variants and $m(\mathbf{s}) \approx 0$ if the phases are distributed almost uniformly in the $[0,2\pi]$ interval. The found m, φ^{best} values may be used to define the probability distribution for the next cycle of random phase generation.

Two phase sets, apparently different, may result in Fourier syntheses that differ only by an origin shift **t** permitted for the given space group (and/or enantiomer choice). For example, this is the case of two syntheses calculated with the same magnitudes $\left\{F^{obs}(\mathbf{s})\right\}$, $\mathbf{s} \in S$ and with two phase sets, $\{\varphi_1(\mathbf{s})\}$ and $\varphi_2(\mathbf{s}) = \varphi_1(\mathbf{s}) + 2\pi(\mathbf{s}, \mathbf{t})$, respectively. These variants must be considered as equivalent ones. At the same time the result of a direct comparison or averaging these sets is unpredictable. To calculate a proper estimate of the closeness of two phase variants a map *alignment* with respect to the choice of the origin and enantiomer must be performed [4, 7]. For example, the translation vector $\mathbf{t}^*$ and the sign $\theta = \pm 1$ must be found such that make the differences $\varphi_1(\mathbf{s}) - [\theta \cdot \varphi_2(\mathbf{s}) - 2\pi(\mathbf{s}, \mathbf{t})]$ for the set $\mathbf{s} \in S$ as small as possible, and the comparison and the averaging must be applied to the phase sets $\{\varphi_1(\mathbf{s})\}$ and $\left\{\varphi_2^*(\mathbf{s})\right\}$ where $\varphi_2^*(\mathbf{s}) = \theta \cdot \varphi_2(\mathbf{s}) - 2\pi(\mathbf{s}, \mathbf{t}^*)$. This procedure of the phase alignment is especially important at first stages of phasing when any phase values may be generated.

A more accurate treatment of a selected population involves methods of cluster analysis. Cluster analysis is a developed branch of applied mathematics aimed to separate a set of points in a multidimensional space into several compact groups of points called *clusters* (or *classes*) so that the points inside a particular cluster are close to each other while different clusters are distanced in space. Methods of cluster analysis use the matrix of point-to-point distances as input information. The mean phase difference in two previously aligned phase sets (as well as other measures for phase sets closeness) may be used to calculate such a matrix in our case. If the cluster analysis shows that the selected population can be divided into several significantly different clusters, then the averaging is performed in each cluster separately, resulting in several alternatives for the solution of the phase problem. This creates a sort of branching of the procedure requiring multisolution strategies.

17.6 Example of Low Resolution *Ab Initio* Phasing: Low Resolution Structure of Na+-NQR

The multi-subunit membrane protein complex (210 kDa, MW) of Na^+-translocating NADH:ubiquinone oxidoreductase from *Vibrio cholerae* generates a Na^+ gradient that is essential for substrate uptake, motility, pathogenicity and efflux of antibiotics. The crystals of Na^+-NQR have low diffractive power and only a few crystals diffracted to 4 Å resolution – a phenomenon known for many other membrane protein crystals. After unsuccessful attempts to apply commonly used phasing methods an *ab initio* phasing procedure was tried to obtain primary structural information [15].

For a crystal with a large unit cell very-low resolution reflections are usually shadowed by a beam-stop making their measurement difficult. To collect these data for Na+-NQR a special experimental setup developed earlier [17] was used. Briefly, the combination of a helium tube and a small beam stop placed at the detector side of the helium tube allows data collection between 300 and 8 Å resolution at the beam line ID14-4 ESRF (Grenoble, France).

The phasing procedure was based on the observation that a high-density region in a low-resolution macromolecular Fourier synthesis is composed of a small number of connected components. For each randomly generated phase set the number of connected components in the high-density region of the corresponding Fourier synthesis was calculated as well as the components' size. A phase set was considered as admissible and stored for further analysis if the corresponding Fourier synthesis had acceptable connectivity properties. The phase sets were generated until the desired number of admissible phase sets (100 in the reported work) has been obtained. The phasing procedure for Na^+-NQR consisted of 21 cycles. In the first cycle 32 reflections in the 36 Å resolution zone were used; the number of reflections was increased stepwise up to 838 reflections in the 12 Å resolution zone at the final cycles. In the initial cycles, the high-density region was defined as 15 % of the grid points possessing the highest values in the Fourier synthesis. This region was extended to 20 % of points in the final cycles. The applied selection rule allowed a high-density region composed of only two connected components of equal volume.

In the phase set generation process, every phase value was generated with its individual probability distribution, which was updated after every phasing step. Generally, the reflections were generated according to the Von Mises distribution. The parameter value was chosen such that the expected value of the cosine of phase

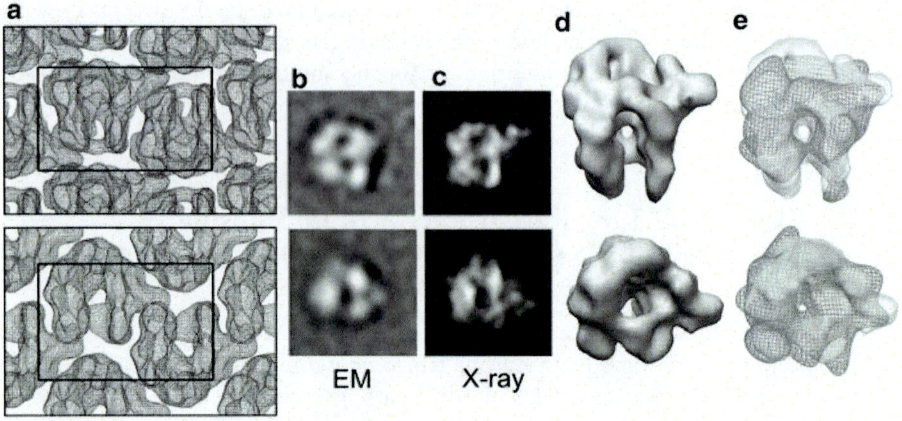

Fig. 17.3 *Ab intio* phasing of Na^+-NQR [15]. (**a**) The projections of an *ab initio* phased 16 Å resolution Fourier synthesis along two crystallographic axes. The unit cell is indicated by a *box*. (**b**) Electron microscopy 2D class averages corresponding to the projections shown at **a**. (**c**) 35 Å filtered X-ray projections. (**d**) A 16 Å resolution envelope for Na^+-NQR particle. (**e**) Superposition of the 16 Å crystal structure (*surface*) and the EM reconstruction (*mesh*) showing similar features and dimensions of the Na^+-NQR complex

deviation from the best phase was equal to the mean deviation found at the end of the previous step. The exception was the case when a new reflection was included for the first time into phasing. In this case, the probability distribution was supposed to be uniform. After the selected admissible phase sets were aligned by permitted space group origin shifts, the best (centroid) phase value and figure of merit were calculated for each reflection.

The effective resolution of the resulting map was estimated as 16 Å being based on the observation [6] that newly added reflections (*e.g.*, from some new resolution shell) improve the quality of the Fourier synthesis if their map correlation coefficient is greater than half the correlation coefficient for previously included reflections. A simple consequence of this rule is that the correlation on newly phased reflections above 0.5 is high enough to obtain an improved map. Similar to some other phasing methods, *ab initio* phasing also needs to choose the correct enantiomer. Currently the obtained resolution of the Fourier syntheses does not allow doing this unambiguously.

The resulting Fourier map revealed the shape of the Na^+-NQR complex and the particles packing in the unit cell (Fig. 17.3). The dimensions of the particle fit well to a protein complex with a molecular mass of 210 kDa.

3D negative-stain electron microscopy (EM) was used as a complementary approach to check results. Three-dimensional structures obtained by EM methods are often considered as perspective preliminary models for consequent crystallographic study. In this study this order was reversed. First, in negative-stain EM, homogenous particles with a diameter of about 100–150 Å were identified. Multivariate statistical analysis and 2D class averaging yielded particles (Fig. 17.3) with features particularly similar to certain views of the *ab initio* calculated molecular envelope, in particular an asymmetric particle with a central cavity. The overall similarity of the EM and X-ray results was further confirmed by calculation of a 3D reconstruction where the X-ray structure (filtered to 35 Å) was used as an initial reference model. The final reconstruction is based on 9,772 particles and has a resolution of 26 Å as estimated by the Fourier shell correlation 0.5 criterion. This reconstruction shows similar features and dimensions (Fig. 17.3) with the exception of three protruding elements which are not present in the EM structure. These protrusions are poorly visible or absent in the individual images of the Na^+-NQR complex and in the 2D class averages. It is likely that these regions of the Na^+-NQR complex are poorly stained by uranyl acetate, or they may exhibit intrinsic flexibility resulting in different conformations. Consequently, the density is not visible in the 3D reconstruction as it is averaged out. The crystal packing suggests that some of these protruding elements are involved in intermolecular contacts and thereby constrained to a single conformation in the crystal.

Acknowledgments VYL and NLL were supported by RFBR grant 10-04-00254-a. The authors thank M.W. Baumstark, G. Fritz, M.S. Casut, K. Knoops, C. Schaffitzel and J. Steuber for highly fruitful collaboration in study of low resolution structure of Na^+-NQR which was describe as an example of phasing in this paper.

References

1. Baker D, Krukowski AE, Agard DA (1993) Uniqueness and the *ab initio* phase problem in macromolecular crystallography. Acta Crystallogr D 49:186–192
2. Bricogne G, Gilmore CJ (1990) A multisolution method of phase determination by combined maximization of entropy and likelihood. I. Theory, algorithms and strategy. Acta Crystallogr A 46:284–297
3. Lunin VY (1993) Electron-density histograms and the phase problem. Acta Crystallogr D 49:90–99
4. Lunin VY, Lunina NL (1996) The map correlation coefficient for optimally superposed maps. Acta Crystallogr A 52:365–368
5. Lunin VY, Urzhumtsev AG (1984) Improvement of protein phases by coarse model modification. Acta Crystallogr A 40:269–277
6. Lunin VY, Woolfson MM (1993) Mean phase error and the map correlation coefficient. Acta Crystallogr D 49:530–533
7. Lunin VY, Urzhumtsev AG, Skovoroda TP (1990) Direct low-resolution phasing from electron-density histograms in protein crystallography. Acta Crystallogr A 46:540–544
8. Lunin VY, Lunina N, Petrova T, Vernoslova E, Urzhumtsev A, Podjarny A (1995) On the *ab initio* solution of the phase problem for macromolecules at very low resolution: the Few Atoms Model method. Acta Crystallogr D 51:896–903
9. Lunin VY, Lunina NL, Petrova TE, Urzhumtsev AG, Podjarny AD (1998) On the ab initio solution of the phase problem for macromolecules at very low resolution. II. Generalized likelihood based approach to cluster discrimination. Acta Crystallogr D 54:726–734
10. Lunin VY, Lunina N, Urzhumtsev A (2000) Topological properties of high density regions and *ab initio* phasing at low resolution. Acta Crystallogr A 56:375–382
11. Lunin VY, Lunina N, Petrova T, Skovoroda T, Urzhumtsev A, Podjarny A (2000) Low resolution *ab initio* phasing. Problems and advances. Acta Crystallogr D 56:1223–1232

12. Lunin VY, Lunina N, Ritter S, Frey I, Keul J, Diederichs K, Podjarny A, Urzhumtsev A, Baumstark M (2001) Low-resolution data analysis for the low-density lipoprotein particle. Acta Crystallogr D 57:108–121
13. Lunin VY, Lunina N, Podjarny A, Bockmayr A, Urzhumtsev A (2002) *Ab initio* phasing starting from low resolution. Z Kristallogr 217:668–685
14. Lunin VY, Urzhumtsev AG, Podjarny A (2012) *Ab initio* phasing of low resolution Fourier syntheses. In: Arnold E, Himmel DM, Rossmann MG (eds) International tables for crystallography F, 2nd edn. Wiley, Chichester, pp 437–442
15. Lunin VY, Lunina NL, Casutt MS, Knoops K, Schaffitzel C, Steuber J, Fritz G, Baumstark MW (2012) Low resolution structure determination of Na+-translocating NADH: ubiquinone oxidoreductase from *Vibrio cholerae* by *ab initio phasing* and electron microscopy. Acta Crystallogr D 68:724–731
16. Petrova T, Lunin VY, Podjarny A (2000) *Ab initio* low-resolution phasing in crystallography of macromolecules by maximization of likelihood. Acta Crystallogr D 56:1245–1252
17. Ritter S, Diederichs K, Frey I, Berg A, Keul J, Baumstark M (1999) Crystallization of human low density lipoprotein (LDL), a large lipid-protein complex: collection of X-ray data at very low resolution. J Crystallogr Growth 196:344–349
18. Sheldrick GM, Gilmore CJ, Hauptman HA, Weeks CM, Miller R, Usón I (2012) Ab initio phasing. In: Arnold E, Himmel DM, Rossmann MG (eds) International tables for crystallography F, 2nd edn. Wiley, Chichester, pp 413–432

Chapter 18
Model-Building and Reduction of Model Bias in Electron Density Maps

Thomas C. Terwilliger

Abstract Model-building is a key element of interpretation of electron density maps. Once a model is built it can then be used to further improve the map and hence improve the quality of a new model. It is helpful in this process to have effective methods for automated model-building and for ensuring that the resulting maps are minimally biased by the model. Many powerful methods for automatic interpretation of macromolecular electron density maps have been developed recently. Here we describe one method based on the identification of regular secondary structure and extension with fragments from known structures. We then describe the use of density modification procedures ("prime-and-switch") to reduce the model bias in maps calculated from models. Finally we describe how these prime-and-switch maps can be used as part of procedures to improve molecular replacement models just after initial placement and how this can extend the range of molecular replacement.

Keywords Model-building • Molecular replacement • Morphing • Prime-and-switch maps • Model bias

18.1 Model-Building at Moderate or High Resolution

Crystallographic model-building is the interpretation of electron density maps in terms of the atomic coordinates of the molecules that are present in a crystal. As these atomic coordinates are typically used in nearly all further analyses of the crystal structure, they are the key result obtained from a crystallographic

T.C. Terwilliger (✉)
Bioscience Division and Los Alamos Institutes, Los Alamos National Laboratory, Los Alamos, NM 87545, USA
e-mail: terwilliger@lanl.gov

R. Read et al. (eds.), *Advancing Methods for Biomolecular Crystallography*, NATO Science for Peace and Security Series A: Chemistry and Biology, DOI 10.1007/978-94-007-6232-9_18,

experiment. In this section, approaches to interpreting electron density maps that make use of the presence of regular secondary structure in macromolecules and that can be applied at resolutions as low as about 3.5 Å will be described.

18.1.1 Model-Building Approaches

In recent years a number of very powerful methods for model-building have been developed. Some of these involve manual interpretation of electron density maps using sophisticated graphical tools (e.g., [10, 12, 13, 20, 24, 41]). Others are fully automatic or nearly so. One automated approach begins with the interpretation of high-resolution maps in terms of the locations of individual atoms, followed by the interpretation of these arrangements of atoms in terms of residues, side-chains and polypeptide (or nucleic acid) chains (ARP/wARP, [16, 28]). Another approach begins with the identification of patterns of density corresponding to secondary structure elements [17, 27, 34]. A third approach consists of identification of Cα positions for proteins or phosphate and base positions for nucleic acids, followed by tracing the remainder of the chains [6, 11, 14, 25, 26]. Additional methods include the use of extensive conformational sampling [7], analysis of low-resolution features in maps [2], and probabilistic analyses of maps [8].

18.1.2 Identifying Short Segments of Regular Secondary Structure in a Map and Using Them in Model-Building

Figure 18.1 illustrates how the identification of secondary structure in a map can be used as part of model-building at moderate resolution (about 2.5 Å in this case). An FFT-based convolution search is used to examine the map shown in Fig. 18.1a, finding all the places in which density typical of a β-strand or of an α-helix are present [1, 34]. The templates used in this search are small density maps calculated from fragments of these secondary structural elements found in the PDB (Protein Data Bank, [3, 4]). When a position and orientation are found that match the map, then the corresponding short fragment is placed in that location, as illustrated in Fig. 18.1a.

A second step in model-building is to extend all these short placed fragments, this time using segments with a wide range of conformations found in high-resolution structures in the PDB (Fig. 18.1b). For proteins, the last residue in a placed fragment is used to position the first residue in a tripeptide taken from the PDB. This determines the positions of the other residues in the tripeptide. Thousands of different tripeptides are considered for each extension, and the one that both fits

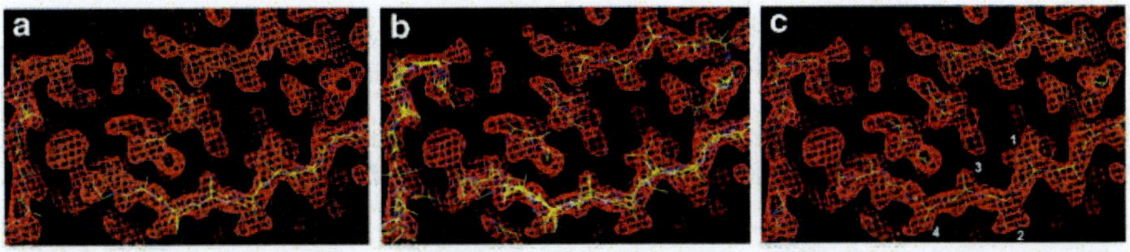

d

#	G	A	S	V	I	L	M	C	F	Y	K	R	W	H	E	D	Q	N	P	T
1	6	5	4	18	18	6	1	1	1	2	6	2	2	1	9	6	1	0	1	4
2	4	11	14	37	5	2	0	2	0	0	2	3	0	0	1	2	0	0	0	6
3	11	23	5	12	5	3	2	0	1	3	7	3	1	0	5	3	2	0	2	2
4	7	9	6	16	8	5	2	0	1	3	8	4	1	0	7	6	2	0	3	4
5	31	7	3	7	4	2	1	0	1	3	5	4	1	0	6	2	2	0	11	1
6	1	3	3	41	14	8	0	0	0	0	2	1	0	0	2	4	0	0	1	9
7	0	0	0	0	0	0	0	0	15	63	1	0	17	1	0	0	0	0	0	0
8	2	3	6	23	10	6	2	1	0	1	4	3	0	0	5	16	1	0	1	6
9	96	0	0	0	0	0	0	0	0	0	0	0	0	0	0	0	0	0	0	0

Fig. 18.1 Interpretation of an electron density map at a resolution of about 2.5 Å. (**a**) Overlapping placement of short fragments of β-structure in the map. (**b**) Iterative extension of fragments with 3-residue segments from the PDB. (**c**) Choosing the best-fitting residues at each position. (**d**) Matrix illustrating relative probabilities of each amino acid at each position (as numbered in **c**) (Figure drawn with Coot Emsley et al. [10])

the density best and can itself be extended again is chosen. The process is then repeated until no new residues can be added that fit the density reasonably well. The result of this model-building is a set of overlapping chains that cover much of the structure.

The third step is to connect chains that overlap and then choose the chains that best fit the electron density and are as long as possible. Chains are connected if two C_α in a row superimpose in the two chains within a small distance (typically 1 Å) and crossing over at this point improves the fit to the density. Once all chains are connected, the longest and best-fitting chain in the entire structure is identified, and all overlapping parts of all other chains are eliminated. Then the next-best-fitting chain is picked and the process is repeated until there are no more chains remaining. The result is shown in Fig. 18.1c, with a single best chain at each position.

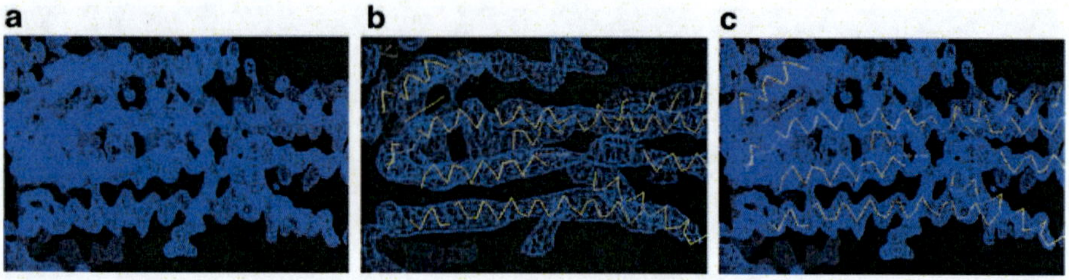

Fig. 18.2 Finding helices in a map at 3.1 Å (PDB entry 1T5S; [32]). (**a**) Electron density map of a region containing helices. (**b**) The same map, calculated at a resolution of 7 Å. (**c**) The map at 3.1 Å with helices placed based on the map in **b** followed by rotation and translation along the helix axis to optimize the fit to density (Figure drawn with Coot [10]. This analysis of helices is described in Terwilliger [38])

18.1.3 Aligning the Sequence to the Backbone Structure

Figure 18.1d illustrates how the sequence of a protein can be matched to the side chain-density in a map once the main-chain has been built. At each position in the chain illustrated in Fig. 18.1c, the density at the position of the side-chain is compared to density expected of each common rotamer of each of the 20 amino acids [35]. Then using this fit and the known sequence composition, the relative probability for each of the 20 amino acids is calculated for that position. These are shown in Fig. 18.1d. Finally, the table in Fig. 18.1d can be used to calculate the relative probability of any sequence alignment. The alignment that is most probable is then chosen (provided the overall confidence is at least 95%).

18.1.4 Knowledge-Based Methods for Identification of Regular Secondary Structure

The overall model-building approach illustrated in Fig. 18.1 is quite effective and only takes a few minutes for a moderate-sized structure, but there are circumstances where it may be useful to have approaches that are even faster. One way to speed up the interpretation of an electron density map is to look in the map for predetermined patterns of density that can be identified quickly. Figure 18.2 illustrates how this approach can be applied to find helices in an electron density map. Figure 18.2a shows a section of a map from a membrane protein at a resolution of about 3.1 Å. The density clearly shows helices but their detailed interpretation as an atomic model is less obvious. One way to quickly identify the locations of helices is to consider the map at lower resolution (7 Å) as shown in Fig. 18.2b. The helices, and in particular, the helix axes, are very clear in this map. As the map is at low resolution, and as the identification of the helices at this resolution only requires specification of the location and direction of the helix axis, this analysis can be carried out very quickly. Then, given each helix axis, the direction and rotational orientation of the corresponding helix can be identified as the one that best fits the density (Fig. 18.2c). This entire process can normally be carried out in a few tens of seconds (typically five residues per second, [38]). A similar procedure can be applied to find β-sheets in electron density maps [39].

18.2 Reducing Model Bias in Electron Density Maps

Once a preliminary model for a crystal structure is obtained, this model can be used along with *2mFo-DFc* $exp(i\varphi_c)$ σ_A-weighted maps ([29]; e.g., [28]) or density-modification techniques (e.g., [36]) to iteratively improve the crystallographic phases and therefore the maps for that crystal structure. This approach is very powerful because building a model for one part of the structure improves the phases, and therefore improves the map for other parts of the structure. Then building these other parts of the structure in turn improve the map in the first part of the structure. In many cases the improvement in map quality through this iterative process can be dramatic.

The use of models in phase calculations does however come with the risk of introducing bias from the model into the electron density maps. This risk is relatively low for model-building at high resolution (<2 Å) and when models are built conservatively from experimental maps. It is somewhat higher for models at lower resolution (>3 Å), particularly those obtained with molecular replacement and refined extensively before close examination. Two methods for reducing potential model bias are prime-and-switch phasing [37] and calculation of iterative-build omit maps [40].

18.2.1 Prime-and-Switch Phasing

One way to reduce model bias resulting from incorrect features in a model being used to calculate phases is to move away from the use of that model as quickly as possible. This can be accomplished with the technique of prime-and-switch phasing [36]. The idea of prime-and-switch phasing is to use the model to obtain initial phases (priming), but then to switch to a different source of information for further phase improvement. If this second source of phase information is both very strong and unrelated to the model used to prime the process, then the resulting phases can be improved and can remove the incorrect features introduced by the model.

To carry out prime-and-switch phasing a source of phase information that is independent of the model is required. A very good source of phase information that is model-independent is density modification. Density modification is a procedure in which our expectations about features of electron density maps are used to improve phases. For example, phases that lead to a flatter solvent region are more likely to be

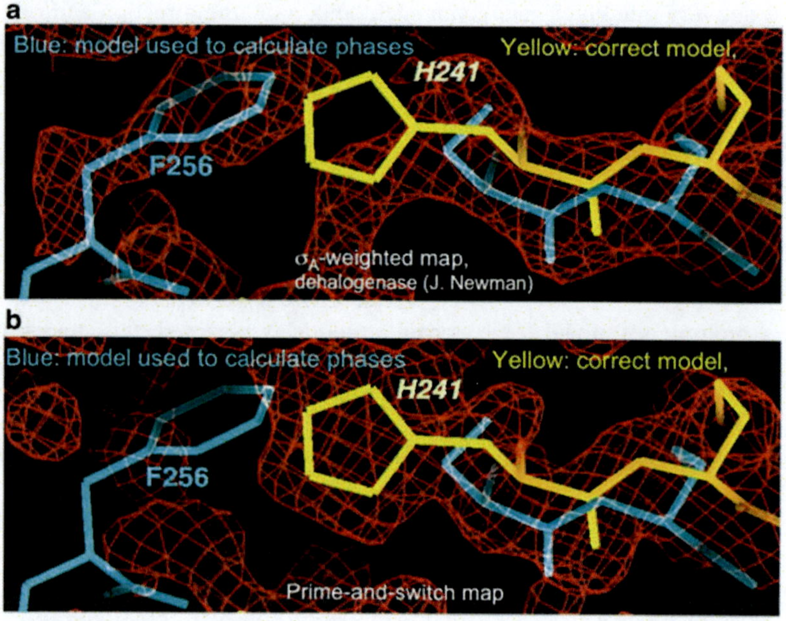

Fig. 18.3 Prime-and-switch phasing. (**a**) *2mFo-DFc exp(iφ_c)* σ_A-weighted map [29] based on PDB entry 1CV2 [19] with data from PDB entry 1BN7 [23]. (**b**) Prime-and-switch map using the same model and data. This prime-and-switch analysis is described in Terwilliger [37]

correct than those that do not. They key element of using density modification as the second step in prime-and-switch phasing is that the phase information coming from density modification is essentially independent of the phase information coming from the model [32]. This means that model bias that is present in the model-based maps can be reduced by prime-and-switch phasing.

Figure 18.3 shows how prime-and-switch phasing can reduce model bias [37]. In Fig. 18.3a a *2mFo-DFc exp(iφ_c)* σ_A-weighted maps [29] is shown based on a model of one dehalogenase enzyme (linB, PDB entry 1CV2, [19]) and data for a second dehalogenase enzyme that differs by an rmsd of about 1.4 Å (dhlA, PDB entry 1BN7, [23]). The map shows features that are very similar to the template dehalogenase linB used to calculate phases and differs substantially from the structure of the dhlA protein in the crystals from which the data were collected. In Fig. 18.3b a prime-and-switch map based on the same model and data is shown. This map has features much more similar to those of the dhlA dehalogenase and less like those of the linB protein used to calculate phases. Prime-and-switch phasing can be used in many situations, but it is most powerful for structures with high solvent content, with data at high resolution, or with non-crystallographic symmetry as all these contribute to the amount of phase information that can be obtained from density modification.

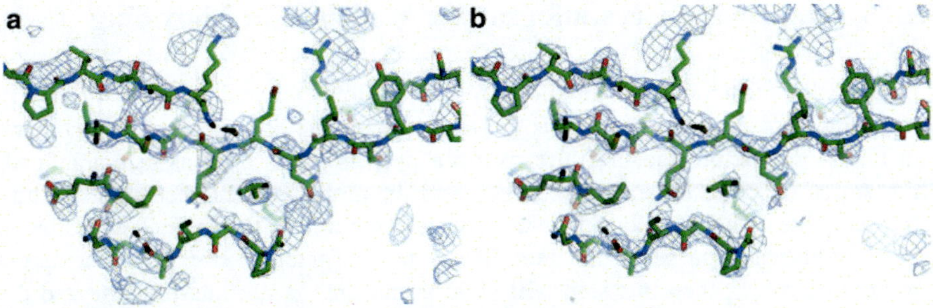

Fig. 18.4 Iterative-build omit map calculation. (**a**) A density-modified map based on SAD phases for 1VQB [31]. (**b**) The same region after calculation of an iterative-build composite omit map (Figure drawn with Coot [10]. This calculation of iterative build omit maps is described in Terwilliger et al. [40])

18.2.2 Iterative-Build Omit Maps

Although prime-and-switch phasing is very effective at removing model bias, an even better way to prevent model bias is to avoid using a model in the first place. It is possible to go through an entire structure determination process yet to create a map with no model bias by calculating an iterative-build omit map [40].

The basic idea of an omit map [5] is to delete all the atoms in a certain region, then to use the resulting model to calculate phases for a map. Then that map is (relatively) unbiased by the model. The caveat for omit maps of this type is that if the structure has been refined, then coordinates of all the other atoms have been adjusted to compensate for any errors that were originally present for the omitted atoms, so the map may retain some model bias.

In an iterative-build omit map, a series of model-building analyses are carried out in parallel, and for each one an omit region is defined. Within this omit region, the occupancies of all atoms are always set to zero. In this way, these atoms never contribute to phase calculations, and therefore they do not bias the maps. The entire process of iterative model-building, density modification and refinement can be carried out with this zeroing of occupancies in one region of the map. As the rest of the structure is being improved, the density map in the omitted region can improve greatly, but with no model bias. A composite omit map can then be created by putting together all the omit regions for all the parallel analyses. Figure 18.4a shows a section of a density-modified map obtained from experimental SAD phasing of gene 5 protein (PDB entry 1VQB, Skinner et al. 1994). The map has many breaks and is somewhat difficult to interpret. Figure 18.4b shows the same region of the composite iterative-build omit map. In this map the density is greatly improved yet it has no model bias.

18.3 Improving Crystallographic Models with Morphing

One of the most powerful methods for determining crystallographic phases in macromolecular crystallography is to use the technique of molecular replacement, in which a structure that has already been determined is used as a template for the structure to be determined. A key step in molecular replacement is to use the template structure, placed in the correct position in the crystal, to estimate crystallographic phases. If the template is not sufficiently similar to the target protein, however, then this step will yield a map that is not interpretable and the procedure can fail. Consequently there has recently been considerable effort to develop methods that can begin refinement or model-building with templates that are very different from the target structure.

New methods that are particularly well-suited to this situation include DEN refinement (deformable elastic network; [30]) and jelly-body refinement [22]. These methods take advantage of the fact that differences among related structures often consist of deformations, so that restraints that allow large-scale motions but restrict small-scale ones can greatly improve refinement. A different kind of approach has been to incorporate sophisticated modeling tools from the structure-modeling field to improve the quality of templates, both before and after placing them in the crystallographic cell [9].

Another approach for improving a model after it has been placed in the crystallographic cell is to systematically deform it based on an electron density map. This can yield a "morphed" structure that is more similar to the target structure and that can be useful in further model-building.

The basic idea of morphing is that in many cases the template structure that is placed in the crystallographic cell is locally correct, but globally distorted. Figure 18.5a illustrates an example of such a situation. The molecular replacement

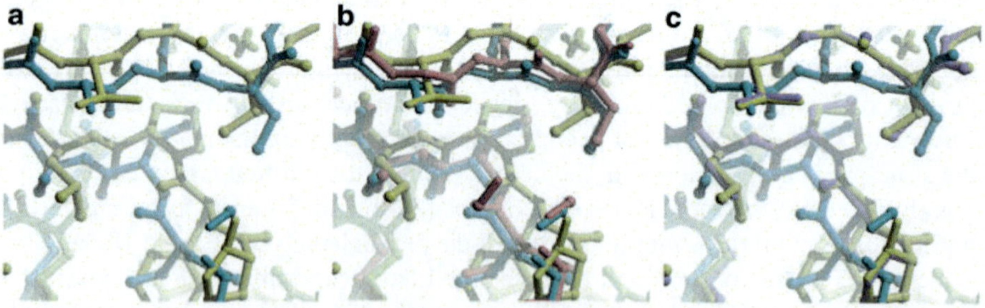

Fig. 18.5 Morphing. (**a**) Molecular replacement template (*blue or upper mid grey if viewed in b/w*) and the final refined structure (*yellow or lower mid grey if viewed in b/w*) for the protease XMRV PR [18]. (**b**) As in **a**, but also showing initial morphed model (*in maroon or middle mid grey if viewed in b/w*). (**c**) As in **b**, but final morphed and refined model after six cycles of morphing (Color figure online; Figure drawn with Coot [10] and Raster3D [21])

solution for the protease XMRV PR [18] based on a related protease structure (PDB entry 2HS1, [15]) was generally correct, but the template and target structure had significant global changes though they are locally similar. Consequently the template (in blue) is somewhat offset from the target structure (in yellow) in this region.

The goal of morphing is to identify the appropriate translation to apply to a local segment of a structure in order to optimize its overlap with the true structure. The way this is done is to choose all atoms in the template within a sphere (typically a radius of 5 Å) of a Cα atom in the template. Then all possible small translations are tested to determine how well the shifted atoms would overlap with the electron density map. The best translations are recorded for each Cα atom in the template, and then these translations are smoothed in a window of 11 residues and applied to all the atoms in the corresponding residues. The result is a new template that is deformed to match the electron density map (Fig. 18.5b). The optimal electron density map for carrying out this process is found to typically be a prime-and-switch map as described above.

After refinement to restore geometry and optimize fit to the data, the procedure can be repeated. Figure 18.5c shows the result after six cycles of morphing. In this case the morphed model is now quite similar to the final refined structure of XMRV PR. Note that as the side chains have not all been changed to match the final sequence the side chains are not fully overlapped, though many of their atoms superimpose.

18.4 Conclusions

Model-building and density modification procedures have become very powerful in recent years. With diffraction data measured to moderate or high resolution and phase information from experiment or molecular replacement, high-quality models can be built and refined in minutes to hours. Automated model-building can be carried out for both proteins and nucleic acids. Methods to ensure that the model-building process does not introduce bias include prime-and-switch phasing and iterative-build omit maps. These approaches can be carried out either during structure determination, to obtain unbiased maps for model-building, or afterwards, to validate that key aspects of a model are correct.

Acknowledgments The author is most grateful to the entire crystallographic community for feedback on methods development and to the other members of the Phenix team for development of the algorithms and software that form a foundation for the methods described here, and to the NIH for generous support of the Phenix project (PI, Paul Adams).

References

1. Adams PD, Afonine PV, Bunkóczi G, Chen VB, Davis IW, Echols N, Headd JJ, Hung L-W, Kapral GJ, Grosse-Kunstleve RW, McCoy AJ, Moriarty NW, Oeffner R, Read RJ, Richardson DC, Richardson JS, Terwilliger TC, Zwart PH (2010) PHENIX: a comprehensive Python-based system for macromolecular structure solution. Acta Crystallogr D 66:213–221
2. Baker ML, Ju T, Chiu W (2007) Identification of secondary structure elements in intermediate-resolution density maps. Structure 15:7–19
3. Berman HM, Westbrook J, Feng Z, Gilliland G, Bhat TN, Wiessig IN, Shindyalov IN, Bourne PE (2000) The protein data bank. Nucleic Acids Res 28:235–242
4. Bernstein FC, Koetzle TF, Williams GJB, Meyer EF Jr, Brice MD, Rodgers JR, Kennard O, Shimanouchi T, Tasumi M (1977) The protein data bank: a computer-based archival file for macromolecular structures. J Mol Biol 112:535–542
5. Bhat TN (1988) Calculation of an OMIT map. J Appl Crystallogr 21:279–281
6. Cowtan K (2006) Buccaneer software for automated model building. Acta Crystallogr D 62:1002–1011
7. DePristo MA, de Bakker PIW, Johnson RJK, Blundell TL (2005) Crystallographic refinement by knowledge-based exploration of complex energy landscapes. Structure 13:1311–1319
8. DiMaio F, Kondrashov DA, Bitto E, Soni A, Bingman CA, Phillips GN Jr, Shavlik JW (2007) Creating protein models from electron-density maps using particle-filtering methods. Bioinformatics 23:2851–2858
9. DiMaio F, Terwilliger TC, Read RJ, Wlodawer A, Oberdorfer G, Wagner U, Valkov E, Alon A, Fass D, Axelrod HL, Das D, Vorobiev SM, Iwaï H, Pokkuluri PR, Baker D (2011) Improving molecular replacement by density- and energy-guided protein structure optimization. Nature 473:540–543
10. Emsley P, Lohkamp B, Scott WG, Cowtan K (2010) Features and development of Coot. Acta Crystallogr D 66:486–501
11. Ioerger TR, Sacchettini JC (2003) TEXTAL system: artificial intelligence techniques for automated protein model building. Methods Enzymol 374:244–270
12. Jones TA, Kjeldgaard M (1997) Electron-density map interpretation. Methods Enzymol 227:173–230
13. Jones TA, Zou J-Y, Cowan SW, Kjeldgaard M (1991) Improved methods for building protein models in electron-density maps and the location of errors in these models. Acta Crystallogr A 47:110–119
14. Keating KS, Pyle AM (2010) Semiautomated model building for RNA crystallography using a directed rotameric approach. Proc Natl Acad Sci 107:8177–8182
15. Kovalevsky AY, Liu F, Leshchenko S, Ghosh AK, Louis JM, Harrison RW, Webber IT (2006) Ultra-high resolution crystal structure of HIV-1 protease mutant reveals two binding sites for clinical inhibitor TMC114. J Mol Biol 363:161–173
16. Langer G, Cohen SX, Lamzin VS, Perrakis A (2008) Automated macromolecular model building for X-ray crystallography using ARP/wARP version 7. Nat Protoc 3:1171–1179
17. Levitt DG (2001) A new software routine automates the fitting of protein X-ray crystallographic electron-density maps. Acta Crystallogr D 57:1013–1019
18. Li M, DiMaio F, Zhou D, Gustchina A, Lubkowski J, Dauter Z, Baker D, Wlodawer A (2011) Crystal structure of XMRV protease differs from the structures of other retropepsins. Nat Struct Mol Biol 18:227–229
19. Marek J, Vevodova J, Smatanova IK, Nagata Y, Svensson LA, Newman J, Takagi M, Damborsky J (2000) Crystal structure of the haloalkane dehalogenase from *Sphingomonas paucimobilis* UT26. Biochemistry 39:14082–14086
20. McRee DE (1999) XtalView/Xfit – a versatile program for manipulating atomic coordinates and electron density. J Struct Biol 125:156–165
21. Merritt EA, Bacon DJ (1997) Raster3D-Photorealistic molecular graphics. Methods Enzymol 277:505–524

22. Murshudov GN, Skubák P, Lebedev A, Pannu NS, Steiner RA, Nicholls RA, Winn MD, Fei L, Vagin AA (2011) REFMAC5 for the refinement of macromolecular crystal structures. Acta Crystallogr D 67:355–367
23. Newman J, Peat TS, Richard R, Kan L, Swanson PE, Affholter JA, Holmes IH, Schindler JF, Unkefer CJ, Terwilliger TC (1999) Haloalkane dehalogenases: structure of a rhodococcus enzyme. Biochemistry 38:16105–16114
24. Oldfield TJ (1994) In: Bailey S, Hubbard R, Waller DA (eds) Proceedings of the CCP4 study weekend from first map to final model. Daresbury Laboratory, Warrington, pp 15–16
25. Oldfield TJ (2002) Pattern-recognition methods to identify secondary structure within X-ray crystallographic electron-density maps. Acta Crystallogr D 58:487–493
26. Oldfield TJ (2003) Automated tracing of electron-density maps of proteins. Acta Crystallogr D 59:483–491
27. Pavelcik F, Schneider B (2008) Building of RNA and DNA double helices into electron density. Acta Crystallogr D 64:620–626
28. Perrakis A, Morris R, Lamzin VS (1999) Automated protein model building combined with iterative structure refinement. Nat Struct Biol 6:458–463
29. Read RJ (1986) Improved Fourier coefficients for maps using phases from partial structures with errors. Acta Crystallogr A 42:140–149
30. Schröder G, Levitt M, Brünger AT (2010) Super-resolution biomolecular crystallography with low-resolution data. Nature 464:1218–1222
31. Skinner MM, Zhang H, Leschnitzer DH, Guan Y, Bellamy H, Sweet RM, Gray CW, Konings RNH, Wang AH-J, Terwilliger TC (1994) Structure of the gene V protein of bacteriophage f1 determined by multiwavelength X-ray diffraction on the selenomethionyl protein. Proc Natl Acad Sci USA 91:2071–2075
32. Sorensen TL-M, Molleer JV, Nissen P (2004) Phosphoryl transfer and calcium ion occlusion in the calcium pump. Science 304:1672–1675
33. Terwilliger TC (2001) Map-likelihood phasing. Acta Crystallogr D 57:1763–1775
34. Terwilliger TC (2003) Automated main-chain model-building by template-matching and iterative fragment extension. Acta Crystallogr D 59:38–44
35. Terwilliger TC (2003) Automated side-chain model building and sequence assignment by template-matching. Acta Crystallogr D 59:45–49
36. Terwilliger TC (2003) Improving macromolecular atomic models at moderate resolution by automated iterative model building, statistical density modification and refinement. Acta Crystallogr D 59:1174–1182
37. Terwilliger TC (2004) Using prime-and-switch phasing to reduce model bias in molecular replacement. Acta Crystallogr D 60:2144–2149
38. Terwilliger TC (2010) Rapid model-building of α-helices in electron density maps. Acta Crystallogr D 66:268–275
39. Terwilliger TC (2010) Rapid model-building of β-sheets in electron density maps. Acta Crystallogr D 66:276–284
40. Terwilliger TC, Grosse-Kunstleve RW, Afonine PV, Moriarty NW, Adams PD, Read RJ, Zwart P, Hung L-W (2008) Iterative-build OMIT maps: map improvement by iterative model building and refinement without model bias. Acta Crystallogr D 64:515–524
41. Turk D (1992) Weiterentwicklung eines Programms fuer Molekuelgraphik und Elektrondichte-Manipulation and Seine Anwendung auf Verschiedene Protein-Strukturaufklerungen, PhD thesis, Technische Universitaet Muenchen, Germany

Chapter 19
Using *Coot* to Model Protein and Ligand Structures Using X-ray data

Paul Emsley

Abstract The software *Coot* is an on-going project to provide graphical tools to assist with the fitting of protein and ligands to X-ray data. Presented here are tools that manipulate side-chain rotamers and main-chain geometry. In addition, ligand tools that interact with CCP4 software and libraries are discussed.

19.1 Introduction

Coot [4] is a molecular graphics application for particularly designed for building and validation of protein models from X-ray data. *Coot* has become popular in the UK community and is gaining popularity in Europe, China and the US both amongst academics and for pharmaceutical companies as part of their drug-development pipeline. It is approaching 10 years since the release of *Coot* and in that time various comments have been made as to the future directions and features to be added - some of which have been taken on board. This presentation will focus on the development of tools to assist model building at lower resolutions and in the presence of non-crystallographic symmetry, including "Backrub rotamers", an optimisation based on the "Backrub" motion described by the Richardsons and co-workers [2].

Recently tools for ligand handling have been introduced. These include 2D→3D, interaction between *Coot* and programs of the CCP4 suite, particularly PRDRG and and JLigand (with on-the-fly link determination). Also a 3D→2D tool has been introduced and represents ligand interactions with the protein residues in the binding pocket.

P. Emsley (✉)
Department of Biochemistry, University of Oxford, Oxford OX1 3QU, UK

MRC Laboratory of Molecular Biology, Hills Road, Cambridge CB2 0QH, UK
e-mail: pemsley@mrc-lmb.cam.ac.uk

R. Read et al. (eds.), *Advancing Methods for Biomolecular Crystallography*,
NATO Science for Peace and Security Series A: Chemistry and Biology,
DOI 10.1007/978-94-007-6232-9_19,

19.2 Tool Description

19.2.1 Backrub Rotamers

"Backrub Rotamers" are based on the formalism for the identification of alternative conformations described by the Richardsons and co-workers [2] and is used in *Coot* to prevent backbone distortion when fitting the rotamer orientations of individual residues. Given a residue of interest with sequence number N, the "backrub vector" is the line between the Cα atom of the previous residue and that of the next one (typically the vector $N - 1 \rightarrow N + 1$). The rotation of the atoms of the central residue and the backbone atoms of the neighbour residues to which they are attached around the backrub vector forms one dimension of the Backrub Rotamer search. Simultaneous to this rotation, the atoms of the individual peptides can be rotated around vectors between the neighbouring Cαs. In *Coot*, this rotation is minimizes the positional variation of the carbonyl oxygen.

The second dimension of the Backrub Rotamer search is the selection of each of the rotamers above a particular probability density value (typically set at 1%). So, each position on the grid provides a different model that can be tested against both the electron density and undesirable interaction with residues of the environment.

The search provides models that have (relatively) high rotamer probability and preserve (to a large extent) the main-chain geometry (which, one hopes at least, was good before the Backrub Rotamer search commenced).

19.2.2 Ramachandran Restraints

At lower resolutions it is occasionally difficult to obtain a good fit to the electron density and have the back-bone geometry (the Ramachandran plot) conform to acceptable notions of high quality. To assist in this dual optimization problem, we optionally add a "Log Ramachandran Plot score" (R) to the target function such that as R improves, the position of the residue on the Ramachandran plot improves (*i.e.* is consistent with high ϕ, ψ probability density).

The log Ramachandran plot provides the value of the Ramachandran plot for a particular ϕ, ψ pair and the derivates with respect to both phi and psi. Using the chain rule, this is combined with the analytical derivatives for torsion angles, thus generating the partial derivative of R with respect to the atomic coordinates contributing to the ϕ, ψ pair.

The technique is to be used with caution. After optimization, one cannot subsequently use the Ramachandran Plot statistics as a measure of the quality of the geometry of the model – it is no longer a "free" statistic.

19.2.3 Ligand Tools

19.2.3.1 Generating Ligands

When generating ligands, the starting point is often a SMILES string or a sketch of the molecule as a 2D MDL molfile. For ligand-fitting, model-building and refinement we typically need a 3D model of the ligand (in a low energy conformer) and a set of restraints that describe the geometry (that is to say, details (including estimated standard deviations) of the bond length, bond angles, torsion angles, planar groups and chiral centres. There are several means (external to *Coot*) to generate these inputs.

19.2.3.2 2D Ligand Sketcher

LIDIA, A JME-work-alike, has been introduced as a 2D molecular editor. This can be used to hand-edit chemical diagrams in a straightforward manner. LIDIA is both built into *Coot* and can be used as a stand-alone application.

19.2.3.3 Working with PRODRG

LIDIA was designed, in the first instance, to work with CCP4's command-line driven version of PRODRG [6] (called cprodrg). The interface has since been generalized so that it is easy to replace the 3D engine with the tool of your choice. The outputs of the 3D engine are a PDB file containing a 3D representation of the ligand and restraints in mmCIF/PDBx format and these are read into *Coot* automatically.

19.2.3.4 Database Searching

Chemical structures can be represented as mathematical graphs, where the graph vertices are the atoms and the edges are the bonds. Using an improved backtracking algorithm in the common subgraph isomorphism search [5] has increased the search speed over more traditional methods. This has enabled a number of ligand comparison and manipulation tools.

SBase contains information similar to that contained in the RCSB's Chemical Component Dictionary. However, SBase has the advantage of being rapidly accessible via a programmatic interface. Using the 2D sketcher one can generate a chemical structure diagram that can be used to find similar structures in the database. Such compounds (in 3D form, together with restraints) can be loaded into *Coot* as desired.

19.2.3.5 Atom Name Matching, Torsion Matching

When comparing similar ligands, it is often convenient that the atom names match – that is to say, structurally equivalent atoms have the same atom name. By graph matching, we can rename the atoms of peer compounds to match a reference set.

By using the same mechanism, we provide tools for ligand overlaying and torsion matching.

19.2.3.6 Ligand Chemistry Representation

Coot uses the bond-order descriptions from the restraints to represent the chemistry of the ligands. *Coot* can parse the common bond types and represent the bonds appropriately. Aromatic ring detection allows aromatic systems to be represented using a ring rather than a number of alternating double and single bond or delocalized bonds.

19.2.3.7 Conformer Generation

It is unlikely, in the general case, that the conformation of the ligand in the crystal structure matches that of the ligand generated by CPRODRG (or some other coordinates-generating software). Thus Coot generates conformers, by variation of the torsion angles based on a probability distribution derived from the restraints. Each conformer undergoes energy minimization before being compared to the density.

19.2.3.8 NCS Ligands

Coot detects non-crystallographic symmetry (NCS) between the chains of the protein model by examination of the sequence. A least-squares fit provides the operator orienting peer molecules relative to a reference molecule. This NCS orientation matrix can then be used to position ligands in the active site of peer molecules.

19.2.3.9 Ligand Validation

Perhaps the most obvious way of measuring the geometric validity of a ligand molecule is to assess the similarities of the actual geometry of the ligand to that of the restraints, comparing actual values of bond length, bond angles and so on, with the ideal values tabulated in the restraints.

However, if the restraint set is incorrect, or has some distortion, then this method may not detect such problems – particularly when using low resolution data.

The program Mogul [1] from the CCDC provides rapid comparison of the fragments of the ligand to a range of small molecule crystal structures. Bond lengths, bond angles and torsion angles can be compared. This analysis is on-going [3].

References

1. Bruno IJ, Cole JC, Kessler M, Luo J, Motherwell WDS, Purkis LH, Smith BR, Taylor R, Cooper RI, Harris SE, Orpen AG (2004) Retrieval of crystallographically-derived molecular geometry information. J Chem Inf Comput Sci 44:2133–2144
2. Davis I, Arendall W, Richardson D, Richardson J (2006) The backrub motion: how protein backbone shrugs when a sidechain dances. Structure 14(5):265–274. doi:10.1016/j.str.2005.10.007
3. Debreczeni JE, Emsley P (2012) Handling ligands with coot. Acta Crystallogr D Biol Crystallogr 68(Pt 4):425–430
4. Emsley P, Lohkamp B, Scott WG, K C (2010) Features and development of coot. Acta Crystallogr D Biol Crystallogr 66:486–501
5. Krissinel EB, Henrick K (2004) Common subgraph isomorphism detection by backtracking search. Software – Practice and Experience 34:591–607
6. Schüttelkopf AW, van Aalten DMF (2004) Prodrg: a tool for high-throughput crystallography of protein-ligand complexes. Acta Cryst D Biol Crystallogr 60(Pt 8):1355–1363. doi:10.1107/S0907444904011679

Chapter 20
Crystallographic Structure Refinement in a Nutshell

Pavel V. Afonine and Paul D. Adams

Abstract The objective of these notes is to provide a general overview of modern crystallographic structure refinement. No specific or technical details are presented, but concepts only. However, references are provided to the relevant literature for those who desire to learn more.

Keywords Crystallographic structure refinement • TLS • ADP • Twinning • Maximum-likelihood • Real-space

20.1 Introduction

Crystallographic structure refinement is a procedure that combines a large number of complex steps (for recent reviews, see for example [6, 52, 55]). The goal is to improve a starting model such that it better agrees with the experimental data and *a priori* knowledge (such as molecule stereochemistry). Any refinement procedure requires decisions be made about *model parameterization*, the *refinement target* and the *optimization method*. These decisions are typically made by the researcher or increasingly the refinement program itself, and are dictated by the experimental data quality (completeness, resolution, crystal specifics such as twinning), data type (X-ray or neutron diffraction experiment) and model quality (number of unmodeled atoms, or current *R*-factors). The variety of possible data and model qualities leads to a range of model parameterizations, refinement targets and optimization methods. Also, the chance for errors always exists. These errors may originate from a lack of

P.V. Afonine (✉) • P.D. Adams
Lawrence Berkeley National Laboratory, One Cyclotron Road, MS64R0121,
Berkeley, CA 94720, USA
e-mail: PAfonine@lbl.gov

R. Read et al. (eds.), *Advancing Methods for Biomolecular Crystallography*,
NATO Science for Peace and Security Series A: Chemistry and Biology,
DOI 10.1007/978-94-007-6232-9_20,

researcher expertize, plain accidents and/or limitations of refinement program. To prevent these errors, the *validation* of inputs and outcomes of refinement should be performed throughout the entire process.

20.2 Model Parameters

Model parameters are the variables that are used to describe the crystal contents and its properties. Macromolecular crystals contain ordered parts (macromolecules and ligands, for example) and bulk-solvent, which is the name for disordered solvent that fills the space between ordered molecules and may constitute from 10 to 90 % of the unit cell volume. Therefore model parameters can be of two categories: (a) those that describe the atomic model (atomic model parameters) and (b) non-atomic model parameters that describe the remainder: bulk-solvent, and crystal properties (twinning, anisotropy).

20.2.1 Atomic Model Parameters

Atoms in crystal models are described by their positions (coordinates), degree of small-scale (harmonic) movements or disorder (atomic displacement parameters, ADPs or B-factors), larger scale disorder (occupancy) and anomalous scattering parameters (f' and f"). These parameters are used to calculate structure factors, $\mathbf{F}_{calc}$, arising from atomic model.

Atomic positions can be parameterized through individual coordinates of atoms in a Cartesian basis with three or less (if an atom is located on a special position) refinable parameters per atom, in torsion angle space with about a 7-fold reduction of refinable parameters [25, 48], or using a rigid-body parameterization with three rotational and three translation parameters per selected group of atoms that are considered to be rigid [4].

The parameterization of ADPs involves describing the different contributions to atomic displacements. The total ADP is the superposition of motions originating from local atomic vibration, constrained group motion (motion about a rotatable bond, for example), residue movement as a whole, domain movement, whole molecule movement and crystal lattice vibrations [20, 43, 49, 56]. If data and model quality permit, the local atomic vibrations can be modeled more accurately by accounting for their anisotropic nature (using 6 or less refinable parameters per atom), as opposed to assuming atoms moving isotropically with only one refinable parameter [5]. Most of ADP parameterizations assume the atomic displacements are small enough to be within a harmonic approximation. An anharmonic approximation can be used to model larger but still not too large disorder [29]. Concerted movements of group of atoms are described using the TLS (translation-libration-screw) parameterization (for review see [54]).

Atomic occupancies are used to model discrete larger-scale disorder beyond the harmonic approximation. Occupancy is the fraction of unit cells in the crystal in which a given atom occupies the position indicated in the model. If all unit cells in the crystal are identical, then occupancies for all atoms are 1. Refinement of occupancy is necessary when the molecule has several distinct conformations or occupies different positions in different unit cells. Refining occupancies provides an estimate of the frequency of alternative conformations.

20.2.2 Non-atomic Model Parameters

Non-atomic model parameters are included as corrections and scales at the level of total model structure factor calculation [3]

$$\mathbf{F}_{\text{model}} = k\,(\mathbf{F}_{\text{calc}} + \mathbf{F}_{\text{bulk}})$$

where k is the Miller-index-dependent scale factor which can account for crystal anisotropy, and $\mathbf{F}_{\text{bulk}}$ is the contribution from the bulk-solvent. In case of hemihedral twinning [28, 42], most typically observed in macromolecules, the square of the total model structure factor is the weighed sum of squared model structure factors corresponding to the twin domains:

$$|\mathbf{F}_{\text{model}}|^2 = \alpha_1|\mathbf{F}_{\text{model.1}}|^2 + \alpha_2|\mathbf{F}_{\text{model.2}}|^2$$

where the weights α_1 and α_2 are the twin fractions – parameters describing a relative contribution of each of the two twin domains into the total intensity.

The total model structure factor $\mathbf{F}_{\text{model}}$ is used virtually everywhere where the model and experimental structure factors are compared, such as R-factors, electron density maps and refinement targets.

20.3 Refinement Target

A crystallographic refinement target is a mathematical function that relates model parameters (expressed through $\mathbf{F}_{\text{model}}$) and the experimental data (amplitudes, F_{obs}, or intensities, I_{obs}, and experimental phases if available). Typically, target functions are defined such that their value decreases as the model improves. This in turn formulates the goal of crystallographic structure refinement as an optimization problem in which the model parameters are modified in order to achieve the lowest possible value of the target function, or in other words, minimization of the refinement target.

Two fundamentally different types of refinement targets are in use: real- and reciprocal-space targets.

A real-space target relates the atomic model directly to the best available map by either minimizing the least-squares difference between the electron density map calculated from the atomic model and the best available density map calculated using the experimental data, or by guiding the atoms towards the nearest density peaks through maximization of electron densities interpolated at atomic centers [15, 18, 19, 30, 31, 38, 39, 53].

A simplest reciprocal-space refinement target is a least-squares target that is a weighed sum of squared differences of observed and calculated structure factor amplitudes or intensities. This kind of target is widely used in small-molecule crystallography and used to be routinely employed in macromolecular crystallography until maximum-likelihood target functions (ML; [32, 33, 44, 45]) were introduced for refinement [1, 10, 38, 40, 41].

The principal reason why the ML function is the target of choice for macromolecular refinement is because crystal structures of macromolecules are never complete (for example, the flat bulk-solvent model is only an approximation, ordered or partially ordered solvent and ligands may not be fully accounted for, all side chain alternative conformations may not be modeled, etc.). A ML target accounts for missing (unmodeled) atoms statistically, while LS functions do not [35]. This is why the use of ML target in macromolecular refinement typically results in better-refined model parameters.

Experimental phase information can be included into both least-squares or maximum-likelihood targets ([34, 36, 37, 40]).

A major advantage of real-space refinement versus reciprocal-space refinement is that it can be performed locally and therefore quickly (for example, for a residue or a stretch of residues; [26]). Also, it has a larger convergence radius. The disadvantage is that it requires an electron density map to refine against. If experimental phases are not available or combined with the model phases, then the target map contains model information and therefore refining against such a map could introduce model bias. However, this bias may not be significant if real-space refinement is performed locally for a small number of atoms.

Both, real- and reciprocal-space refinement methods are complimentary and are used in various refinement protocols together [6].

20.4 Optimization Method

Methods to optimize the refinement target can be gradient-driven minimization, simulated annealing, grid (or systematic) searches and interactive model manipulation using a graphical environment. These methods vary in speed, scalability, convergence radius, and applicability to current model parameters [52]. The type and number of refinable parameters, as well as current model quality dictate the choice of optimization method.

Gradient-driven minimization is employed in refinement of most types of parameters. However, its convergence radius is rather small since it doesn't allow parameter shifts against the gradient, and therefore a model trapped in a local minimum cannot escape from it (see references below). Simulated Annealing (SA) refinement extends the convergence radius of refinement of coordinates. Grid searches can be used when there are a few (typically one to three) parameters to optimize and the range of possible parameter values is known. An example may be the optimization of bulk-solvent [3, 22] or bulk-solvent mask parameters [11].

20.5 Validation

Central to successful model refinement is the validation of what goes into refinement and what comes out of it. Three entities need validation: data, model and model-to-data fit. While validation of the data should be performed during the diffraction experiment or at least right after it, the model and model-to-data fit need to be validated throughout the refinement process, and not only at the very end of structure solution [17, 47].

The part of reflection data validation that is pertinent to refinement includes checking the reflection intensity statistics for the presence of twinning, translational NCS and obtaining the Wilson *B*-factor [57].

Validation of the model includes making sure that it makes sense physically and chemically. This includes analysis of crystal packing, steric contacts (all-atom clashscore), molecular geometry (correctness of bond and angle values, flatness of ring systems, such as Tyr or Phe residues, absence of unexplained Ramachandran plot outliers and C_{beta}-deviations for proteins, correct chirality and energetically favorable choices of rotamers). It is essential that such a validation is performed locally, that is per atom or group of atoms involved in the analyzed metric (as opposed to only analyzing overall values averaged over the whole structure).

Finally, there are a number of metrics to quality assess model-to-data fit. These include R_{work} and R_{free} factors, overall or calculated in resolution bins, local assessment of model to electron density fit through analysis of real-space (map) correlation coefficients, and electron density values for both, 2mFo-DFc and residual maps, parameters describing crystal anisotropy and bulk-solvent.

20.6 Refinement Workflow

A refinement protocol typically consists of three parts: (1) *initialization*, (2) *macro-cycle*, and (3) *output* of refinement results.

The initialization step includes processing of input data and making various decisions about the refinement procedure. The input data are: a structural model typically supplied as a PDB [8, 9] file, reflection data (intensities or amplitudes of measured reflections, and experimental phase information, if available) and parameters defining the refinement run. Automatic decisions include choosing the model parameterization (for example, whether to use isotropic or anisotropic ADPs), selecting the refinement target (phased ML target if experimental phases available, or twin-target in case of twinning, or switching to use a least-squares target if the number of free-*R* flags is insufficient), whether or not to update ordered solvent (water), detection of NCS and deciding how to use it, grouping occupancies for constrained refinement of alternative conformers and more. Various restraints are built at this step as well [21, 24, 27]. Input reflection data manipulations, if requested or necessary, can happen at this step too. This may be filtering reflections by sigma or resolution, converting intensities to amplitudes [23], and removing outliers [46].

A refinement protocol (*macro-cycle*) typically consists of multiple steps repeated iteratively. Each step is specifically tailored to the refinement of particular parameters. The required number of such steps should be such that the refinement converges, which means that any additional iteration of the refinement process does not change the model parameters significantly. Convergence of the particular refinement run depends on the data quality and model quality. The rationale behind this iterative procedure is based on the following:

1. The refinement target function has many local minima. Gradient-driven minimization can reach only the nearest local minimum, therefore sophisticated optimization algorithms, such as local real-space searches [26, 39] or simulated annealing [1, 12–14] may need to be applied.
2. Some model parameters are highly correlated. For example: isotropic displacement parameters and the overall exponential scale factor, ADPs and occupancies [16], components of the total ADP [2], parameters of overall anisotropic and bulk-solvent scales [22, 51].
3. Different minimization methods imply different convergence speed and radii for different types of model parameters [7, 50].
4. As the model improves during refinement, a different model parameterization may be more appropriate (for example, switch from using isotropic ADPs to anisotropic for all or selected atoms). Also, model content may need to be updated too; for example, new water molecules may be added or/and erroneous ones removed.

Each refinement typically produces three kinds of output information: (a) a refined model, (b) various electron density maps and (c) a log file.

Electron density maps output by refinement programs are the Fourier syntheses calculated with various weights to best represent the model, such as a 2mFo-DFc σ_A weighted map [44], and model errors, such as a residual (or difference) map, mFo-DFc. Since missing (unmeasured) reflections can cause mild to severe map distortions, often they are modeled with some non-zero values [6] obtained from the current model. This helps to alleviate map errors due to the data incompleteness but also introduces the risk of bias w.r.t. the model.

The log file contains information about the data, model and model-to-data fit, as well as some components of the validation report that do not require an interactive analysis (for example, using a graphics program).

References

1. Adams PD, Pannu NS, Read RJ, Brünger AT (1997) Cross-validated maximum likelihood enhances crystallographic simulated annealing refinement. Proc Natl Acad Sci 94:5018–5023
2. Afonine PV, Urzhumtsev A (2007) On determination of T matrix in TLS modelling. CCP4 newsletter on protein crystallogrophy 45. Contribution 6
3. Afonine PV, Grosse-Kunstleve RW, Adams PD (2005) A robust bulk-solvent correction and anisotropic scaling procedure. Acta Crystallogr D 61:850–855
4. Afonine PV, Grosse-Kunstleve RW, Urzhumtsev A, Adams PD (2009) Automatic multiple-zone rigid-body refinement with a large convergence radius. J Appl Crystallogr 42:607–615
5. Afonine PV, Urzhumtsev A, Grosse-Kunstleve RW, Adams PD (2010) Atomic Displacement Parameters (ADPs), their parameterization and refinement in PHENIX. Computational crystallography newsletter. 1: 24–31. (http://www.phenix-online.org/newsletter)
6. Afonine PV, Grosse-Kunstleve RW, Echols N, Headd JJ, Moriarty NW, Mustyakimov M, Terwilliger TC, Urzhumtsev A, Zwart PH, Adams PD (2012) Towards automated crystallographic structure refinement with phenix.refine. Acta Crystallogr D 68:352–367
7. Agarwal RC (1978) A new least-squares refinement technique based on the fast Fourier transform algorithm. Acta Crystallogr A 34:791–809
8. Berman HM, Westbrook J, Feng Z, Gilliland G, Bhat TN, Weissig H, Shindyalov IN, Bourne PE (2000) The Protein Data Bank. Nucleic Acids Res 28:235–242
9. Bernstein FC, Koetzle TF, Williams GJB, Meyer EF, Brice MD, Rodgers JR, Kennard O, Shimanouchi T, Tasumi M (1977) The Protein Data Bank: a computer-based archival file for macromolecular structures. J Mol Biol 112:535–542
10. Bricogne G, Irwin J (1996) Maximum-likelihood structure refinement: theory and implementation within BUSTER-TNT. Proceedings of the CCPD Study Weekend, Daresbury Laboratory, Warrington, England, pp 85–92
11. Brünger AT (2007) Version 1.2 of the Crystallography and NMR system. Nat Protoc 2: 2728–2733
12. Brünger AT, Adams PD (2002) Molecular dynamics applied to X-ray structure refinement. Acc Chem Res 35:404–412
13. Brünger AT, Kuriyan J, Karplus M (1987) Crystallographic R factor refinement by molecular dynamics. Science 235:458–460
14. Brünger AT, Adams PD, Rice LM (2001) Enhanced macromolecular refinement by simulated annealing. International tables for crystallography, vol F. Crystallography of biological macromolecules. Kluwer Academic Publishers, Dordrecht, pp 375–381
15. Chapman MS (1995) Restrained real-space macromolecular atomic refinement using a new resolution-dependent electron-density function. Acta Crystallogr A 51:69–80
16. Cheetham JC, Artymiuk PJ, Phillips DC (1992) Refinement of an enzyme complex with inhibitor bound at partial occupancy. Hen egg-white lysozyme and tri-N-acetylchitotriose at 1.75 A resolution. J Mol Biol 224:613–628
17. Chen VB, Arendall WB III, Headd JJ, Keedy DA, Immormino RM, Kapral GJ, Murray LW, Richardson JS, Richardson DC (2010) MolProbity: all-atom structure validation for macromolecular crystallography. Acta Crystallogr D 66:12–21
18. Deisenhofer J, Remington SJ, Steigemann W (1985) Experience with various techniques for the refinement of protein structures. Methods Enzymol 115B:303–323
19. Diamond R (1971) A real-space refinement procedure for proteins. Acta Crystallogr A 27: 436–452
20. Dunitz JD, White DNJ (1973) Non-rigid-body thermal-motion analysis. Acta Crystallogr A 29:93
21. Engh RA, Huber R (1991) Accurate bond and angle parameters for X-ray protein structure refinement. Acta Crystallogr A 47:392–400
22. Fokine A, Urzhumtsev A (2002) Flat bulk-solvent model: obtaining optimal parameters. Acta Crystallogr D 58:1387–1392

23. French S, Wilson K (1978) On the treatment of negative intensity observations. Acta Crystallogr A 34:517–525
24. Grosse-Kunstleve RW, Afonine PV, Adams PD (2004) cctbx news. Newsletter of the IUCr commission on crystallographic computing 4: 19–36
25. Grosse-Kunstleve RW, Moriarty NW, Adams PD (2009) Torsion angle refinement and dynamics as a tool to aid crystallographic structure refinement. Proceedings of the ASME 2009 International Design Engineering Technical Conferences & Computers and Information in Engineering Conference, IDETC/CIE 2009, Aug 30–Sept 2, 2009, San Diego, California, USA
26. Headd JJ, Immormino RM, Keedy DA, Emsley P, Richardson DC, Richardson JS (2009) Autofix for backward-fit sidechains: using MolProbity and real-space refinement to put misfits in their place. J Struct Funct Genomics 10(1):83–93
27. Headd JJ, Echols N, Afonine PV, Grosse-Kunstleve RW, Chen VB, Moriarty NW, Richardson DC, Richardson JS, Adams PD (2012) Use of knowledge-based restraints in phenix.refine to improve macromolecular refinement at low resolution. Acta Crystallogr D 68:381–390
28. Helliwell JR (2008) Macromolecular crystal twinning, lattice disorders and multiple crystals. Crystallogr Rev 14:189–250
29. Johnson CK, Levy HA (1974) Thermal motion analysis using Bragg diffraction data. International tables for X-ray crystallography, vol IV, ed. Ibers JA, Hamilton WC, pp 311–335. Birmingham: The Kynoch Press
30. Jones TA, Zou J-Y, Cowan SW, Kjeldgaard M (1991) Improved methods for building protein models in electron density maps and the location of errors in these models. Acta Crystallogr A 47:110–119
31. Korostelev A, Bertram R, Chapman MS (2002) Simulated-annealing real-space refinement as a tool in model building. Acta Crystallogr D 58:761–767
32. Lunin VY, Skovoroda TP (1995) R-free likelihood-based estimates of errors for phases calculated from atomic models. Acta Crystallogr A 51:880–887
33. Lunin VY, Urzhumtsev A (1984) Improvement of protein phases by coarse model modification. Acta Crystallogr A 40:269–277
34. Lunin VY, Urzhumtsev A (1985) Program construction for macromolecule atomic model refinement based on the fast Fourier transform and fast differentiation algorithms. Acta Crystallogr A 41:327–333
35. Lunin VY, Urzhumtsev AG (1999) Maximal Likelihood refinement. It works, but why?. CCP4 newsletter on protein crystallography 37:14–28
36. Lunin VY, Urzhumtsev A (1989) FROG – high-speed restraint-constraint refinement program for macromolecular structure. J Appl Crystallogr 22:500–506
37. McCoy AJ, Storoni LC, Read RJ (2004) Simple algorithm for a maximum-likelihood SAD function. Acta Crystallogr D 60:1220–1228
38. Murshudov GN, Vagin AA, Dodson EJ (1997) Refinement of Macromolecular Structures by the Maximum-Likelihood Method. Acta Crystallogr D 53:240–255
39. Oldfield TJ (2001) A number of real-space torsion-angle refinement techniques for proteins, nucleic acids, ligands and solvent. Acta Crystallogr D 57:82–94
40. Pannu NS, Read RJ (1996) Improved Structure Refinement Through Maximum Likelihood. Acta Crystallogr A 52:659–668
41. Pannu NS, Murshudov GN, Dodson EJ, Read RJ (1998) Incorporation of Prior Phase Information Strengthens Maximum-Likelihood Structure Refinement. Acta Crystallogr D 54:1285–1294
42. Parsons S (2003) Introduction to twinning. Acta Crystallogr D 59:1995–2003
43. Prince E, Finger LW (1973) Use of Constraints on Thermal Motion in Structure Refinement of Molecules with Librating Side Groups. Acta Crystallogr B 29:179
44. Read RJ (1986) Improved Fourier coefficients for maps using phases from partial structures with errors. Acta Crystallogr A 42:140–149
45. Read RJ (1990) Structure-factor probabilities for related structures. Acta Crystallogr A 46: 900–912

46. Read RJ (1999) Detecting outliers in non-redundant diffraction data. Acta Crystallogr D 55:1759–1764
47. Read RJ, Adams PD, Arendall WB III, Brunger AT, Emsley P, Joosten RP, Kleywegt GJ, Krissinel EB, Lütteke T, Otwinowski Z, Perrakis A, Richardson JS, Sheffler WH, Smith JL, Tickle IJ, Vriend G, Zwart PH (2011) A new generation of crystallographic validation tools for the protein data bank. Structure 19:1395–1412
48. Rice LM, Brunger AT (1994) Torsion angle dynamics: reduced variable conformational sampling enhances crystallographic structure refinement. Proteins 19:277–290
49. Sheriff S, Hendrickson WA (1987) Description of overall anisotropy in diffraction from macromolecular crystals. Acta Crystallogr A 43:118–121
50. Tronrud DE (1994) From First Map to Final Model. Proceedings of the CCP4 Study Weekend, edited by S. Bailey, R. Hubbard & D Waller, Warrington: Daresbury Laboratory, pp. 111–124
51. Tronrud DE (1997) TNT refinement package. Methods Enzymol 277:306–319
52. Tronrud DE (2004) Introduction to macromolecular refinement. Acta Crystallogr D 60: 2156–2168
53. Urzhumtsev AG, Lunin VY, Vernoslova EA (1989) FROG – high-speed restraint-constraint refinement program for macromolecular structure. J Appl Crystallogr 22:500–506
54. Urzhumtsev A, Afonine PV, Adams PD (2011) TLS for dummies. Computational crystallography newsletter 2: 42–84. (http://www.phenix-online.org/newsletter)
55. Watkin DJ (2008) Structure Refinement: Some Background Theory and Practical Strategies. Appl Crystallogr 41:491–522
56. Winn MD, Isupov MN, Murshudov GN (2001) Use of TLS parameters to model anisotropic displacements in macromolecular refinement. Acta Crystallogr D 57:122–133
57. Zwart PH, Grosse-Kunstleve RW, Adams PD (2005) Characterization of X-ray data sets. CCP4 newsletter, Winter, Contribution 7

Chapter 21
Crystallographic Maps and Models at Low and at Subatomic Resolutions

Alexandre G. Urzhumtsev, Pavel V. Afonine, and Vladimir Y. Lunin

Abstract Crystallographic studies at both extremes of the resolution interval, low and subatomic, are less common in macromolecular crystallography and have their own specific features. Ignoring these features may complicate structure solution or lead to errors in crystallographic Fourier maps and in their interpretation.

Keywords Low resolution • Subatomic resolution • Crystallographic Fourier maps • Macromolecular models

21.1 Introduction

The *electron density distribution* of an ideal crystal $\rho_{exact}(x,y,z)$ is a periodic function of three space coordinates and can be calculated as a Fourier series containing an infinite number of Fourier coefficients $F_{hkl}\exp(i\varphi_{hkl})$, called, in crystallography, *structure factors*. In practice, the Fourier series contain only a finite number of terms corresponding to measured (or calculated) diffraction data. The integer indices (h,k,l) are the coordinates of a point in some basis and characterize the direction

A.G. Urzhumtsev (✉)
IGBMC, CNRS-INSERM-UdS, 1 rue Laurent Fries, B.P.10142, 67404 Illkirch, France

Physics Department, Faculté des Sciences et des Technologies, Université de Lorraine, B.P. 239, 54506 Vandoeuvre-lès-Nancy, France
e-mail: sacha@igbmc.fr

P.V. Afonine
Lawrence Berkeley National Laboratory, One Cyclotron Road, BLDG 64R0121, Berkeley, CA 94720, USA

V.Y. Lunin
Institute of Mathematical Problems of Biology, Russian Academy of Sciences, Pushchino 142290, Russia

R. Read et al. (eds.), *Advancing Methods for Biomolecular Crystallography*, NATO Science for Peace and Security Series A: Chemistry and Biology, DOI 10.1007/978-94-007-6232-9_21, © Springer Science+Business Media Dordrecht 2013

of the diffracted wave. In crystallography, the inverse of the distance from the point (h,k,l) to the origin, $d(h,k,l) = |(h,k,l)|^{-1}$, is called the *resolution of the corresponding reflection*. The set S of reflections inside a sphere of radius D, i.e. such that $|(h,k,l)| \leq D$, forms the *complete data set of resolution* $d_{high} = D^{-1}$. In what follows, we consider only complete data sets $\{F_{hkl}\}$, $hkl \in S$, i.e. we assume that the Fourier series is calculated over all reflections inside this sphere.

The Fourier sum $\rho(x,y,z)$ is called *a Fourier synthesis* corresponding to a given set S of reflections and it is different from the electron density distribution $\rho_{exact}(x,y,z)$. In crystallography, Fourier syntheses are visualized as a set of isosurfaces $\rho(x,y,z) = \mu$ where the choice of the constant μ depends on specific tasks. Such isosurfaces are often called *crystallographic maps*.

In some instances crystallographic studies at low resolution ($d_{high} > 6$–10 Å or lower) may be the only opportunity to obtain structural information. Studies using very-high-resolution data ($d_{high} < 0.8$–0.9 Å) may provide unique structural information not accessible at lower resolutions or by utilizing other methods (see for example [5], and references therein). In this article we briefly overview the key features of Fourier syntheses and structure modeling at these extreme resolutions.

21.2 Resolution Cut-Off, Fourier Maps and *R*-Factors

Calculating a Fourier synthesis from a truncated set of reflections results in a loss of detail and accuracy of the image: the lower the resolution (the larger d_{high}), the less terms are included into the Fourier summation, and the less detailed the image is (Fig. 21.1). Another effect of a resolution cut-off is an oscillation of the synthesis values, called Fourier truncation ripples. The ripples are most pronounced in the regions of sharp variation of the function, in particular around its peaks (which correspond to atomic centers in the case of electron density).

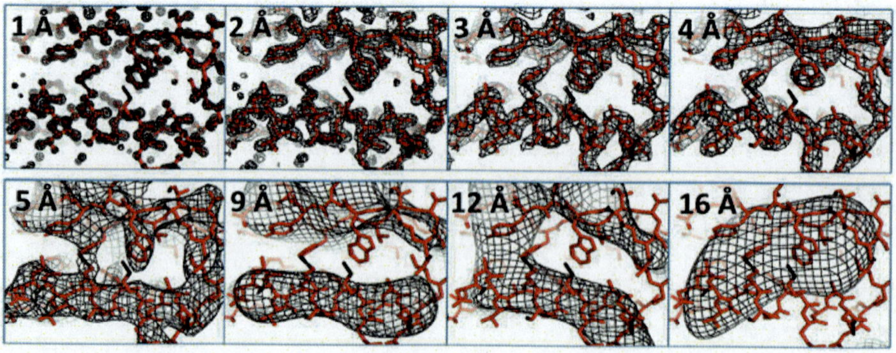

Fig. 21.1 An example of protein Fourier maps at different resolutions. The corresponding syntheses were obtained with the structure factors calculated from an atomic model (superposed)

Degradation of map quality due to limited data resolution d_{high} complicates building of atomic models. However *a priori* structural knowledge makes it possible to build and refine atomic models at resolutions lower than atomic resolution, such as 1.0–1.2 Å, where individual atoms are visible. These procedures have become more and more routine and automated (see, for example, [1]).

Since the maps become less interpretable with decreasing resolution, one may expect that models built using these maps become less accurate as well. *R*-factor statistics for all PDB [8, 9] structures reveals that the most frequent values of the *R*-factors grow with the resolution d_{high} [38]. Comparison of these values with the *R*-factors for the best models at the same resolution indicates two particular points, one at approximately 6 Å and the second at 0.8 Å. These resolution limits are the focus of discussion below.

21.3 Low Resolution and Bulk Solvent

A plot of mean observed structure factor amplitudes, $\langle F_{hkl}^{obs} \rangle$, and those calculated from an atomic model, $\langle F_{hkl}^{calc} \rangle$, as a function of resolution, shows two diverging curves starting around 6 Å (Fig. 21.2a; see also [29]). Not surprisingly, the *R*-factor increases as well at this resolution (Fig. 21.2b). This divergence is because macromolecules in crystals are surrounded by unstructured solvent, called bulk solvent. The amount of bulk solvent can vary from 10 to 90 % of the unit cell volume making its contribution to the X-ray scattering quite significant. The electron density in this region is rather featureless. According to general properties of the Fourier transform (the flatter the function, the sharper its Fourier transform, and *vice versa*),

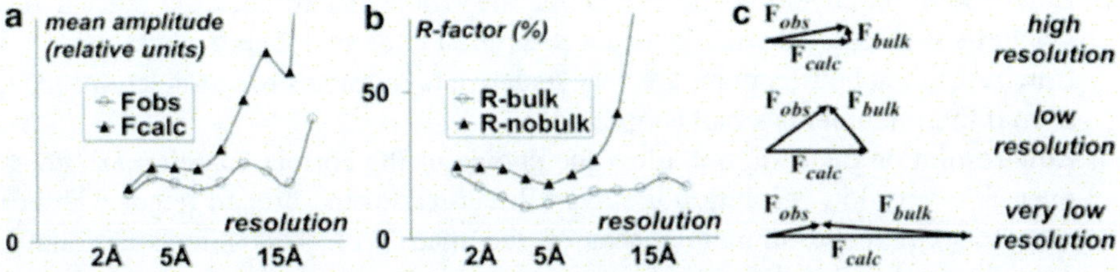

Fig. 21.2 Bulk-solvent contribution to the diffraction data. (**a**) Average amplitude shown in resolution shells for experimental structure factors (*circles*) and those calculated from a macromolecular atomic model (*triangles*). (**b**) *R*-factor value shown in resolution shells: data calculated from an atomic model with (*circles*) and without (*triangles*) bulk-solvent contribution. (**c**) Schematic relation between structure factors (as complex numbers) calculated from a macromolecular atomic model, F^{calc}, and from bulk solvent, F^{bulk}. They are of the same order of magnitude at resolutions lower than ~5–6 Å (*middle diagram*) but $F^{bulk} << F^{calc}$ at higher resolutions (*upper diagram*). At resolutions below 20–30 Å (*bottom diagram*) the amplitudes of F^{bulk} and F^{calc} are proportional to each other, and the phases differ by ~180° [37]

its Fourier coefficients F_{hkl}^{bulk} are much smaller than F_{hkl}^{calc} at resolutions 1–3 Å but are comparable at 5–6 Å and lower (Fig. 21.2c). We call this resolution limit *low resolution.*

The flatness of the density in the solvent region is the basic assumption for commonly used models, in particular that by Phillips [29] and by Jiang and Brunger [19]. Various improvements of this model have been reported over the last decade (see, for example, [16], and references therein).

21.4 Low- and Very-Low Resolution

Comparison of two last images in Fig. 21.1 shows a typical situation arising at resolutions 15–20 Å or lower, which we call *very-low resolution*. Here the macromolecular Fourier maps give scarce, but important, structural information, such as the shape of the molecule and its position in the crystal. The shape of an envelope may also be obtained from other methods, such as electron microscopy or SAXS, and its position may be determined by molecular replacement procedures [36], as was done for the T50S particle [6]. However, a direct determination or analysis of low-resolution envelopes [28] may be important when no extra information is available, or when an independent result is needed. Some key features of very-low resolution crystallographic studies are given below.

1. Low-resolution Fourier syntheses represent molecular envelopes at relatively low map contouring levels. There is a common belief that they indicate the centers of the molecules when this level is high. In fact, the peaks at such syntheses are often shifted from corresponding molecular centers toward regions of close intermolecular contacts (see for example Fig. 21.3).
2. Low-resolution envelopes do not cover the whole macromolecule even if the synthesis is calculated with the exact structure factor values (Fig. 21.1).
3. Low-resolution Fourier maps are very sensitive to missing strong reflections [35], even if their number is small (Fig. 21.4).
4. Low-resolution maps do not allow the choice of the correct enantiomer. Moreover, at a very-low resolution the overall features of the flipped (sign-inverted) map -$\rho(\mathbf{r})$ may be similar to that of the map $\rho(\mathbf{r})$. This may additionally complicate *ab initio* phasing [28].
5. Due to a relatively small number of reflections and their low resolution, it is difficult to distinguish between the space-group symmetry, twinning or local (non-crystallographic) symmetry (see for example, [26]).
6. Very-low resolution maps calculated with relatively few reflections may show a superposition of images corresponding to different choices of the origin [25].
7. Increasing the resolution of the low-resolution Fourier maps does not always help to interpret them better; they may stop showing the molecular envelope but do not yet reveal secondary structure elements (Fig. 21.1, last two images).

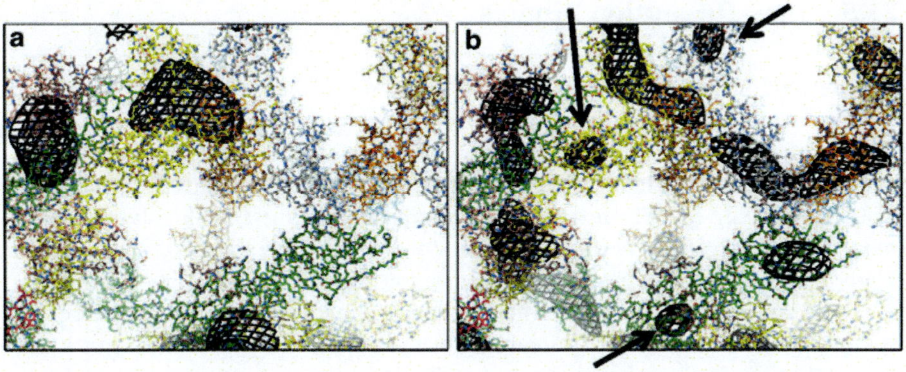

Fig. 21.3 Titin model [40] consisting of six quasi identical molecules per asymmetric unit, superposed with the maps of the resolution of 22 Å shown with a high cut-off level. The maps shown are (**a**) the best possible 'experimental' map [Fourier coefficients are $F^{obs}\exp(i\varphi^{model})$ where the phases take into account the contribution of both the macromolecule and bulk solvent] and (**b**) the best possible model map [where Fourier coefficients $F^{calc}\exp(i\varphi^{calc})$ were calculated from the atomic model only]. In both maps the peaks do not correspond to the centers of individual molecules, with a few exceptions indicated by arrows. The same is true for the maps of both lower and higher resolutions (not shown), and for peaks at higher and at lower cut-off levels (not shown)

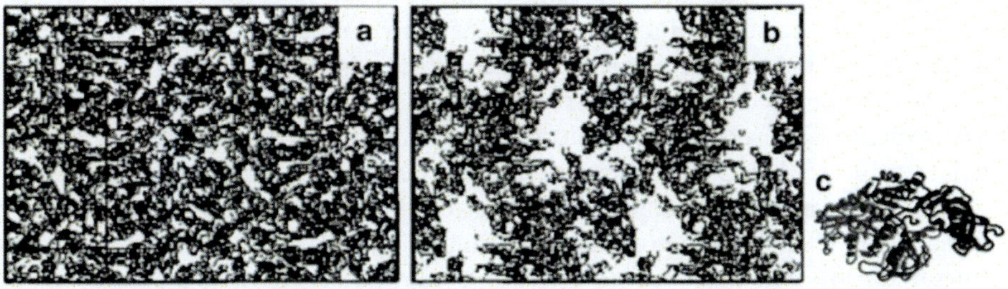

Fig. 21.4 8-Å resolution maps for elongation factor G [12]. (**a**) Synthesis calculated with experimental amplitudes and SIR-phases, approximately 1,700 reflections in total. (**b**) The same data but 29 reflections of the resolution below 30 Å were restored and added. (**c**) An atomic model [2] in an orientation approximately corresponding to the left bottom molecule at (**b**)

8. The standard deviation σ used to define the synthesis cut-off level μ becomes very sensitive to various factors making map comparison difficult. Other units may be used instead, for example a volume per residue or a selected percent of the unit cell volume (see for example [23]).
9. At very-low resolutions the bulk-solvent structure factors $F_{hkl}^{bulk}\exp\left(i\varphi_{hkl}^{bulk}\right)$ become quasi proportional to those $F_{hkl}^{calc}\exp\left(i\varphi_{hkl}^{calc}\right)$ calculated from a macromolecular model (Fig. 21.2); this is also true for their sum. As a result, at such resolutions an explicit modeling of bulk solvent becomes less critical. For example, molecular replacement with an atomic model at ~20 Å resolution does not need modeling of bulk solvent while it may be crucial at 8–10 Å.

21.5 Low-Resolution Models

Low-resolution crystallographic studies may be the only option in a number of practical situations, for example when no high-resolution intensities are measured or when usual phasing methods do not yield a solution. While phasing of low-resolution maps is challenging [28], the interpretation of these maps and the building of corresponding models also have specific problems. First, bulk solvent shall be modeled and taken into account when low-resolution data are used. Also different kinds of models may be used for the macromolecule itself.

The geometry of some structural elements (amino acids, nucleotides, α-helices) is known, and the structure of domains or even molecules themselves may be known from either previous or alternative studies. This allows one to fit atomic models even into low-resolution crystallographic maps using a kind of molecular replacement procedures. A direct interpretation of such maps may lead to ambiguous map interpretation and erroneous structures [11].

Non-atomic models, such as Gaussian spheres for individual residues [7, 34] or for the whole molecule (for example [31]), may be used instead. Lunin et al. [24] showed that modeling crystal content by a few Gaussian scatterers is ambiguous and that quite different models may result in structure factor amplitudes equally similar to the experimental data. Moreover, the centers of such large Gaussian spheres do not necessarily correspond to the centers of individual components, domains, or molecules, and may be shifted towards the interface of such components (see feature 1 of the previous Section). Therefore although these models are useful for phasing their structural interpretation should be done with care. Other geometric objects such as cylinders [21, 33] or spherical shells (e.g., [20, 26]) may be used as well.

Finally, at very-low resolutions the electron density in the protein region may be modeled by a flat continuous function, similarly to bulk solvent. Low-resolution structure factors calculated from such a model may be very close to the exact values [27] indicating that the crystal electron density can be approximated by a macromolecular envelope. The border of this envelope, a two-dimensional surface that separates the macromolecule and solvent, can be parameterized and refined [16].

21.6 Subatomic Resolution

Crystallographic maps at subatomic (also called sub-angstrom, ultra-high) resolutions show macromolecules more accurately and reveal extra details that are not available at typical macromolecular resolutions (∼2 Å). Computational artifacts are also common. Figure 21.5 shows crystallographic Fourier maps with circular peaks around interatomic bonds, which are the Fourier truncation ripples and do not bear any structural information [10]. The first positive spurious peaks superimpose at the points approximately at a distance $0.9d_{high}$ from the atomic centers. Such points form the circles around the middle of interatomic bonds; the smaller the d_{high}, the smaller the radius of the circle.

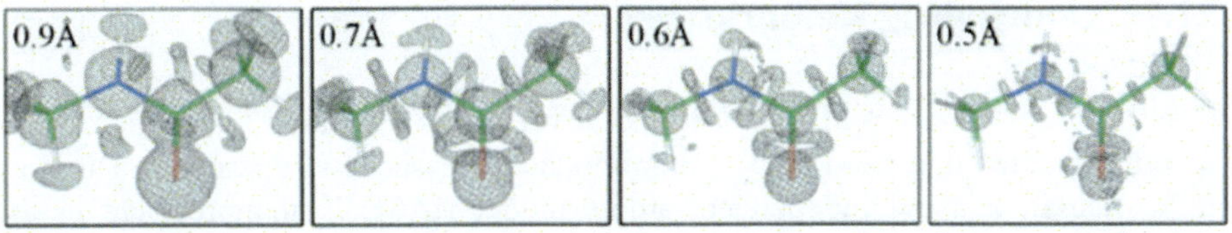

Fig. 21.5 Fourier ripples forming rings around interatomic bonds [10]. Note the variation of the rings with resolution

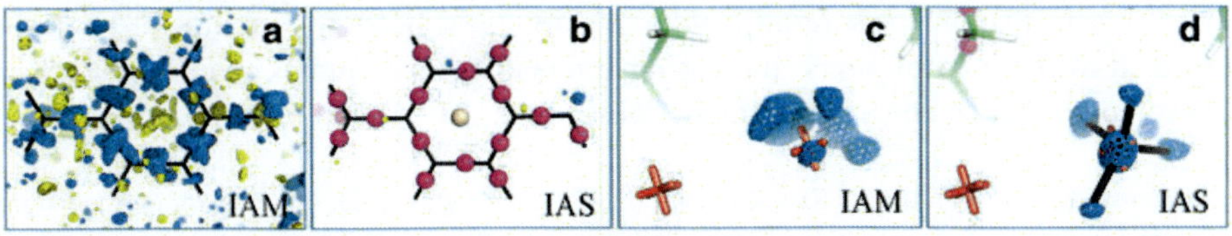

Fig. 21.6 Difference maps calculated with a conventional (IAM) model (**a**,**c**), and with IAS model (**b**,**d**) [4]. (**a**,**b**) tripeptide YGG at 0.43 Å [30]; cut-off levels 0.20 and 0.10 eÅ-3, respectively. (**c**,**d**) antifreeze protein at 0.62 Å [22], cut-off levels 0.40 and 0.25 eÅ-3, respectively. IAS are shown by spheres. In (**b**) there is a positive IAS at each interatomic bond and a negative one in the middle of the aromatic ring; note that the new difference map has only four tiny residual peaks. In (**d**) the closest IAS is very far from the region shown; SO4 molecule could be unambiguously identified in the improved map instead of a water molecule as it is in the deposited model

This indicates that unbalanced Fourier syntheses should be used with care in such cases, and difference maps are preferable, showing less Fourier truncation artifacts. At subatomic resolution these maps would typically show positive and negative density peaks around atoms and bonds (Fig. 21.6a), reflecting the redistribution of the electron density due to atomic interaction and bond formation [3]. Not modeling these densities may results in incorrect values of refined displacement parameters [3] of atoms and overall noisier maps (Fig. 21.6c). Also, a large number of experimental diffraction amplitudes compared to the number of atoms allows the introduction of more detailed models which are capable of describing extra features visible at this resolution.

An example of such a model is the multipolar model [17]. While routinely used in small-molecule crystallography, its straightforward application to macromolecules [18] is impractical due to the many parameters that this model introduces and runtime considerations. Other possible techniques have been later suggested, such as invariomes [14], transferable libraries [39] and Cartesian Gaussian multipoles [32].

A simple alternative solution is to model deformation peaks directly with Gaussian interatomic scatterers, the IAS model ([3], and references therein). This modeling improves refined atomic displacement parameters and maps (Figs. 21.6a, b). This in turn may allow interpretation and modeling of details otherwise hidden in map noise (Figs. 21.6c, d).

The high quality of data and models at ultra-high resolution allows map calculation on an absolute scale while working in σ becomes inconvenient.

21.7 Concluding Remarks

Crystallographic studies at very-low and at very-high resolution have a number of specific features that need to be accommodated for successful structure solution. In particular, a usual independent-atoms model (IAM) is inappropriate in the case of low resolution and insufficient in the case of high resolution. At these extreme resolutions, the Fourier ripples can severely interfere with the map quality and therefore with the map interpretation. A conventional definition of the map contouring level μ expressed in standard deviations σ is inconvenient in both cases.

To date, relatively few (compared to all structures in PDB) structures have been studied at these resolutions. Increasing interest in such studies requires a further development of corresponding methodologies.

Acknowledgment VL thanks RFBR 10-04-00254-a grant for financial support. PA acknowledges the NIH (grant GH063210) and the Phenix Industrial Consortium for support of the Phenix project. *PyMol* [13] and *coot* [15] were used for illustrations. The authors thank all persons contributed to different parts of the relevant projects and A. McEwen for careful reading and correcting the text

References

1. Adams PD, Afonine PV, Bunkóczi G, Chen VB, Davis IW, Echols N, Headd JJ, Hung L-W, Kapral GJ, Grosse-Kunstleve RW, McCoy AJ, Moriarty NW, Oeffner R, Read RJ, Richardson DC, Richardson JS, Terwilliger TC, Zwart PH (2010) PHENIX: a comprehensive Python-based system for macromolecular structure solution. Acta Crystallogr D 66:213–221
2. Aevarsson A, Brazhnikov E, Garber M, Zheltonosova J, Chirgadze Y, Al-Karadaghi S, Svensson LA, Lilias A (1994) Three-dimensional structure of the ribosomal translocase: elongation factor G from *Thermus thermophilus*. EMBO J 13:3669–3677
3. Afonine PV, Lunin VY, Muzet N, Urzhumtsev A (2004) On the possibility of observation of valence electron density for individual bonds in proteins in conventional difference maps. Acta Crystallogr D 60:260–274
4. Afonine PV, Grosse-Kunstleve RW, Adams PD, Lunin VY, Urzhumtsev A (2007) On macromolecular refinement at subatomic resolution with interatomic scatterers. Acta Crystallogr D 63:1194–1197
5. Afonine PV, Mustyakimov M, Grosse-Kunstleve RW, Moriarty NW, Langan P, Adams PD (2010) Joint X-ray and neutron refinement with phenix.refine. Acta Crystallogr D 66:1153–1163
6. Ban N, Freeborn B, Nissen P, Penczek P, Grassucci RA, Sweet R, Frank J, Moore PB, Steitz TA (1998) A 9 Å resolution X-ray crystallographic map of the large ribosomal subunit. Cell 93:1105–1115
7. Bentley GA, Lewit-Bentley A, Finch JT, Podjarny AD, Roth M (1984) Crystal structure of the nucleosome core particle at 16 A resolution. J Molec Biol 176:55–75
8. Berman HM, Westbrook J, Feng Z, Gilliland G, Bhat TN, Weissig H, Shindyalov IN, Bourne PE (2000) The protein data bank. Nucleic Acids Res 28:235–242
9. Bernstein FC, Koetzle TF, Williams GJ, Meyer EF, Brice MD, Rodgers JR, Kennard O, Shimanouchi T, Tasumi M (1977) The protein data bank: a computer-based archival file for macromolecular structures. J Mol Biol 112:535–542

10. Bochow A, Urzhumtsev A (2005) On the Fourier series truncation peaks at subatomic resolution. CCP4 Newsletter on Protein Crystallography 42: http://www.ccp4.ac.uk/newsletters/newsletter42/content.html
11. Chang G, Roth CB, Reyes CL, Pornillos O, Chen YJ, Chen AP (2006) Retraction. Science 314:1875
12. Chirgadze YN, Brazhnikov EV, Garber MB, Nikonov SV, Fomenkova NP, Lunin VY, Urzhumtsev A, Chirgadze NY, Nekrasov YV (1991) Crystal structure of ribosomal factor G from bacteria *Thermus thermophilus* at low resolution. Dokl Acad Nauk SSSR 320:488–491
13. DeLano WL (2002) The PyMOL Molecular Graphics System, DeLano Scientific, San Carlos, CA, USA. http://www.pymol.org
14. Dittrich B, Hübschle CB, Messerschmidt M, Kalinowski R, Girnt D, Luger P (2005) The invariome model and its application: refinement of D, L-serine at different temperatures and resolution. Acta Crystallogr A 61:314–320
15. Emsley P, Lohkamp B, Scott WG, Cowtan K (2010) Features and development of coot. Acta Crystallogr D 66:486–501
16. Fenn TD, Schnieders MJ, Brunger AT (2010) A smooth and differentiable bulk-solvent model for macromolecular diffraction. Acta Crystallogr D 66:1024–1031
17. Hansen NK, Coppens P (1978) Testing aspherical atom refinements on small-molecule data sets. Acta Crystallogr A 34:909–921
18. Jelsch C, Pichon-Pesme V, Lecomte C, Aubry A (1998) Acta Crystallogr D 54:1306–1318
19. Jiang JS, Brünger AT (1994) Protein hydration observed by X-ray diffrcation. Solvation properties of penicillopepsin and neuraminidase crystal structures. J Mol Biol 243:100–115
20. Johnson JE, Akimoto T, Suck D, Rayment I, Rossmann MG (1976) The structure of southern bean mosaic virus at 22.5 resolution. Virology 75:394–400
21. Kalinin DI (1980) Use of a cylindrical model of a protein to determine the spatial structure of the rhombic modification of leghaemoglobin. Sov Phys Crystallogr 25:307–313
22. Ko TP, Robinson H, Gao YG, Cheng CHC, DeVries AL, Wang AHJ (2003) The refined crystal structure of an Eel pout type III antifreeze protein RD1 at 0.62-A resolution reveals structural microheterogeneity of protein and solvation. Biophys J 84:1228–1237
23. Lunin VY (1988) Use of information on electron density distribution in macromolecules. Acta Crystallogr A 44:144–150
24. Lunin VY, Lunina NL, Petrova TE, Vernoslova EA, Urzhumtsev A, Podjarny AD (1995) On the ab initio solution of the phase problem for macromolecules at very low resolution: the Few Atoms Model method. Acta Crystallogr D 51:896–903
25. Lunin VY, Lunina N, Urzhumtsev A (1999) Seminvariant density decomposition and connectivity analysis in very low resolution macromolecular phasing. Acta Crystallogr A 55:916–925
26. Lunin VY, Lunina N, Ritter S, Frey I, Keul J, Diederichs K, Podjarny A, Urzhumtsev A, Baumstark M (2001) Low-resolution data analysis for the low-density lipoprotein particle. Acta Crystallogr D 57:108–121
27. Lunin VY, Urzhumtsev A, Bockmayr A (2002) Direct phasing by binary integer programming. Acta Crystallogr A 58:283–291
28. Lunin VY, Urzhumtsev A, Podjarny AD (2012) *An initio* phasing of low-resolution Fourier syntheses. In: Himmel DM, Rossmann MG, Arnold E (eds) International tables for crystallography, vol F. Wiley, Chichester, pp 437–442
29. Phillips SEV (1980) Structure and refinement of oxymyoglobin at 1.6 Å resolution. J Mol Biol 142:531–554
30. Pichon-Pesme V, Lachekar H, Souhassou M, Lecomte C (2000) Electron density and electrostatic properties of two peptide molecules: tyrosyl-glycyl-glycyne monohydrate and glycyl-aspartic acid dehydrate. Acta Crystallogr B 56:728–737
31. Podjarny AD, Rees B, Thierry JC, Cavarelli J, Jesior JC, Roth M, Lewitt-Bentley A, Kahn R, Lorber B, Ebel JP, Giegé R, Moras D (1987) Yeast tRNAAsp –aspartyl-tRNA synthetase complex: low resolution crystal structure. J Biomol Struct Dyn 5:187–198

32. Schnieders MJ, Fenn TD, Pande VS, Brunger AT (2009) Polarizable atomic multipole X-ray refinement: application to peptide crystals. Acta Crystallogr D 65:952–965
33. Strop P, Brzustowicz MR, Brunger AT (2007) Ab initio molecular-replacement phasing for symmetric helical membrane proteins. Acta Crystallogr D 63:188–196
34. Svergun DI, Petoukhov MV, Koch MHJ (2001) Determination of domain structure of proteins from X-ray solution scattering. Biophys J 80:2946–2953
35. Urzhumtsev A (1991) Low-resolution phases: their influence on SIR-syntheses and retrieval with double-step-filtration. Acta Crystallogr A 47:794–801
36. Urzhumtsev A, Podjarny AD (1995) On the solution of the molecular-replacement problem at very low resolution: application to large complexes. Acta Crystallogr D 51:888–895
37. Urzhumtsev A, Podjarny AD (1995) On the problem of solvent modelling in macromolecular crystals using diffractional data: 1. The low resolution range. Joint CCP4 and ESF-EACBM Newsletter on Protein Crystallography, 31: 12–16
38. Urzhumtsev A, Afonine PV, Adams PD (2009) On the use of logarithmic scales for analysis of diffraction data. Acta Crystallogr D 65:1283–1291
39. Volkov A, Messerschmidt M, Coppens P (2007) Improving the scattering-factor formalism in protein refinement: application of the University at Buffalo Aspherical-Atom Databank to polypeptide structures. Acta Crystallogr D 63:160–170
40. von Castelmur E, Marino M, Svergun DI, Kreplak L, Labeit D, Ucurum-Fotiadis Z, Konarev PV, Urzhumtsev A, Labeit S, Mayans O (2007) A regular pattern of Ig super-motifs defines segmental flexibility as the elastic mechanism of the titin chain. Proc Natl Acad Sci 105:1186–1191

Chapter 22
Recent Advances in Low Resolution Refinement Tools in *REFMAC5*

Robert A. Nicholls, Fei Long, and Garib N. Murshudov

Abstract This contribution deals with some aspects of low resolution refinement and map calculation, as implemented in the crystallographic refinement program – *REFMAC5*. Refinement at low resolution is considered as a regularisation problem. Regularisers for application in both real space and reciprocal space have been implemented. In real space, regularisers are applied as interatomic distance restraints. There are two types of real space regularisers: those with a target, and those without a target. External restraints to reference structures belong to the class with a target, where targets are calculated using corresponding interatomic distances in the reference structure(s). Such reference structures may arise from homologous structures, secondary structural elements, generic or specific structural fragments (including self-restraints), and other sources of generically derived information, e.g. hydrogen bonds. Such interatomic distance restraints are generated using the conformation-independent protein structure comparison and analysis program – *ProSMART*. Jelly-body restraints belong to the real space targetless class of regularisers, where interatomic distance self-restraints are recalculated at every cycle. This regulariser has the power to stabilise refinement without imposing externally-derived information. Importantly, it is not dependent on initial values, and does not change the extrema of the target function. Regularisers in reciprocal space are designed to sharpen (deblur) the electron density map, whilst inhibiting the amplification of noise. These regularisers do not affect refinement. Rather, they are applied at the end of a refinement session, with the intention of aiding subsequent map interpretation.

Keywords Low resolution refinement • Regularisers • External restraints • Map sharpening • Refmac5 • Prosmart

R.A. Nicholls • F. Long • G.N. Murshudov (✉)
Structural Studies Division, MRC Laboratory of Molecular Biology, Cambridge CB2 0QH, UK
e-mail: nicholls@mrc-lmb.cam.ac.uk; flong@mrc-lmb.cam.ac.uk; garib@mrc-lmb.cam.ac.uk

R. Read et al. (eds.), *Advancing Methods for Biomolecular Crystallography*, NATO Science for Peace and Security Series A: Chemistry and Biology,
DOI 10.1007/978-94-007-6232-9_22, © Springer Science+Business Media Dordrecht 2013

22.1 Introduction

The intrinsic high mobility of molecules, subunits and domains inside unit cells, coupled with long-range lattice disorder in large macromolecular crystals, often causes poor quality diffraction from such crystals. This causes the observation of weak, anisotropic and noisy data, resulting in the obtained data having very poor information content. However, the structures of individual components of a complex may have been independently determined at a higher resolution. Such information might then be used to aid the refinement of the lower-resolution structure.

There are other factors that can reduce the information content of macromolecular crystallographic data thus reducing effective resolution. These include crystal growth peculiarities such as twinning and order-disorder. In these cases, although the nominal resolution may be high, not all of the observations are independent. For example, in the case of perfect hemihedral twinning, the number of independent observations is decreased by a factor of two, corresponding to a resolution reduction by a factor of $2^{\frac{1}{3}} = 1.26$. Therefore, in the limit, the quality of the electron density map in the presence of perfect hemihedral twinning at 2 Å would correspond to that of at 2.52 Å in the single crystal case. The refinement of models against data from twinned crystals is now routine [1–3]. However, statistics after refinement against such data should be interpreted with care [4]. It is important to remember that R-factors and other overall statistics are dependent on the statistical properties of the data, and therefore comparison of R-factors from different crystals may give the wrong impression about the comparative quality of the models.

There are many problems that need to be tackled in order to make low resolution structure analysis routine, two of which are considered here:

1. The use of chemical and structural information as restraints to increase consistency of the derived atomic models with available prior knowledge. The use of chemical information in the form of bond lengths, bond angles, and torsion angles has always been routine. For details on the organisation and use of chemical knowledge in refinement, see Ref. [5], for example. Recent years have seen an explosion of approaches towards utilising structural information [1, 3, 6–8]. This demonstrates the importance of finding a (and the lack of a unique) solution to the problem of exploiting structural information.
2. Calculation of electron density to aid the reduction of errors introduced during manual and automatic model building. Data from low resolution crystals usually exhibit high isotropic and anisotropic B-values. This contributes to observing smeared regions of electron density, with vanishing side chains, secondary structural elements, and even domains. Were this effect removed, the electron density map may reveal more features. Current approaches use only one B-value for crystal map sharpening. However, the problem is complicated by the non-negligible influence of contributing factors such as anisotropic diffraction, rigid body oscillation of individual structural units, and correlated motion of whole chains.

Many tools have been developed to aid crystallographic refinement at medium and higher resolutions over the past few decades. One of the current challenges is to develop complementary approaches for dealing with cases where only low resolution data are available (lower than around 3 Å). One of the sources of available information is the 3D structures of macromolecules deposited in the Protein Data Bank [9]. Structural information may be utilised in various forms, such as secondary structure restraints, homologous reference structures, and homology models, by various modern refinement software packages including *REFMAC5* [1,10] of *CCP4* [11], *BUSTER-TNT* [12], *phenix.refine* [2, 13], *SHELX* [3], and *CNS* [7, 14].

The concept of calculating an electron density map showing more features, e.g. side chains, has been proposed by many authors. Notably, Brunger et al. [15, 16] suggest a procedure known in the field of image processing [17] as inverse filtering. However, it is known that such filters can amplify noise, thus masking out real signal. Unfortunately, the electron density always contains noise, which stems from several sources:

1. Noise due to variations in the experimental data;
2. Noise due to errors in the model (e.g. atomic coordinates, model incompleteness, misparameterisation, *B*-factors, scale factors), and thus in calculated phases. Such noise correlates with the "true" electron density, and is consequently very hard to address;
3. Noise due to Fourier series termination. When data are collected to the crystal diffraction limit and no map sharpening is used, such noise usually dies out approaching the high resolution limit. However, when map sharpening is used as an inverse filter then the effect of series termination becomes pronounced.

In general, we consider refinement at low resolution as a regularisation problem [18]. In the context of parameter optimisation using the maximum likelihood method, information derived from external sources can be used to regularise the problem. Injecting regularisers in the form of restraints reduces the effective number of parameters, thus increasing the effective residual degrees of freedom, resulting in refinement stabilisation and improved resistance to overfitting. Also, regularisers can be used in order to find a stable solution to the ill-posed problem of map sharpening, which is considered as an inverse deblurring problem.

Various regularisers have been designed, each of which are suitable in different situations. Regularisers for application in both real space and reciprocal space have been implemented. In real space, regularisers are applied as interatomic distance restraints, of which there are two types: those with a target, and those without a target.

Here, we first describe the use of external structural information, e.g. from reference homologous structures, generic or specific structural fragments, and other sources of generically derived information, e.g. hydrogen bonds. We also describe regularisers for use in the absence of appropriate external information, specifically jelly-body restraints. Then, we describe anisotropic regularised map sharpening. We also provide examples of practical usage.

22.2 External Structural Restraints

Information from external sources can be incorporated during refinement using a Bayesian framework, where the distribution of interatomic distances serves as prior knowledge. Restraints generated using external structural information should help the macromolecule under refinement to adopt a conformation that is more consistent with previous observations. This is similar to the use of geometry terms in refinement, which helps local structure adopt chemically reasonable conformations.

The minus log posterior distribution target used in *REFMAC5* [10], may be expressed:

$$f_{\text{total}} = f_{\text{geom}} + w f_{\text{xray}} \tag{22.1}$$

where f_{geom} and f_{xray} are the contributions of geometry terms (-prior distribution) and experimental data(-loglikelihood), and w weights their relative contributions. The geometry component is a linear combination of various factors (effectively equivalent to the assumption that these contributors are independent), including any available external structural information:

$$f_{\text{geom}} = f_{\text{other}} + w_{\text{ext}} \sum_{\substack{(d,r,\sigma)\in R \\ r\leq r_{\max}}} f_{\text{ext}}\left(\tfrac{d-r}{\sigma}, \kappa\right) \tag{22.2}$$

where $f_{\text{ext}}(\hat{r}, \kappa)$ is the unweighted contribution of an external interatomic distance restraint $(d, r, \sigma) \in R$ to the target function, where R is the list of external restraints, and the function $f_{\text{ext}}(\hat{r}, \kappa)$ depends on the normalised residual $\hat{r} = \frac{d-r}{\sigma}$ and parameter κ. The parameter w_{ext} adjusts the weight of the external restraints relative to the other geometry components, and f_{other} represents the contribution of all other prior information [1]. An interatomic distance restraint comprises the current distance d between two atomic positions, the objective value r, and standard uncertainty σ. The mechanism used for application of external restraints in *REFMAC5* is described by Mooij et al. [19]. The effect of an external restraint during crystallographic refinement is illustrated in Fig. 22.1.

Here, we stipulate that the objective value r of an external restraint should be lower than some threshold $r_{\max}$, typically 4.2 Å, so that only reasonably short range restraints are utilised. External restraints are designed to be longer-range than chemical bond and angle restraints, whilst being sufficiently short to allow resistance to differences in global conformation. This allows potential for external restraints to be used even when the target and reference structures are, for example, in different bound states, or from different crystal forms.

In *REFMAC5*, the Geman-McClure [22] robust estimation function is used for external restraints:

$$f_{\text{ext}}(r, \kappa) = \frac{r^2}{1 + \kappa^2 r^2} \tag{22.3}$$

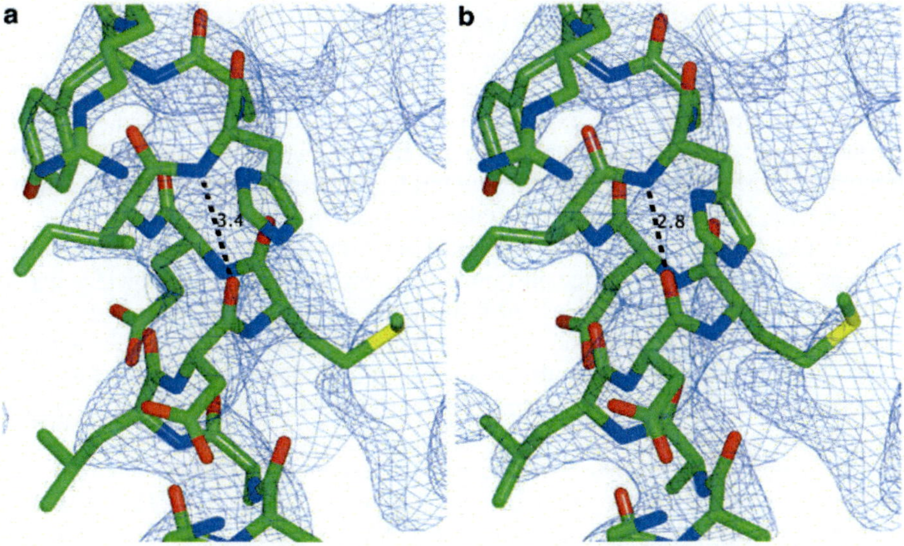

original structure. re-refined with an external restraint.

Fig. 22.1 Illustration of the effect of an external restraint. Subfigure (**a**) depicts a region of helical structure from 3v4w [20] taken from the PDB [9]. Subfigure (**b**) shows the same region after refinement using an external restraint between a nitrogen and an oxygen atom (visually identified by a *dashed line*), which corresponds to a single α-helical generic h-bond restraint, with objective value set to 2.8 Å. Before refinement, the interatomic distance is 3.4 Å. During refinement, the interatomic distance is gradually pulled towards the restraint objective value. At the end of refinement, the interatomic distance is reduced to 2.8 Å, and the helical structure appears more regular (Images were produced using *CCP4mg* [21])

This function, which is equivalent to least-squares for $\kappa = 0$, helps to reduce the influence of outliers, and sensitivity to conformational changes.[1]

Various criteria have been used for optimisation of the X-ray weight w, notably R_{free} [23], $-LL_{\text{free}}$ [24, 25], and geometry statistics. Similarly, the appropriate selection of external weight w_{ext} is not obvious. It should be acknowledged that using R_{free} as an optimisation criterion may reduce the usefulness of R_{free} as an independent indicator of refinement quality, and similarly for other statistics. Therefore, careful consideration should be given to selection of the weight for external structural information w_{ext}, and other parameters such as κ. For example, suitable parameters for external restraints to homologous structures might generally be $w_{\text{ext}} = 10$ and $\kappa = 0.02$. However, suitable parameter values will depend on the context of application. For example, if wanting to force a particular region to adopt specific local structure using a given reference fragment during earlier stages of refinement when density is poor, then using a higher w_{ext} and lower κ may be more appropriate in order to enforce compliance.

[1]Note that a similar formulation is used for local NCS restraints. This allows the NCS-related chains to adopt different conformations if justified by the X-ray data.

22.2.1 Selection of External Structural Information

External structures should be selected on the basis of their reliability and similarity to the current model. For example, suitable reference structures may include sequence-identical, homologous, or structurally similar models solved at a higher resolution, or generically derived structural information from non-homologous sources (e.g. secondary structure restraints obtained from an ideal α-helix).

The use of external restraints may, in some cases, be justified by any resultant increase in reliability of atomic positions. However, it should be acknowledged that such an approach introduces bias; the influence of such bias may result in the model adopting a conformation less consistent with the observed data. The use of external restraints might make a particular model adopt a conformation very similar to a high-resolution homologue, assuming it is appropriate to do so, and ideally result in an improved model.

We suggest that external restraints should only be used if the benefits of any improvement in reliability are deemed to outweigh the negative effects. Indeed, this may well be the case for data of poor quality collected at low resolution. For example, refinement of a model might cause some regions of very poor electron density to adopt an incorrect conformation. Increasing the weight of geometry terms may help the structure to adopt a more chemically-reasonable configuration, but the region may still be incorrectly modelled due to the effect of the misleading density; geometry restraints operate at a very high level of structural resolution. However, external restraints can operate at a lower level of structural resolution, as desired (e.g. by increasing the maximum restraint distance $r_{\max}$).

22.2.2 Generation of External Restraints to Homologous Structures

Here, we refer to the chain that is to be refined as the *target* chain, and to the chain that is to be used to generate the restraints as the *external* or *reference* chain.

External restraints for use in refinement by *REFMAC5* may be generated using the tool *ProSMART* [26]. Amongst various other functionalities, *ProSMART* can generate interatomic distance restraints utilising structural information. Whilst reference structures would generally be near-identical in sequence, the approach allows the alignment of, and subsequent restraint-generation using, any reference chain(s) regardless of sequence homology. However, it is not implied that there would be any utility of using external restraints based on dissimilar structures; a high degree of local structural conservation would generally be required for the successful application of external restraints. In general, we assume that the target and external reference structures are sufficiently similar, although at present such decisions should ultimately be made by the user.

The alignment approach adopted by *ProSMART* is independent of global conformation, being instead concerned with the net conservation of local structure, at a chosen level of structural resolution. Indeed, restraints generated by *ProSMART* allow great global flexibility, rather than rigidly pulling the target structure towards the same global conformation. Consequently, it is not necessary for the external reference chain to adopt the same global conformation as the target, e.g. structures in different bound states can be used. It is, however, necessary for local structure to be sufficiently well conserved along the chain so that the effect on refinement is positive.

22.2.2.1 General Approach

Suppose that we want to generate an interatomic distance restraint between two atoms in the target structure, given knowledge of their positions and thus the distance d between them. Given an external reference structure, and a residue alignment between the target and reference structures, it is possible to find the distance r between the corresponding atoms in the reference structure. The distance r is the objective value of the restraint.

If the target and external chains share a high degree of structural similarity, then we might expect for d to be approximately equal to r, with some error. Consequently, the restraint distances r, with appropriate distributional assumptions, can be used as prior information during crystallographic refinement. Since we want to maintain a degree of global conformational independence between the target and reference structures, it is undesirable to generate restraints between atoms that are far apart. Therefore, restraints are only generated whose objective values are less than some threshold $r_{\max}$. This parameter represents the structural resolution of the restraints; lower thresholds allow better conformational independence, whilst higher thresholds provide more information about the surrounding structural environment.

The adopted procedure of external restraint generation thus involves: the identification of the lists of corresponding intrachain atom-pairs in both the target and reference structures; filtering these lists in order to identify only those atom-pairs suitable for restraint generation (based on interatomic distance criteria); identification of corresponding atom-pairs between the target and reference structures; and finally estimation of restraint distributions.

22.2.2.2 Identification of Atom-Pairs to Be Restrained

Knowledge of an alignment between residues allows the direct inference of an atomic correspondence between target and reference structures. Such a correspondence may include both main and side chain atoms (providing aligned amino acids are the same), or only main chain N, C^{α}, C and O atoms (allowing main chain structural restraints to be generated even for residues of different amino acid type).

The alignment may also be filtered according to conservation of local main chain and/or side chain structure in an attempt to avoid the generation of potentially unsuitable restraints.

Given an alignment of atoms, it is then necessary to identify the list of sufficiently close atom-pairs, independently for each of the two structures. Various methods for near-neighbour searching have been developed. Here, in order to efficiently reduce the search space, we use a cell technique [27] previously used in biology [28], which involves the uniform partitioning (voxelisation) of space into cubic cells with edge length $r_{\max}$ (the value of $r_{\max}$ is chosen to be 1.5 times greater for the target structure than for the reference). This approach allows the efficient identification of all atoms with positions $\mathbf{x_i}$ and $\mathbf{x_j}$ such that their interatomic distance satisfies the criteria: $r_{\min} \leq |\mathbf{x_i} - \mathbf{x_j}| \leq r_{\max}$.

Using the achieved atomic correspondence, we may then calculate the list of all pairs of corresponding atom-pairs, only considering those identified as being sufficiently close. The quantities of interest directly follow, namely the interatomic distance $d_{ij} = |\mathbf{x_i}^{\text{target}} - \mathbf{x_j}^{\text{target}}|$ between atoms i and j in the target structure, and the distance $r_{ij} = |\mathbf{x_i}^{\text{ref}} - \mathbf{x_j}^{\text{ref}}|$ between corresponding atoms in the reference structure.

Finally, distances between atom-pairs that are already tightly restrained by standard geometry terms are removed from the list of external restraints. In particular, we remove any short restraints separated by only one or two chemical bonds. This removal is particularly vital when attempting to estimate restraint standard deviations. Note that variability of longer-range restraints is generally very different to that of short restraints separated by only few bonds.

22.2.2.3 Form of the Restraint Distributions

Suppose the distributions of the positions of two atoms are $X_1 \sim N(\mathbf{x}_1, \sigma_1^2)$ and $X_2 \sim N(\mathbf{x}_2, \sigma_2^2)$, where $\mathbf{x}_i$ is the coordinate corresponding to atom i. Since we are generally interested in low-resolution structures, we assume spherical Normality; the variance terms are scalar to emphasise this point. Note that B-factors are closely related to the variabilities of these distributions, which are usually chosen to be isotropic for low-resolution structures.

The distribution of vectors from the first atom to the second is given by:

$$\Delta X = X_2 - X_1 \sim N\left(\mathbf{x}_2 - \mathbf{x}_1, \sigma_1^2 + \sigma_2^2 - 2\text{cov}(X_1, X_2)\right) \tag{22.4}$$

If the atoms are close, then their positions are likely to be positively correlated, which will reduce the variability of the distance between them. Conversely, if the atoms are far apart, then it is reasonable to surmise that their positions would be more independent, and thus the variability of their interatomic distance would be larger.

Given that, under assumption of independence of atomic positions, $\sqrt{\sum_{k=1}^{3}\left(\frac{\Delta X_k}{\sigma}\right)^2}$ follows a noncentral chi distribution with 3° of freedom with noncentrality parameter $\sqrt{\sum_{k=1}^{3}\left(\frac{E(\Delta X_k)}{\sigma}\right)^2}$, we deduce that the interatomic distance $D = \sqrt{\sum_{k=1}^{3}(\Delta X_k)^2}$ is related to the noncentral chi distribution; specifically, $D\sigma^{-1} \sim \chi_3'$ where $\sigma^2 = \text{var}(\Delta X)$. However, we use the approximation of Normally distributed interatomic distances[2]

$$D \sim N(\mu, \sigma^2) \tag{22.5}$$

which constitutes the restraint to be used in refinement. Given knowledge of external structural information, we estimate the mean as $\mu = r$, the distance between the corresponding atoms in the reference structure. Appropriate selection of the standard deviation σ is less obvious; currently used approaches are described below.

22.2.2.4 Estimation of Restraint Standard Deviations

Restraint standard deviations (SDs) may be estimated utilising the observed distribution $P(d|r)$ of interatomic distances in the target structure given corresponding distances r in the reference structure. For example, estimation of uniform SDs would allow restraints to be automatically weighted according to the overall agreement between interatomic distances in the two structures. In this trivial case, all SDs are estimated using:

$$\sigma^2 = \frac{1}{n-1}\sum_i (d_i - r_i)^2 \tag{22.6}$$

This would result in higher SDs (lower weights) assigned to all external restraints when the reference structure is less similar to the target. Due to the distance-dependence of the variability of $|d - r|$, using a higher distance threshold $r_{\max}$ would also result in higher SDs. It follows that the restraints would naturally be weighted down if the target and reference structures exhibit conformational differences.

Another choice would be to allow the SDs to increase with the mean in order to account for the distance-dependence of the observed distribution of restraints. This would allow restraints with small objective values (r) to have higher weights, whilst naturally weighting down the longer-range restraints. For example, the restraint variance could be allowed to increase linearly with restraint distance, that is:

$$\sigma^2(\mathbf{k}) = \mathbf{k}_1 + \mathbf{k}_2 r \tag{22.7}$$

[2]The chi distribution with degrees of freedom greater than two can be approximated by a Normal distribution. This approximation is reasonable for the purpose of parameter estimation. However, in other applications that are sensitive to the tails of the distribution, such as statistical testing, such an approximation may give unreliable overoptimistic results.

where the parameters **k** depend on the particular chain-pair. This could be justified by the observation that any signalling causing correlation in atomic position would generally become weaker as restraint distance increases. Furthermore, peculiar behaviour may be observed when there are multiple rigid substructures (e.g. domains) present, the effect of which would be exacerbated when the maximum restraint threshold $r_{\max}$ is large. The presence of multiple domains, or indeed any deterministic conformational changes, would tend to cause a systematic increase in observed restraint error for higher distances r.

Note that this approach allows more complicated functional forms of $\sigma^2(\mathbf{k})$. For example, further to the restraint objective value (r), it would be possible to allow dependency on factors such as B-values or reliability of atomic position (if available).

Given a functional form for the restraint variances $\sigma^2(\mathbf{k})$, we then use maximum likelihood estimation to optimise the parameters **k**. Optimisation is performed using a quasi-Newton method, in which an approximation of the Hessian matrix is updated after each procedural iteration. Specifically, we use the BFGS formula for updating the inverse Hessian approximation, and a line search algorithm for selecting trial parameter values as described by Nocedal and Wright [29].

If experimental data corresponding to the external prior information are available, then we estimate individual atomic uncertainties σ_i for atoms in the reference structures using the using the procedure of Murshudov and Dodson [30], which allows for dependencies on atomic B-value, crystallographic resolution, model and data completeness, and data quality. Determination of restraint SDs is then reduced to the problem of estimating the correlation between atomic positions. Unfortunately, this information is lost during the experiment, and so is either set to a fixed value or estimated as above.

Alternatively, attempts to estimate restraint SDs may be bypassed, instead allowing the weight of external restraints to be controlled using only the weighting terms in the refinement program.

22.2.3 Generic Interatomic Distance-Based Restraints

Further to using a reference structure, *ProSMART* is able to generate restraints based on individual structural units. Specifically, these include generic bond-like restraints and structural fragment restraints, each of which are useful in different circumstances. Such restraints might be used when a suitable reference structure is not available, or when the reference chain is itself not sufficiently well-refined.

Both generic bond and fragment-based restraints may have broad application, most notably in the generation of restraints for secondary structural elements. Usage may be varied in practical application. For example, these restraints can be used to stabilise refinement in later stages, or temporarily force the maintenance of sensible conformations during earlier stages of the model building/refinement process.

22.2.3.1 Generic Bond Restraints (Hydrogen Bond Restraints)

Generic restraints representing specific atomic interactions may be generated using *ProSMART*. By default, these are configured to represent backbone hydrogen bonds, and as such can be used to help stabilise secondary structure. However, the approach is general and thus is extensible to custom applications.

All generic bonds have a detection range (e.g. 2–3.5 Å) and a target value (e.g. 2.8 Å). Rules may be applied that govern which atoms are allowed to interact, e.g. only O–N atom-pairs for helices; both N–O and O–N atom-pairs for sheets and loops. Also, specific residue separations between the atoms may be required, e.g. 3 residues for 3_{10}-helices; 4 for α-helices; 5 for π-helices. Furthermore, other criteria may be used in order to ensure chemical sense, for example a nitrogen atom may only form one hydrogen bond restraint, and an oxygen atom may form a maximum of two hydrogen bond restraints; where necessary the most favourable configuration is always selected.

22.2.3.2 Fragment-Based Restraints

External restraints may be generated using a sequential n-residue fragment representing a particular local conformation. For example, an "ideal" α-helix may be used to generate restraints that will keep helical structures intact. Such helical restraints are different to generic hydrogen bond helical restraints, since they include restraints between all sufficiently close backbone atoms. Also, the fragment based helical restraints do not require strict compliance with ideal secondary structure conformation in order to be detected, unlike with conventional secondary structure detection methods. This is particularly relevant at low resolution, where secondary structure may not be sufficiently well-formed to be detected from predicted hydrogen bonding patterns.

The suitability of other general in-sequence fragment-based restraints, such as for β-strands, is less obvious due to their comparatively high degrees of flexibility, and the fact that hydrogen-bonding occurs between, not within, β-strands. Another potential application would be when it is desired for a particular region to adopt a known conformation (e.g. if a specific small portion of conserved structure is found between the target and a reference chain); the suitability of such an approach would have to be carefully considered for the particular case. In principle, any structures may be used as reference fragments.

Since aligned fragments may overlap (e.g. consecutive helical fragments), it is possible for a particular atom-pair to be restrained to several atom-pairs in the reference fragment. For example, in a helical fragment, the distances between main chain atoms in residues i and j may be very similar to those in residues $i + 1$ and $j + 1$. In such cases, restraints for a target atom-pair in a helix might be generated using corresponding atoms from residues i and j, or those from residues $i + 1$ and $j + 1$. Consequently, it is necessary to decide which residues to use for restraint generation. More generally, any restraint between atoms from residues

i and j may result from several fragment alignments. Specifically, the reference fragment, which has residue range $[1, n]$, may be aligned to any of the residue ranges $[j-n+1, j], \ldots, [i, i+n-1]$ in the target structure, whilst still implying correspondences for residues i and j (where $j-n < i < j$). Therefore, ignoring heterogeneities and boundary conditions, there may be up to $i-j+n$ potential alignments of residues i and j with some residues in the reference fragment.

The list of potential residue correspondences is reduced by fragment score criteria, since we only want to generate fragment-based restraints for regions of structure sufficiently similar to the reference fragment; only configurations with associated Procrustes dissimilarity (local RMSD) scores below some threshold are included. Of the remaining potential residue-pair alignments, if any, the one with the most favourable associated fragment Procrustes score is selected for restraint generation.

22.3 Jelly-Body Restraints

Similarly to external structural restraints, jelly-body restraints are real space regularisers in interatomic distance space. However, jelly-body restraints belong to a special class of regularisers that do not have a target value. These "restraints" do not impose any externally-derived prior information, and may in principle be applied to any structures.

Jelly-body restraints take the following form:

$$w_{\text{jelly}} \sum_{d_{\text{current}} \leq d_{\max}} (d - d_{\text{current}})^2 \tag{22.8}$$

for all atom-pairs from the same chain separated by no greater than some maximum distance $d_{\max}$ (default 4.2 Å) so that all sufficiently-close atom-pairs are conceptually restrained to their current values. However, the objective value is updated at every step, so that for all atom-pairs we always have $d = d_{\text{current}}$. In effect, the current state of the system is the prior information utilised by jelly-body restraints.

Note that jelly-body restraints inherently make zero contribution to the target function and its gradient. Consequently, jelly-body restraints have the desirable property that, in the absence of any external information, they do not alter the minima of the target function. However, jelly-body restraints do contribute to the second derivative of the target function, and so change the refinement search direction. These restraints thus act as regularisers, improving the effective parameter to observation ratio and thus stabilising refinement, without introducing external bias.

This technique has strong similarity to the elastic network model calculations [31]. It should be noted that a similar (although notably different) technique has been implemented in CNS, termed DEN [7].

22.4 Practical Usage

Here we present an example of re-refinement of a low resolution structure taken from the Protein Data Bank [9]. Where appropriate, the model was re-refined using 40 iterations of refinement by *REFMAC5*, using main and side chain external structural restraints generated by *ProSMART*.

It should be noted that the examples of the re-refinement of deposited models presented here may not represent typical application, since external restraints may also be applied during both earlier and latter stages of the refinement process in order to help models adopt more reliable conformations. Here, automated *REFMAC5* refinement was performed using largely default settings (e.g. no TLS groups), and no attempt is made to achieve "good" final models. Rather, in order to demonstrate practical usage, our example effectively amounts to considering a snapshot during the model building/refinement process.

22.4.1 Automated Re-Refinement of Ovotransferrin 1ryx

We consider the re-refinement of the low-resolution 3.5 Å structure 1ryx [32] using external restraints from the higher-resolution 2.8 Å homologue 2d3i [33]. Illustrations of the models are displayed in Fig. 22.2. Consideration of this case demonstrates that external restraints can have a positive effect on refinement even when the target and reference structures are from different crystal forms, the models are in different global conformations, and exhibit regions of substantially different local structure. This indicates the approach has the potential to be both flexible and robust with respect to these factors.

We should clarify that, in presenting this example, we do not mean to criticise the original work of the authors of the low-resolution structure. Rather, we are demonstrating the power of these recently-developed refinement techniques that were not available at the time of original refinement.

As can be seen in Fig 22.3, default re-refinement of 1ryx resulted in a substantial increase in both R_{free} and $\Delta R = R_{\text{free}} - R$, indicating a high degree of overfitting. However, the use of jelly-body restraints stabilises refinement, yielding small reductions in both R and R_{free}, and importantly stabilising ΔR. The use of external restraints from 2d3i results in further reductions in both R and R_{free}, whilst maintaining stability of ΔR. Note that, according to the refinement statistics, the jelly-body restrained refinement path is smooth (smoothly approaches the closest local minima of the existing likelihood function), whilst that corresponding to external restraints is more heterogeneous. This is because the use of external restraints alters the likelihood function, and thus the model may undergo more substantial changes in search of the local minima. Such heterogeneity will depend on the agreement between the reference structure and the current state of the target

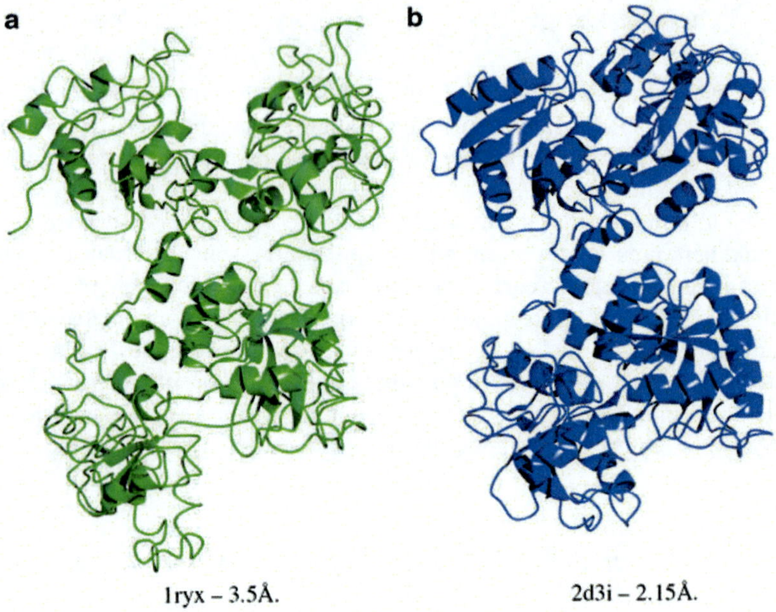

Fig. 22.2 Depictions of ovotransferrin structures 1ryx and 2d3i, taken from the Protein Data Bank [9]. Both structures comprise one chain in the asymmetric unit. 1ryx was processed assuming the space group $P4_32_12$, whilst 2d3i was processed in $P2_1$. The two models adopt different global conformations (consider the relative orientations of the upper two domains). Also, there are regions that exhibit clear structure dissimilarity, which could be due to genuine differences between the crystals or due to model incorrectness (consider the lower-left domain). Note also that the lower resolution structure has fewer regular well-defined secondary structure elements than the higher resolution homologue (Images were produced using *CCP4mg* [21])

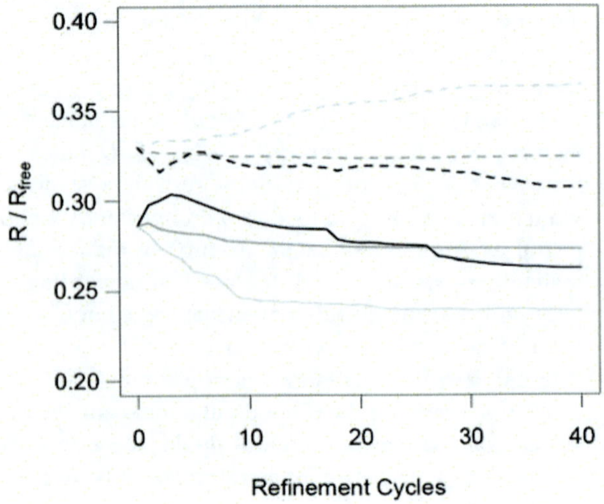

Fig. 22.3 Refinement statistics corresponding to the re-refinement of 1ryx, which had original R/R_{free} of 0.286/0.330. After 40 cycles, default refinement resulted in R/R_{free} of 0.239/0.365 (*light grey lines*), with jelly-body restraints 0.265/0.322 (*dark grey lines*), and with main and side chain external restraints from 2d3i 0.263/0.307 (black lines). R-values are shown as *solid lines*, and R_{free} as *dashed lines* (The image was produced using *R* [34])

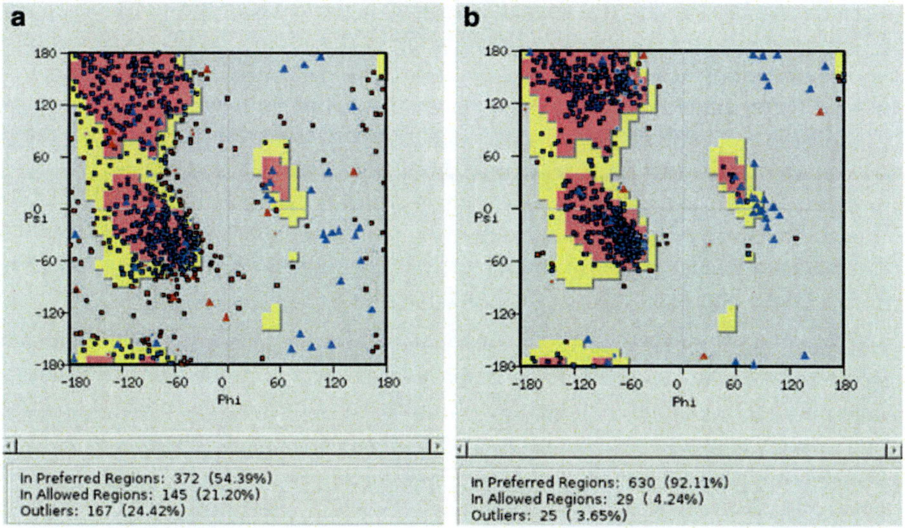

1ryx original structure. 1ryx re-refined with external restraints.

Fig. 22.4 Ramachandran plots corresponding to the model 1ryx. Plots are shown corresponding to (**a**) the original model, and (**b**) the model after re-refinement with main and side chain external restraints from 2d3i. Note that backbone torsion angles are not explicitly restrained by the external restraints. Rather, the general improvements in their values are a consequence of stabilisation of local structure, which is performed in interatomic distance space (Plots were produced using *COOT* [35])

model, the suitability and quality of the reference structure, and on the degree to which the external restraints alter the phases (thus electron density) during the refinement process.

However, it is important not to rely solely on refinement statistics when attempting to qualify or quantify model improvement. Various validation tools are available for accessing model reliability given prior knowledge. Figure 22.4 displays Ramachandran plots corresponding to the model before and after automatic re-refinement with external restraints. Overall, we see that the use of external restraints results in greatly improved backbone geometry, indicating a more reasonable model.

22.4.2 Procedural Refinement Strategies. Example: Localised Conformation Change due to External Restraints

Further to looking at global statistics and geometry validation, it is important to investigate how well the localised regions of the model agree with the electron density. Such manual inspection may reveal parts of the model that:

- can be easily fixed by real space refinement, e.g. side chain flips;
- are spurious, perhaps incorrectly modelled or disordered, which require careful consideration;
- have been pulled out of a sensible conformation into an incorrect one by the external restraints, in which case external restraints should be regenerated excluding these identified regions, and refinement should be repeated;
- have been affected by the use of external restraints, changing interpretation of the electron density map, ultimately allowing potential for subsequent improvement by manual real space refinement;
- have been improved by use of external restraints.

Note that, since external restraints may have a positive effect on some regions of structure and a negative effect on other regions, overall global statistics may be misleading. For example, in the case of improved refinement statistics, it could be that external restraints negatively effect some regions, which would need to be fixed. Equally, in the case of worse refinement statistics, it may be that external restraints positively effect some regions, and that the worsened statistics are only due to the application of external restraints in certain localised regions. Whilst the use of the Geman-McClure function aims to allow robustness to outliers, this behaviour may be unavoidable in some circumstances.

For example, Fig. 22.5 considers a localised region of 1ryx that undergoes substantial conformational change due to the use of external restraints. The region shown on the left (around Asn473) exhibits plausible improvements in both model and density after refinement with external restraints (Fig. 22.5b). However, it is evident that there is scope for further model improvement in this region. Consequently, minor real space refinement is performed (Fig. 22.5c), before subsequent refinement with jelly-body restraints (Fig. 22.5d). Note that in the final round of refinement, the external restraints are removed, and only jelly-body restraints are used. The jelly-body restraints regularise refinement, making it sufficiently stable so that the noisy data does not cause the model to lose the sensible conformation imposed by the external restraints. At the same time, the removal of the external restraints means that the model is free to refine into the density, rather than continue to have strong bias towards the reference structure. This multi-stage process – refine with external restraints; manual real space refinement; refine with jelly-body restraints – can thus form a powerful procedure when refining at low resolution.

The density corresponding to the trace shown on the right side of Fig. 22.5 disappears following application of external restraints. This could be due to a number of reasons, e.g. incorrectly built trace, model bias, disorder, insufficiently refined phases, or inappropriate application of external restraints. In practice, careful remodelling of this region would be considered, although this is beyond the scope and purpose of this example.

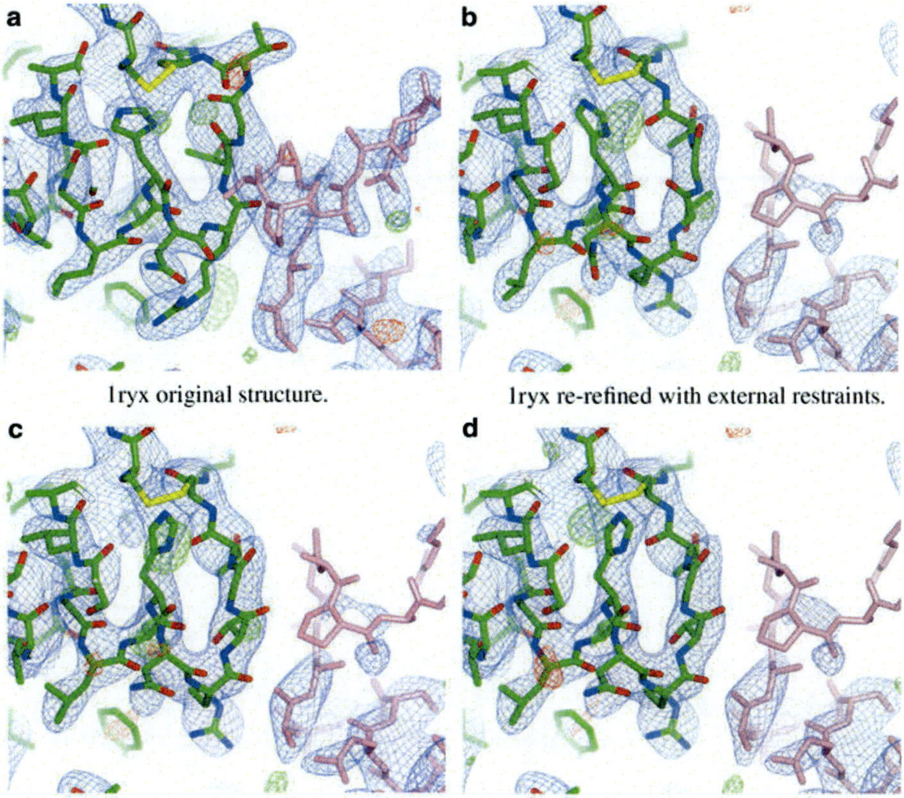

a: 1ryx original structure.

b: 1ryx re-refined with external restraints.

c: 1ryx re-refined with external restraints, then modified.

d: 1ryx re-refined with external restraints, modified, then re-refined using jelly-body.

Fig. 22.5 Example of the re-refinement of 1ryx using main and side chain external restraints, focussing on a region (around Asn473) that undergoes dramatic conformational change during refinement. After automated refinement with external restraints (**b**), the model was manually altered (**c**) by real space refinement of His472–Thr475, before subsequent refinement without external restraints but with jelly-body restraints (**d**). Models corresponding to subfigures (**a**), (**b**) and (**d**) had R/R_{free} 0.286/0.307, 0.263/0.307 and 0.253/0.304, respectively. Electron density map contours are shown at 1.3σ. The trace shown on the right corresponds to a symmetry-related copy. Images were produced using *CCP4mg* [21]

22.4.3 Refinement Strategy May Affect Biological Conclusions. Example: Differences in Density Map Interpretation

The use of external restraints can in some cases result in new or altered features appearing in the electron density map. For example, Fig. 22.6 depicts the N-lobe in 1ryx, which contains the two Tyr residues (Tyr92 and Tyr191) involved in liganding

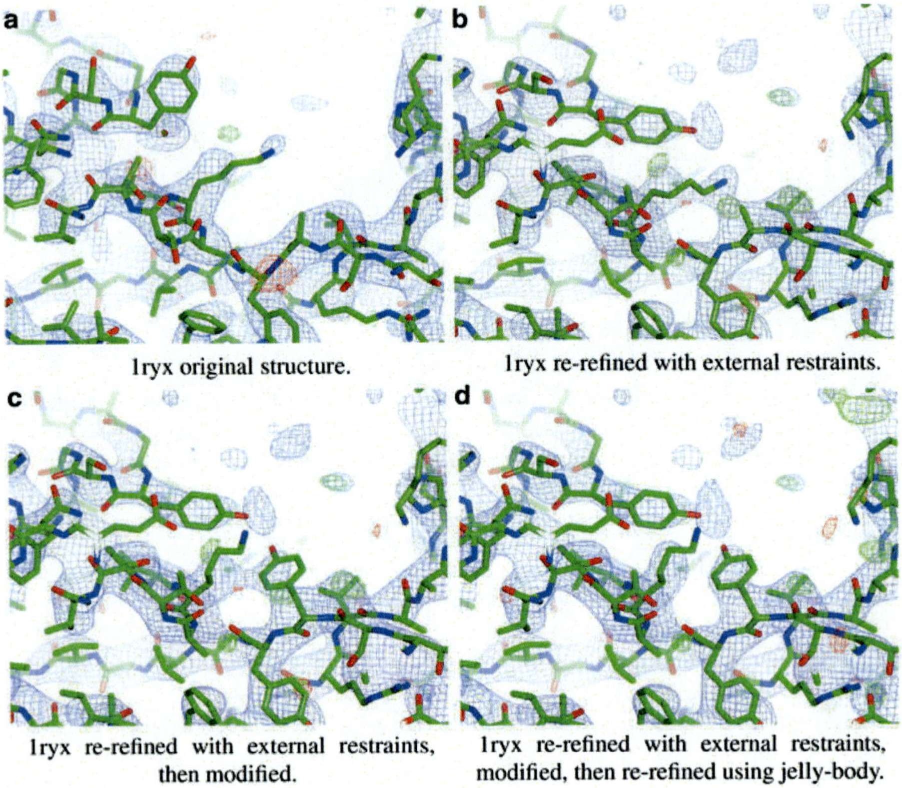

Fig. 22.6 Example of the re-refinement of 1ryx using main and side chain external restraints, focussing on a region (around Lys209) that exhibits changes in the electron density after re-refinement with external restraints. After automated refinement with external restraints (**b**), the model was manually altered (**c**) by building the side chain of Tyr92 and adjusting Lys209, before subsequent refinement without external restraints but with jelly-body restraints (**d**). Models corresponding to subfigures (**a**), (**b**) and (**d**) had R/R_{free} 0.286/0.307, 0.263/0.307 and 0.252/0.307, respectively. Electron density contour shown at 1.3σ. Images were produced using *CCP4mg* [21]

Fe^{3+} ions [32]. Note that the side chain of Tyr92 is unmodelled. However, after re-refinement with external restraints (Fig. 22.6b), density appears for the Tyr92 side chain. Furthermore, the Tyr191 side chain changes its rotameric state. Also note that extra density appears in the N-lobe, which could hint at residual metal binding – in practice, whether or not this is the case would have to be further investigated following subsequent rounds of refinement and model improvement.

After manually building the Tyr92 side chain and adjusting the position of Lys209 (Fig. 22.6c), we re-refine the structure using jelly-body restraints (without external restraints, using a similar procedure to above). The resultant electron density map (Fig. 22.6d) gives a very different picture of the N-lobe in comparison with the original structure. Note that there may be opportunity for significant model improvement in this case. In fact, in this case, refining the structure with external restraints reveals opportunities for flipping and building many other side chains (not shown), as well as significant portions of backbone that would need to be considered carefully (e.g. as can be seen in Fig. 22.5). As such, further rounds of model building and refinement may have lead to further clarity regarding the structure of the N-lobe and other regions of functional relevance (although that is not the purpose of this example, and so is not further explored here).

This example demonstrates that recently developed techniques for refining at low resolution can aid refinement, allowing new features to be revealed in the map. It is evident that improving the model in this way can result in dramatically different interpretation of the electron density, and in some cases may lead to different conclusions being drawn regarding biologically-relevant regions.

22.5 Anisotropic Regularised Map Sharpening

The map sharpening problem can be written in the general form:

$$\rho = g(\rho_0, k, n) \tag{22.9}$$

where ρ_0 is the underlying signal we would like to observe (actual electron density), ρ is observed signal (model of electron density, from observation), g is a process through which blurring operates on the signal, k is a blurring function that changes the signal (ρ_0) before observation is carried out, and n is noise. However, this formulation is too general to be practical. In order to make the problem manageable, we must make assumptions regarding the functional forms of g and k, and assume a model for the noise n. Therefore, for simplicity, we assume that noise is additive and the blurring function is linear:

$$\rho(x) = \int k(x, y)\rho_0(y)dy + n(x) \tag{22.10}$$

If there were no noise then the problem would be a linear equation. This problem is ill-posed, especially when k is near singular, i.e. small perturbations in input parameters may cause large variations in output. For example, the effects of small noise addition, an incorrectly defined blurring function, or Fourier series termination, may result in an uninterpretable "deblurred" electron density map. It should be noted that in crystallography we always deal with limited, noisy data, and that Fourier series termination is always present. Even if there were no noise and we had knowledge of the exact blurring function $k(x, y)$, solving Eq. (22.10) would still not be straightforward. The numbers of equations and parameters to be estimated are equal to the number of grid points in the electron density, which can be very large.

The problem becomes manageable, whilst not completely reflecting reality, when we make the further assumption that the blurring function is independent of position. This simplification essentially means that the whole content of the asymmetric unit oscillates as a unit, with no rotational component, resulting in the blurring function having the property $k(x, y) = k(x - y, 0)$. Using the notation $k(x) = k(x, 0)$, Eq. (22.10) becomes:

$$\rho(x) = \int k(x - y)\rho_0(y)dy + n(x) \tag{22.11}$$

Since the problem is ill-posed, we can approach its solution utilising ideas from the field of regularisation [36]. Under the assumption of white noise, our ill-posed problem may be replaced by the minimisation problem:

$$\left\| \int k(x - y)\rho_0(y)dy - \rho(x) \right\|^2 + \gamma f(\rho_0) \rightarrow \min \tag{22.12}$$

where $||.||$ denotes the L_2 norm, f is a regularisation function, and γ is a regularisation parameter to be selected. Usually, regularisers are chosen so that the resultant function obeys certain conditions. For purposes of example, we shall consider two popular conditions: (1) the function should be small; and (2) the first derivatives of the function should be small (i.e. the function should vary slowly). For the first case we have:

$$f_1(\rho) = ||\rho(x)||^2 = \int \rho^2(x)dx \tag{22.13}$$

and for the second case:

$$f_2(\rho) = \int \sum_i \left(\frac{\partial}{\partial x_i} \rho(x) \right)^2 dx \tag{22.14}$$

which is known as a first order Sobolev norm. Since ρ is a periodic function, we can write:

$$f_2(\rho) = -(\Delta\rho, \rho) = -\int \Delta\rho(x)\rho(x)dx \tag{22.15}$$

where Δ is the Laplace operator ($\Delta = \sum_i \frac{\partial^2}{\partial x_i^2}$), and (.,.) it denotes the scalar product in Hilbert space.

Now the problem is reduced to finding the minimum of the functional:

$$\left\| \int k(x - y)\rho_0(y)dy - \rho(x) \right\|^2 + \gamma(L\rho_0, \rho_0) \rightarrow \min \tag{22.16}$$

where $L = I$ (identity operator) for L_2 type regularisers (first case), and $L = -\Delta$ for Sobolev type regularisers (second case).

Using Plancherel's theorem, the convolution theorem, and the fact that the Fourier transformation of the Laplacian is proportional to the negative squared length of the reciprocal space vector, we can rewrite the problem as:

$$\frac{1}{2}\sum_{hkl}(\mathscr{F}(k(x))F_{0hkl} - F_{hkl})^2 + \frac{1}{2}\alpha t(|s|)F_{0hkl}^2 \rightarrow \min \tag{22.17}$$

where F_{hkl} is the structure factor before sharpening (e.g. $2mFo - DFc$ type maps) and F_{0hkl} is that after sharpening, $|s| = 2\sin\theta/\lambda$ is the length of the reciprocal space vector, with $t(s) = 1$, $\alpha = \gamma$ for regularisation function f_1, and $t(s) = s^2$, $\alpha = (2\pi)^2\gamma$ for f_2. This minimisation problem has a very simple solution:

$$F_{0hkl} = \frac{\mathscr{F}(k(x))}{\mathscr{F}(k(x))^2 + \alpha t(|s|)}F_{hkl} \tag{22.18}$$

When $k(x)$ is Gaussian then the equation has the especially simple form, since $K(s) = \mathscr{F}(k(x)) = \exp(-s^T B_{\text{deblur}} s/4)$, with B_{deblur} as an anisotropic de-blurring B-value.

Unfortunately, in reality neither B-values nor α are known. Whilst there are several techniques to find an "optimal" value for α when the blurring function is known [18], in our implementation such an approach did not give consistent results. Therefore, we used the following ad hoc procedure for selection of the regularisation parameter. Denoting $K_\alpha(s) = \frac{K(s)}{K^2(s)+\alpha t(|s|)}$ and $A_\alpha(s) = K_\alpha(s)K(s)$, we see that A_α is similar to the hat function used in regression analysis [37]. We can define the degrees of freedom of errors[3] (the number of observations minus the effective number of parameters) as:[4]

$$n_{\text{df}} = \text{tr}(I - A_\alpha) = \sum_{hkl}(1 - A_\alpha(s)) \tag{22.19}$$

Note that when $\alpha = 0$, $n_{\text{df}} = 0$, and when $\alpha \rightarrow \infty$ then n_{df} is equal to the number of observations. We select α so that n_{df} is equal to 10–20 % of the number of observations. Since we do not know the exact values of B and α, we also perform ad hoc integration using an empirically-derived distribution of these parameters. The necessary integral may then be written:

$$F_{0hkl}^{\text{int}} = \int_{\alpha,B} P(B,\alpha)F_{0hkl}(\alpha, B)d\alpha dB \tag{22.20}$$

[3]This is the simplest way of defining effective degrees of freedom of errors. Another formula is: $n_{\text{df}} = \text{tr}(I - A_\alpha)^2$.

[4]The equation has a simple form in reciprocal space and position independent blurring function when sharpening matrices are diagonal.

$$= \int_{\alpha,B} P(B,\alpha)K_\alpha(B)F_{hkl}d\alpha dB \tag{22.21}$$

$$= F_{hkl}\int_{\alpha,B} P(B,\alpha)\frac{e^{-s^T(B_{\text{aniso}}+B)s/4}}{e^{-s^T(B_{\text{aniso}}+B)s/2}+\alpha t(|s|)}d\alpha dB \tag{22.22}$$

where B_{aniso} reflects anisotropy of the data, and is calculated during scaling of the calculated structure factors relative to the observed ones, under the conditions that it obeys crystal symmetry, and $\text{tr}(B_{\text{aniso}}) = 0$.

The joint probability distribution of B and α can be written:

$$P(B,\alpha) = P(B)P(\alpha; B) \tag{22.23}$$

The mean value of the isotropic part is taken to be equal to the median value (B) of coordinates (although it may be better to use Wilson's B-value estimated using the intensity curves derived by Popov and Bourenkov [38]). We approximate the distribution of the isotropic part of the B-values using a Gaussian distribution centred at B_{sharp}, with standard deviation equal to $\frac{B_{\text{sharp}}}{10}$. For each B-value, we select α so that n_{df} is 10–20 % of the number of observations, and the standard deviation of the distribution of α is taken to be $\frac{\alpha_B}{10}$.

Note that Eqs. (22.16) and (22.17) suggest a class of regularisers. They can be selected to use particular knowledge about the electron density in real and reciprocal space. For example, if it is desired to suppress of effect of ice rings, then one can select $t(|s|)$ so that the corresponding reflections are weighted down.

22.5.1 *Implementation and Example*

We have implemented anisotropic sharpening with L_2 and Tikhonov-Sobolev regularisers with and without integration over the ad hoc joint probability distribution of B and α using probability distribution (22.23). We have also implemented the regularisation function $t(s) = 1 + s^2$. These are available from *REFMAC5* version 5.7. In our tests, all regularisation functions have given similar results. This is not surprising, as the major problem is that the blurring function is not position independent. Before finding accurate regularisers, the problem of modelling position dependent blurring functions should be dealt with. All results presented here were achieved using the L_2 type regulariser.

Map sharpening was tested for many cases using data sets from the PDB [9] with resolution below 3 Å. The best results were obtained for PDB code 2r6c [39]. For any low resolution data taken from the PDB, before map calculation we generally try jelly-body, local NCS (if present) and external reference structure (if applicable) restrained refinement and take the best refined results for further analysis. For 2r6c,

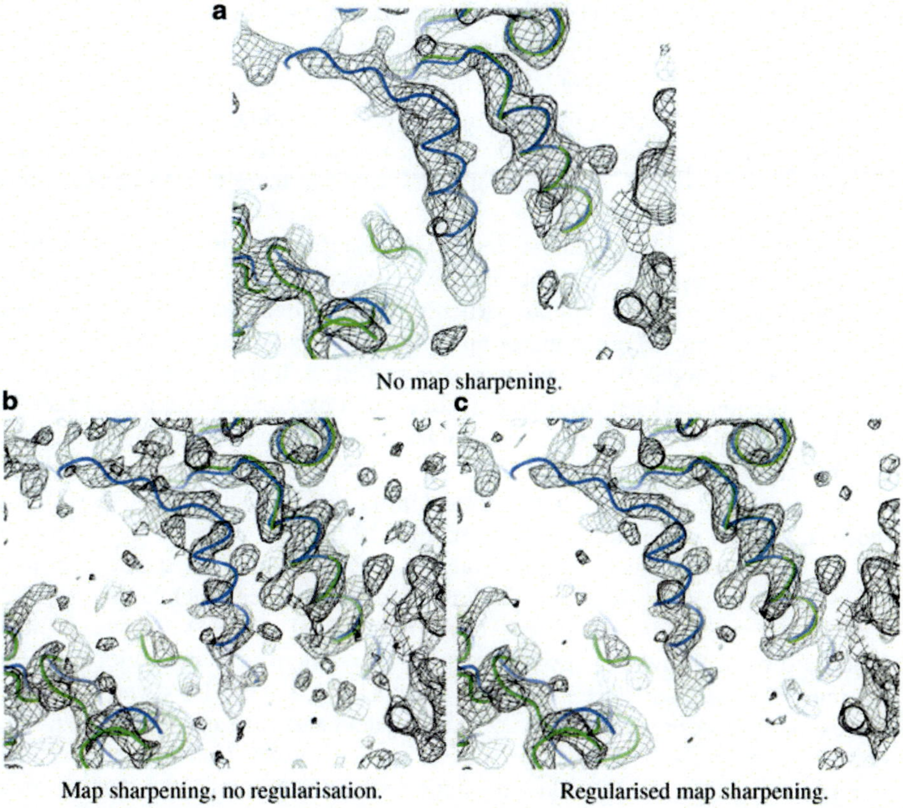

Fig. 22.7 Visual effects of map sharpening on electron density. This example was taken from PDB code 2r6c. Images show the map with: (**a**) no map sharpening; (**b**) map sharpening using the inverse filter (no regularisation); (**c**) regularised sharpened map using the L_2 type Tikhonov regulariser, with sharpening coefficients integrated over B and α, as described in the text. The backbone trace of 2r6c chain C is shown in green (light grey in monochrome). The homologous structure 2r6a chain A is shown in blue (dark grey in monochrome), superposed using residues 270–287 from 2r6a(A). The image shows unmodelled density in 2r6c that corresponds to a helix present in 2r6a. Both sharpened maps show more features than the unsharpened map, with the regularised map giving more connectivity. Images were produced using *CCP4mg* [21]

the original $R/R_{\rm free}$ statistics reported in the PDB were 32.1/34.4. After refinement these values became 24.0/30.0. Figure 22.7 shows an illustration of the maps after refinement with and without unregularised and regularised map sharpening. It is apparent that, in this case, more features (possibly side chains) and connectivity can be seen when using regularised map sharpening. Whilst this example shows regularisation using the L_2 type regulariser, it should be noted that the Sobolev type regulariser gave similar results.

22.6 Conclusions and Future Directions

We have presented tools to aid low resolution refinement, each of which are types of regularisers. Specifically, these include external structural restraints, jelly-body restraints and anisotropic regularised map sharpening. These new techniques are intended to allow model improvement that would not otherwise be possible, and can lead to new map features being revealed.

The use of external restraints derived from reference structural information gives promising results. Sometimes improvements are subtle, whilst in other cases improvements may be dramatic. In particular, we have demonstrated how improved models can be achieved using an example of the externally-restrained re-refinement of a deposited model. We have also shown that, following the application of external restraints, subsequent refinement (without external restraints) can then be stabilised by jelly-body restraints, which are very effective in providing resistance to overfitting.

At low resolution, affects such as model bias cause difficulties in qualifying any model improvement or impairment during the refinement process. We commonly rely on refinement statistics (e.g. R-values) to determine model quality, but they are not always conclusive. It is important to complement such measures with independent validation, e.g. from considering model geometry. However, such statistics are not always conclusive – it is often possible to have worse global scores, but improved local structure in some regions, and vice versa. However, as seen in our examples (see Figs. 22.5 and 22.6), the calculated electron density map may not always be reliable. Indeed, unoptimal refinement can lead to incorrect map interpretation. This results in low resolution structure determination having great potential for error.

Since the use of external restraints will alter global geometry validation statistics, such results should be interpreted accordingly, and the integrity of local structure should always be considered. Indeed, it is important to always manually inspect the electron density to check that the model agrees reasonably well with the data, thus ensuring local suitability of the use of external restraints, despite any apparent improvement or degeneracy in overall statistics. If there are any serious artefacts that arise due to bias towards the reference structure, it may be appropriate to re-attempt externally-restrained refinement, excluding particular residues from external restraint generation.

In some cases, better results can be achieved by utilising information from multiple reference structures, the difficulty often being that this requires the existence and availability of multiple structures suitable as references. Our implementation allows the generation of external restraints based on multiple reference structures; currently, the restraints most consistent with the target model are selected for use during refinement, although other strategies are possible.

For practical application, we anticipate external restraints to also be of particular use during earlier stages of model building/refinement, for stabilising local structure and helping to achieve sensible model geometry. Of course, the degree of any improvement due to external restraints will be limited by the suitability and quality of the reference structural information. Certainly, there is an immediate need for ways to automatically validate the suitability of reference structures, most importantly at the local level, so that destructive restraints are not generated, or are appropriately weighted down (whilst down-weighting is already effectively performed by using robust estimators in our implementation, other complementary approaches would be desirable). For example, in application it may be sensible to attempt re-refinement of any reference homologous structures before restraint generation, in an attempt to improve the quality of the prior information. This might be performed manually or automatically, e.g. using the protocol of *PDB_REDO* [40]. In some cases, manual model re-building and refinement of reference structures may be necessary/appropriate, and thus should ideally always be considered. Such approaches may reduce any error propagation from reference to target models.

We have also implemented generic DNA/RNA basepair restraints based on interatomic distances, torsion angles, and chirality; testing is currently in progress. Parameters for these restraints have been taken from Neidle [41]. For accurate refinement of DNA/RNA, it is necessary to use sugar puckering as well as base stacking restraints. Whilst it is relatively simple to implement sugar puckering restraints, e.g. using the elegant method presented by Cremer and Pople [42], determining appropriate distributional parameters will take some time. Designing restraints for base stacking is a much more challenging problem, for which we do not currently have any satisfactory approaches.

There is much room for improvement and future exploration in the generation and application of external structural information. Some of the more notable features include:

1. Further investigation into the cooperative usage of external restraints from multiple reference structures. Specifically, this would involve expanding the approach of restraint generation and SD estimation to better utilise situations where multiple reference structures are available. In the current implementation, all restraints are pooled, or alternatively only the "best" restraints are selected. A more sophisticated solution would be to more appropriately describe the distribution of each interatomic distance restraint. This would result in the assignment of bespoke restraints for each individual atom-pair that more closely represent reality, being based on observed intraclass flexibility. However, this would require an appropriate array of reference structures, which may include: different forms/models of the same protein; classes of structurally similar proteins; structure ensembles resulting from other experimental (NMR) or theoretical (MD) techniques.
2. Consideration of generic restraints derived from considering the density of fragment conformation space. This may allow the expansion and generalisation of the presented fragment-based approach into an automated method, which is currently only recommended for α-helical restraints and for cases afforded special manual consideration.

3. Assessment and identification of structures appropriate for use as external references, given a target. Currently, reference structures are manually identified, and suitability manually assessed. It would be desirable for such decisions to be reliably automated, e.g. using *BALBES* [43].
4. Ability to generate external restraints to homologous DNA/RNA structures.
5. Multicrystal refinement, whereby multiple datasets are used to achieve a single model (as with multicrystal averaging). Each model would be a refinement target, as well as being used as a reference structure for all other models. Successful implementation of this is an important future prospect for low resolution refinement.
6. Further generalisations to regularised map sharpening. The implemented method of regularised map sharpening uses the assumption that the blurring function is position independent. However, this assumption may not always be valid – it is expected for the oscillation of molecules within a crystal to be more complex, and crystal disorder to be more anisotropic. One natural extension to map sharpening would be to use TLS parameters [44] as a blurring function. However, we are not aware of a simple solution to this problem. Another problem with the current approach is that we assume that noise and signal are uncorrelated, and that the noise is white noise. This may not reflect reality, especially when atomic model errors are dominating contributors to the noise. For density modification, the problem may become even more complicated.

Acknowledgements Part of this work was carried out whilst the authors were at the Structural Biology Laboratory, Department of Chemistry, University of York, during which time RAN was funded by a BBSRC Ph.D. Studentship, and FL and GNM were funded by the Wellcome Trust. Part of this work was supported by the Medical Research Council (grant number: MC_US_A025_0104).

References

1. Murshudov G, Skubák P, Lebedev A, Pannu N, Steiner R, Nicholls R, Winn M, Long F, Vagin A (2011) REFMAC5 for the refinement of macromolecular crystal structures. Acta Crystallogr D Biol Crystallogr 67(4):355
2. Adams P, Afonine P, Bunkoczi G, Chen V, Davis I, Echols N, Headd J, Hung L, Kapral G, Grosse-Kunstleve R, McCoy A, Moriarty N, Oeffner R, Read R, Richardson D, Richardson J, Terwilliger T, Zwart P (2010) PHENIX: a comprehensive Python-based system for macromolecular structure solution. Acta Crystallogr D Biol Crystallogr 66(2):213
3. Sheldrick GM (2008) A short history of SHELX. Acta Crystallogr A A64(1):112
4. Murshudov GN (2011) Some properties of Crystallographic Reliability index – Rfactor: Effect of Twinning. Appl Computat Math 10:250
5. Vagin AA, Steiner RA, Lebedev AA, Potterton L, McNicholas S, Long F, Murshudov GN (2004) REFMAC5 dictionary: organization of prior chemical knowledge and guidelines for its use. Acta Crystallogr D Biol Crystallogr 60(12(1)):2184
6. Schröder G, Brünger A, Levitt M (2007) Combining efficient conformational sampling with a deformable elastic network model facilitates structure refinement at low resolution. Structure 15(12):1630

7. Schröder G, Levitt M, Brünger A (2010) Super-resolution biomolecular crystallography with low-resolution data. Nature 464(7292):1218
8. Smart O, Womack T, Flensburg C, Keller P, Paciorek W, Sharff A, Vonrhein C, Bricogne G (2012) Exploiting structure similarity in refinement: automated NCS and target-structure restraints in BUSTER. Acta Crystallogr D Biol Crystallogr 68(4):368
9. Berman H, Battistuz T, Bhat T, Bluhm W, Bourne P, Burkhardt K, Feng Z, Gilliland G, Iype L, Jain S, Fagan P, Marvin J, Padilla D, Ravichandran V, Schneider B, Thanki N, Weissig H, Westbrook J, Zardecki C (2002) The protein data bank. Acta Crystallogr D Biol Crystallogr 58(6):899
10. Murshudov G, Vagin A, Dodson E (1997) Refinement of macromolecular structures by the maximum-likelihood method. Acta Crystallogr D Biol Crystallogr 53(3):240
11. Winn M, Ballard C, Cowtan K, Dodson E, Emsley P, Evans P, Keegan R, Krissinel E, Leslie A, McCoy A, McNicholas S, Murshudov G, Pannu N, Potterton E, Powell H, Read R, Vagin A, Wilson K (2011) Overview of the CCP4 suite and current developments. Acta Crystallogr D Biol Crystallogr 67(4):235
12. Blanc E, Roversi P, Vonrhein C, Flensburg C, Lea SM, Bricogne G (2004) Refinement of severely incomplete structures with maximum likelihood in BUSTER-TNT. Acta Crystallogr D Biol Crystallogr 60:2210
13. Afonine P, Grosse-Kunstleve R, Adams P (2005) A robust bulk-solvent correction and anisotropic scaling procedure. Acta Crystallogr D Biol Crystallogr 61(7):850
14. Brünger A, Adams P, Clore G, DeLano W, Gros P, Grosse-Kunstleve R, Jiang J, Kuszewski J, Nilges M, Pannu N, Read R, Rice L, Simonson T, Warren G (1998) Crystallography & NMR system: A new software suite for macromolecular structure determination. Acta Crystallogr D Biol Crystallogr 54(5):905
15. Brünger AT, DeLaBarre B, Davies JM, Weis WI (2009) X-ray structure determination at low resolution. Acta Crystallogr D Biol Crystallogr 65(2):128
16. DeLaBarre B, Brünger AT (2006) Considerations for the refinement of low-resolution crystal structures. Acta Crystallogr D Biol Crystallogr 62(8):923
17. Gonzalez RC, Woods RE (2002) Digital image processing. Prentice Hall, Upper Saddle River
18. Vogel R (2002) Computational methods for inverse problems. SIAM, Philadelphia
19. Mooij W, Cohen S, Joosten K, Murshudov G, Perrakis A (2009) "Conditional Restraints": Restraining the Free Atoms in ARP/wARP. Structure 17(2):183
20. Bollati M, Barbiroli A, Favalli V, Arbustini E, Charron P, Bolognesi M (2012) Structures of the lamin A/C R335W and E347K mutants: Implications for dilated cardiolaminopathies. Biochem Biophys Res Commun 418(2):217–221
21. McNicholas S, Potterton E, Wilson KS, Noble MEM (2011) Presenting your structures: the CCP4mg molecular-graphics software. Acta Crystallogr D Biol Crystallogr 67(4):386
22. Geman S, McClure D (1987) Statistical methods for tomographic image reconstruction. Bull Int Stat Inst LII:5
23. Brünger A (1997) FreeR value: Cross-validation in crystallography. Methods Enzymol 277:366
24. Bricogne G (1997) Bayesian statistical viewpoint on structure determination: Basic concepts and examples. Methods Enzymol 276:361
25. Tickle I (2007) Experimental determination of optimal root-mean-square deviations of macromolecular bond lengths and angles from their restrained ideal values. Acta Crystallogr D Biol Crystallogr 63(12):1274
26. Nicholls R (2011) Conformation-independent comparison of protein structures. Ph.D. thesis, Department of Chemistry, University of York
27. Bentley J (1975) Survey of techniques for fixed radius near neighbor searching. Technical report, Stanford Linear Accelerator Center, Menlo Park
28. Levinthal C (1966) Molecular model-building by computer. Sci Am 214:42
29. Nocedal J, Wright S (1999) Numerical optimization. Springer, New York
30. Murshudov G, Dodson E (1997) Simplified error estimation a la Cruickshank in macromolecular crystallography. CCP4 Newsl Protein Crystallogr 33:31

31. Trion MM (1996) Large Amplitude Elastic Motions in Proteins from a Single-Parameter, Atomic Analysis. Phys Rev Lett 77(9):1905
32. Thakurta P, Choudhury D, Dasgupta R, Dattagupta J (2004) Tertiary structural changes associated with iron binding and release in hen serum transferrin: a crystallographic and spectroscopic study. Biochem Biophys Res Commun 316(4):1124
33. Mizutani K, Mikami B, Aibara S, Hirose M (2005) Structure of aluminium-bound ovotransferrin at 2.15 A resolution. Acta Crystallogr D Biol Crystallogr 61(12):1636
34. R Development Core Team (2011) R: a language and environment for statistical computing. R Foundation for Statistical Computing, Vienna, http://www.R-project.org
35. Emsley P, Cowtan K (2004) Coot: model-building tools for molecular graphics. Acta Crystallogr D Biol Crystallogr 60(12(1)):2126
36. Tychonoff A, Arsenin V (1977) Solution of ill-posed problems. Winston & Sons, Washington, DC
37. Stuart A, Ord K, Arnold S (2009) Kendall's advanced theory of statistics, volume 2A: classical inference. Wiley, New York
38. Popov A, Bourenkov G (2003) Choice of data-collection parameters based on statistic modelling. Acta Crystallogr D Biol Crystallogr 59:1145
39. Bailey S, Eliason W, Steitz T (2007) Structure of hexameric DnaB helicase and its complex with a domain of DnaG primase. Science 318(5849):459
40. Joosten R, Salzemann J, Bloch V, Stockinger H, Berglund A, Blanchet C, BongcamRudloff E, Combet C, Da Costa A, Deleage G, Diarena M, Fabbretti R, Fettahi G, Flegel V, Gisel A, Kasam V, Kervinen T, Korpelainen E, Mattila K, Pagni M, Reichstadt M, Breton V, Tickle I, Vriend G (2009) PDB_REDO: automated re-refinement of X-ray structure models in the PDB. J Appl Crystallogr 42(3):376
41. Neidle S (2008) Principles of nucleic acid structures. Elsevier, London
42. Cremer D, Pople J (1975) A General Definition of Ring Puckering Coordinates. J Am Chem Soc 97(6):1354
43. Long F, Vagin AA, Young P, Murshudov GN (2008) BALBES: a molecular-replacement pipeline. Acta Crystallogr D Biol Crystallogr 64(1):125
44. Winn MD, Isupov MN, Murshudov GN (2001) Use of TLS parameters to model anisotropic displacements in macromolecular refinement. Acta Crystallogr D Biol Crystallogr 57(1):122

Chapter 23
High Resolution Macromolecular Crystallography

Mariusz Jaskolski

Abstract Atomic resolution is achieved when diffraction data extend beyond 1.2 Å. Structure refinement at this resolution allows anisotropic ADPs, reliable interpretation of static disorder, solvent structure and H atoms. Stereochemical restraints can be relaxed or removed, providing unbiased high-quality information about macromolecular stereochemistry, which in turn can be used to define improved conformation-dependent libraries. The surplus of data allows estimating least-squares uncertainties in the derived parameters, analogously to small-molecule standards. Atomic resolution data provide the most reliable information about macromolecular structure, especially important for validating new discoveries or resolving subtle issues of molecular mechanisms. At ultrahigh resolution it is possible to study charge density distribution by multipolar refinement of electrons in non-spherical orbitals. The current limit for macromolecular crystal X-ray diffraction is 0.55 Å for nucleic acids (Z-DNA) and 0.48 Å for proteins (crambin).

Keywords Atomic resolution • Stereochemical restraints • Conformation-dependent stereochemical library • H atoms • Multipolar refinement • Charge density • Standard uncertainties

---

M. Jaskolski (✉)
Department of Crystallography, Faculty of Chemistry, Adam Mickiewicz University, Poznan, Poland

Center for Biocrystallographic Research, Institute of Bioorganic Chemistry, Polish Academy of Sciences, Poznan, Poland
e-mail: mariuszj@amu.edu.pl

R. Read et al. (eds.), *Advancing Methods for Biomolecular Crystallography*,
NATO Science for Peace and Security Series A: Chemistry and Biology,
DOI 10.1007/978-94-007-6232-9_23,

23.1 Introduction

The criteria for high resolution in macromolecular crystallography are drifting in the desired direction (i.e. towards smaller d_{min} values) with time; a 2.5 Å data set that would be considered "high-resolution" 30 years ago is medium resolution, at best, today. Also different techniques have different criteria: a cryo-EM image reconstruction at 6 Å would be proudly classified as high-resolution while in X-ray crystallography it's almost beyond the limit of even low resolution. Although NMR spectroscopists sometimes borrow this term to describe their models, this is actually misleading because NMR-derived models simply do not have a clearly defined resolution. Even in crystallography, strictly speaking, "resolution" refers to the diffraction data and electron density maps (as their Fourier transform) and not to models, which are only interpretations of electron density (and could be even incorrect!). Crystallographic resolution is defined as the minimum d-spacing in Bragg's Law ($\lambda = 2d_{min}\sin\theta_{max}$) corresponding to the maximum glancing angle θ_{max} at which statistically significant reflections are still observed. With rigorous application of optical principles, it can be demonstrated that this d_{min} limit corresponds almost exactly to the minimal separation of two points that can be still distinguished in electron density maps generated by Fourier transformation. To add to the terminological confusion, we note that even crystallographers are not always precise in their vocabulary, as illustrated by the contradictory term "super-resolution with low-resolution data" and its controversial [42] elaboration that "a structure derived from low-resolution diffraction data can have quality similar to a high-resolution structure" [30].

On the somewhat arbitrary scale of resolution intervals (Fig. 23.1), the point at $d_{min} = 1.2$ Å has been defined by Sheldrick [31] as atomic resolution. This definition was later supported by a thorough mathematical argument [8] but its usefulness is obvious as at this level all non-hydrogen atoms should be resolved, including those in the shortest (1.2 Å) C=O bond. In this article, we will limit discussion of high-resolution macromolecular crystallography to atomic-resolution studies satisfying the Sheldrick criterion. By "ultrahigh" resolution we will mean at least 0.8 Å, roughly corresponding to a full sphere of Cu $K\alpha$ data. Atomic resolution macromolecular crystallography has been recurring in systematic reviews. The present article is based on the previous reviews (e.g. [12, 33]) but also discusses the most recent developments.

23.2 Experimental Aspects

Although personal preferences and case-to-case requirements may differ (e.g. demanding high redundancy to enhance weak anomalous signal or, conversely, reduced exposure to minimize radiation damage), if a single advice is to be given, it would be: always get the highest resolution during your diffraction experiment. This will ease all subsequent steps, will help reduce model bias, and will

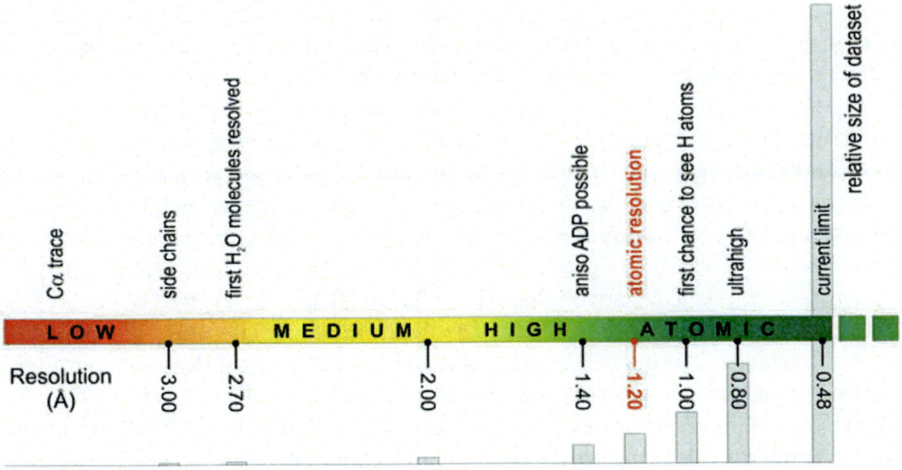

Fig. 23.1 Division of resolution into descriptive intervals. Only the criterion of atomic resolution has precise definition (1.2 Å). The annotations above the resolution bar indicate the allowed level of interpretation. The *histogram* illustrates, on a relative scale, the growth of the data set with increasing resolution

authenticate any unusual features discovered in the structure. With high-brilliance third-generation synchrotron sources and modern highly sensitive, fast and reliable detectors, achieving 2 Å resolution or better, is now almost a norm in routine studies, although difficult cases, especially of large macromolecular complexes, will continue to challenge the frontiers of macromolecular crystallography.

However, collecting meaningful high-resolution data is not equivalent to "visiting" high-order *hkl* indices in the reciprocal space for which no statistically significant intensity signal can be detected. As a rule of thumb, one should expect the average signal-to-noise ratio ($<I/\sigma(I)>$) in the highest resolution shell to be at least 2. Setting $<I/\sigma(I)>$ at 2 is typically equivalent to having about 50 % of the data in the last resolution shell with $I > 2\sigma(I)$, i.e. with statistically significant non-zero intensity. The above criteria are rather conservative but should guarantee a high-quality data set. A resolution-oriented approach could, however, push d_{min} to the extreme limit where adding more observations does not add information [15]. Statistically, this would correspond to a correlation coefficient CC_{true} between the experimental data and ideal noise-free signal of about 0.4. Diederichs and Karplus [21] have shown that CC_{true} can be estimated by $CC_{1/2}$, which measures correlation between two half-sets and should be acceptable down to ~0.1. It is also good to have the last resolution shell as complete as possible. (Parenthetically we note that at very high resolution, e.g. with $d_{min} < 1$ Å, diffraction data statistics should be analyzed in more shells of resolution, e.g. 20, especially for large unit cells.) It does not mean, however, that one should artificially truncate high-resolution data only to see high completeness in the last resolution shell. On the contrary, all reflections are precious and should always be included, particularly at high resolution. If completeness in the last resolution shell is really poor (for instance, as a consequence of a rectangular detector, which captures high resolution only at the corners), it is possible to estimate effective resolution (as opposed to the nominal resolution of the sparsely populated last shell), by finding d_{eff} at which a reciprocal lattice sphere with radius $1/d_{eff}$ would be filled completely with the available data points. Alternatively, one can estimate optical resolution d_{opt} by a Gaussian analysis of the Patterson map based on optical principles, as proposed by Vaguine et al. [36] and implemented in SFCHECK. Another reason for the rapidly deteriorating data completeness at high resolution is the rejection by most data processing programs (with default setting) of reflection intensities with large negative values ($I < -3\sigma(I)$). This practice, which is equivalent to outlier rejection, is acceptable, but the threshold ($-3\sigma(I)$) should not be manipulated (raised) to "improve" (the appearance of) the data set. Likewise, no σ-cutoff should be applied to select reflections for refinement. However, in algorithms that use $|F_o|$ for refinement, the data will be effectively truncated at $0\sigma(I)$, by eliminating negative intensities during the $|F_o| = \sqrt{I}$ conversion. From this point of view, refinement programs utilizing reflection intensities, such as SHELXL [7], are preferred.

It should be noted that the $\sigma(I)$ values, estimating the errors of intensity measurements, obtained from area detectors are typically not very accurate, as there is no good way to estimate them from signal-accumulating measurements. Consequently, the statistics based on $\sigma(I)$ values could be flawed.

The second most important parameter of a good data set is high redundancy, which always improves data quality but has to compete with radiation damage. In addition to reducing random errors, multiple observations can be used to estimate standard deviations of intensity measurements.

R_{merge} as a resolution-limiting criterion is not recommended (although in accurate high-resolution studies high-quality data characterized by low R-values are desirable). R_{merge} is naturally higher at high symmetry and with high redundancy, although multiple observations are certainly improving data quality. Better, redundancy-independent parameters (e.g. R_{rim}) have been proposed by Diederichs and Karplus [13] and Weiss and Hilgenfeld [37, 38].

Most (16-bit) detectors do not have sufficient dynamic range to reliably record very strong and very weak data at the same time. It may be thus necessary to measure the strongest, low-resolution data in a quick additional pass. At ultrahigh resolution, three runs may be necessary, e.g. ∞–2.0, 2.4–1.0 and 1.5–0.7 Å, with relative 1:10:100 exposure, to eliminate detector oversaturation in the respective resolution ranges. The quick low-resolution pass should be recorded first, before extensive radiation damage takes place.

23.3 Application of Direct Methods at Atomic Resolution

Although this chapter is mainly concerned with high-resolution refinement, it may be appropriate to mention that at atomic resolution, solution of the phase problem by direct methods, either in their classic form or as dual-space recycling ("shake and bake"), is possible. Indeed, the formulation of the 1.2 Å criterion was motivated by the applicability of direct methods [31]. Although originally deemed limited to small-molecule structures (100 independent non-H atoms), direct methods can now solve routinely macromolecular structures of 1,000 atoms, provided accurate atomic-resolution data are available. The list of successful phase determinations includes difficult cases (e.g. with unexpected special-position heavy atom; [34]) and is crowned with a protein structure (T4 lysozyme) comprised of over 1,300 non-H atoms [25].

23.4 Model Refinement at High Resolution

Historically, high order refinements were carried out with careful gradual extension of resolution. This strategy, originally dictated by limited computer power, is no longer necessary if the starting model is very good. However, when only an approximate model is available, starting the refinement at about 2 Å, and even inclusion of a rigid-body refinement step may be indicated to increase the radius of convergence.

The first round of refinement at full resolution is done with isotropic atomic displacement parameters (ADPs), and is followed by model adjustment in electron density maps and inclusion of the most evident solvent molecules. Switching from isotropic to anisotropic model at this stage more than doubles the number of model parameters and brings about a dramatic decrease of the R factors (up to 0.05). Anisotropic ADPs are used for the macromolecule and solvent atoms and must not be combined with modeling of anisotropic rigid-body motion (TLS parameters). The subsequent steps of the refinement protocol are listed in Table 23.1.

Table 23.1 Protocol of macromolecular refinement at atomic resolution

Step	Action
1	Full resolution
2	Isotropic ADPs
3	Correction of model errors, evident solvent molecules
4	Bulk solvent correction
5	Anisotropic ADPs
6	Modeling of disorder
7	Riding H atoms
8	Partial water molecules
9	Refinement/adjustment of occupancies
10	Relax/remove restraints
(11)	(H-atoms refined)
12	All reflections
(13)	(Multipolar refinement)
14	Full-matrix least-squares

Steps listed in parentheses are only possible at ultrahigh resolution

As a rule, the proportion of fragments modeled in dual (or sometimes triple) conformation increases, as resolution permits distinction between closely spaced alternate occupancies. This only applies to static disorder. Dynamic disorder can be reduced by collecting diffraction data at the lowest possible temperature. Fractional occupancies of light atoms (C/N/O) are considered from ca. 0.2, or exceptionally from 0.1 at ultrahigh resolution, i.e. from electron density contribution equivalent to an H atom.

At 0.9 Å resolution or better, stereochemical restraints of well-ordered fragments can be gradually relaxed and even removed altogether at ultrahigh resolution. Poorly ordered or multiple-conformation fragments should remain restrained as they are poorly defined by diffraction data.

Usually, refinement at very high resolution is carried out in SHELXL, which uses conventional (accurate) structure-factor summations [32]. Test calculations with minimization of least-squares targets seem to indicate that on convergence, the results of SHELXL and Refmac [26] are similar (S. Krzywda, personal communication). However, newer versions of programs such as Refmac or phenix.refine [1] allow refinement against maximum-likelihood targets, not available in the least-squares oriented SHELXL algorithm, and it is yet to be seen if this offers any benefit at high resolution. SHELXL does have, however, very strong advantages, which include (i) refinement on intensities rather than structure factor amplitudes and (ii) the possibility to calculate the least-squares variance-covariance matrix providing estimations of the uncertainties of individual parameters, and for these reasons should be recommended as the program of choice for meticulous high-resolution refinement.

23.5 Use and Validation of Stereochemical Restraint Libraries

At lower resolution, the use of stereochemical restraints is absolutely necessary simply to improve the data/parameter (d/p) ratio. At 1.2 Å, d/p is about 3 even for anisotropic models and approaches 5 at 1.0 Å, making restraints dispensable from the mathematical point of view. However, while they can be relaxed in well ordered segments, other areas, such as flexible side chains, still need to be restrained. At ultrahigh resolution, for well ordered structures, the refinement is highly overdetermined and stereochemical restraints can be eliminated altogether, as illustrated by the structure of Z-DNA at 0.55 Å resolution, where the macromolecule was refined without any restraints whatsoever [10]. Under strict control of stereochemical restraints at lower resolution, model deviations from the target values should not exceed the uncertainties of the target estimates. In the case of protein bond lengths [14], this is on the order of 0.015–0.020 Å [19]. At very high resolution, the results are dominated by the diffraction terms and the root-mean-square deviations (r.m.s.d.'s) from the target values are likely to reflect errors in the targets. Deviations as high as 0.02–0.03 Å could be still acceptable.

The target values were compiled from an analysis of small-molecule databases about 20 years ago, for protein structures by Engh and Huber [14] and for nucleic acids by Parkinson et al. [27]. Although they are largely correct, some adjustments seem to be necessary. For instance, the protein dictionary entry for the peptide C-N bond may need re-evaluation [19] and the peptide group planarity is most certainly enforced too strictly, deforming in many cases the Ramachandran geometry [2]. The nucleic-acid parameters for the phosphate group and the valence angles at the guanine glycosidic bond certainly should be re-examined [10]. The situation is now very interesting because not only is the small-molecule CSD database [5] about 10 times larger than when originally used for target generation, but we also have a subset of ultrahigh resolution structures in the PDB [7], with minimal target bias, from which the targets can be derived independently.

In addition to covalent geometry, other model parameters, such as ADPs or non-bonded contacts, are also restrained. Main-chain torsion angles should be left unrestrained to guarantee bias-free model validation via Ramachandran plots.

23.6 Conformation-Dependent Stereochemical Restraints

Macromolecular models refined at ultrahigh resolution are largely independent of stereochemical targets, even if restraints are included, and can be used for their validation and improvement. It has been noted in a number of studies that some (especially angular) parameters of such models have surprisingly wide spread that could be correlated with conformation and other characteristic features (e.g. hydrogen-bonding patterns) of macromolecules [19]. For instance, the N–Cα–C angle of the polypeptide backbone has a wide spread [2] and is correlated not only with residue type but also with the local φ/ψ backbone conformation [6]. By modeling main-chain bond distances and angles in proteins characterized at 1 Å resolution or better, as functions of the φ/ψ torsion angles, Tronrund and Karplus [35] were able to create a conformation-dependent stereochemical library (CDL) that allows achieving better models at lower resolution and, when applied at higher resolution, does not distort the models from the diffraction-driven target (r.m.s.d. for bonds ~0.007–0.010 Å) but, indeed, improves the results.

23.7 Restraint-Free Refinement and Disorder

Although restraint relaxation is practiced at atomic resolution and restraint-free refinement is mathematically possible at ultrahigh resolution, from the point of view of the d/p ratio it is somewhat contradictory that the degree of discrete (static) disorder that can be modeled by fractional-occupancy conformations increases with resolution. As demonstrated with BPTI, even in the same crystal structure the percent of disordered residues increases with resolution [2, 11] and reaches 21 %

at 0.86 Å. In the 0.66 Å crystal structure of human aldose reductase, one-third of all residues were modeled in multiple conformations [18]. This makes the improvement of d/p less spectacular (many parameters have to be "wasted" in poorly defined fragments) and requires the retention of stereochemical restraints in multiple-conformation areas. The disorder is usually visible in the macromolecule and in the solvent region, and it is often found to form correlated "networks". Such networks should be identified and refined with common occupancy parameters. There are, however, exceptions from the "more disorder at higher resolution" rule, as illustrated by the structure of Z-DNA [10], where there is no detectable disorder in the macromolecule even at 0.55 Å resolution. On the other hand, only 15 % of the water sites have full occupancy in that structure, which may be perceived as a very rigid macromolecule immersed in a highly fluid milieu.

23.8 Treatment of H Atoms

The X-ray scattering power of the H atom (in particular in polarized X–H bonds) is very low and, therefore, H atoms are omitted in modeling macromolecular crystal structures. Although at high angles the scattering cross section diminishes further, paradoxically H atoms can be better visualized using high-resolution data because the disproportion to C, N or O scattering is less drastic. Besides, only at better than 1 Å resolution is it possible to resolve H atoms, as typical X–H covalent bonds are 0.9–1.1 Å long. Even if H atoms are not fully resolved by the diffraction data, it is recommended to include their contribution to F_c (even at 2 Å resolution), simply to improve the $|F_o|-|F_c|$ agreement and to remove bias from the location of their parent atoms X, which otherwise are placed at the centroid of the X–H electron cloud, i.e. form an "expanded" skeleton. The positions of most H atoms in proteins and nucleic acids are easily generated from the skeleton of the remaining atoms, and their contribution at atomic resolution typically decreases the R factor by ca 0.01. H atoms in $-NH_3^+$ and $-CH_3$ groups, in the ambiguously protonated His residue, –OH groups and (possibly) carboxylic groups cannot be generated automatically and have to be analyzed individually, usually based on logical H-bonding circuits. Generation of H atoms with fractional occupation is not a sensible proposition, especially that dealing with geometry of multiple-conformation groups easily leads to errors in H atom placement.

Generation of H atoms in water molecules cannot be done fully automatically, although there are algorithms that claim to challenge even neutron scattering data (C. Lecomte, personal communication). Considering the high proportion of water molecules with fractional occupancy, it is doubtful if *en bloc* generation of water H atoms is very meaningful. Those special cases where water H atoms are important and are clearly defined in electron density (Fig. 23.2) can be dealt with manually.

With the overwhelming overdeterminacy of ultrahigh resolution structures, full refinement of H atom parameters (*x*,*y*,*z*,Biso) is possible as in small-molecule

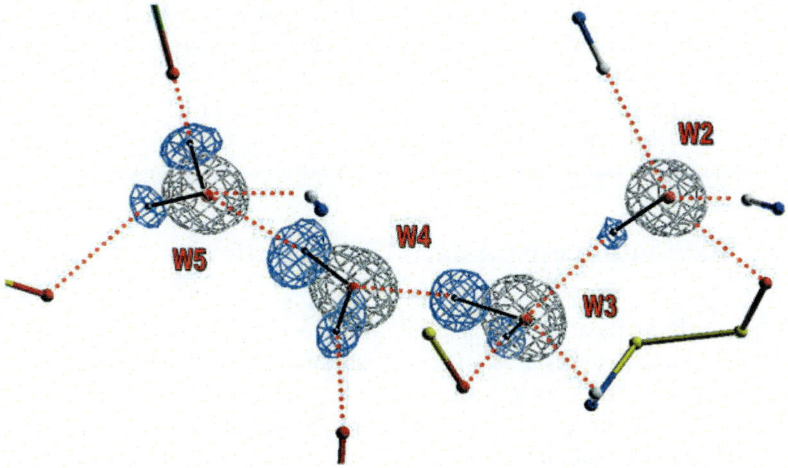

Fig. 23.2 A chain of water molecules linked by homodromic hydrogen bonds in the 0.86 Å structure of BPTI [2]. The positions of H atoms (not included in the model) are clearly marked by the (smaller) spheres of F_o–F_c electron density, contoured at the 2.9σ level. The $2F_o$–F_c electron density, centered on the O atoms, is contoured at the 2.0σ level

crystallography. Such tests have been carried out but the minimal gain in convergence (e.g. R factor drop from 0.0768 to 0.0764 in the case of Z-DNA; M. Gilski, personal communication) does not justify the massive effort needed to verify the results. It is thus concluded that even at very high resolution, conventional refinement should include riding H atoms, and if necessary only the key H atoms should be refined individually.

Some protocols, especially applied in ultrahigh charge density studies, place or shift H atoms along the X–H bonds to neutron distances [4]. While this procedure yields a geometrically correct model, it is not necessarily compatible with X-ray refinement of spherical atoms. Moreover, normalization of H atoms in very short (and thus of key importance) hydrogen bonds may be entirely unjustified.

23.9 Use of R_{free} for Validation

Calculation of R_{free} [9] is now a common method for validating crystallographic models and the process through which they are generated. Although atomic resolution refinement is typically not frustrated with profound strategic ambiguities, some decisions are clearly validated by reference to R_{free}. It is usually enough to set aside 1,000–2,000 test reflections, rather than applying the 5–10 % rule, which could be very wasteful concerning the large data set size at high resolution. One should ensure, however, that the test reflections are selected at random from thin slices of resolution, covering the entire range, i.e. including the highest resolution shell as well. When the model has been completed, the test reflections should be included in the working data set for a final round of refinement and for the generation of final electron density maps. This will further improve the final d/p ratio and reduce series termination errors in the Fourier transform, i.e. will lead to better results, which is the ultimate goal of any high-resolution study.

23.10 Estimation of Standard Uncertainties

Most of the refinement cycles in SHELXL are done *via* the conjugate-gradient algorithm, which, in the interest of speed, circumvents the inversion of the normal-equation matrix. The last refinement cycle (for diagnostic purposes, without application of parameter shifts) should be calculated in the full-(or blocked) matrix least-squares mode to estimate standard uncertainties (s.u.) in the atomic parameters. This is done for all reflections but without restraints and usually for positional parameters only (to obtain s.u.'s of geometrical parameters). If the problem is prohibitively large (over 100 residues), the matrix can be blocked into 50-residue segments (with 5-residue overlap) that will be refined in alternating cycles. Accurately estimated s.u.'s are a treasure trove because they allow meaningful interpretation of model geometry. For instance, it is possible to gauge significant *vs* insignificant geometry differences, or even evaluate potential errors in stereochemical standards. At ultrahigh resolution, s.u.'s in bond lengths, for example, are as low as, or lower than, in small-molecule crystallography. In the 0.55 Å structure of Z-DNA, the σ(bond) values are 0.002–0.004 Å, while in the 0.86 Å structure of BPTI they are on the order of 0.005–0.02 Å.

23.11 Electron Density Maps at High and Ultrahigh Resolution

Work with electron density maps at better than atomic resolution is very gratifying and indeed pleasant because they show most of the atoms in a structure as well resolved spheres. Electron density maps can use $2F_o–F_c$ coefficients, or $3F_o–2F_c$ coefficients as recommended by Lamzin and Wilson [23], but the difference is not very obvious. At very high resolution even F_o maps can be used as series termination effects are negligible (Fig. 23.3). For difference maps, σ_A-derived coefficients are usually used [28]. At ultrahigh resolution difference electron density maps will not only unambiguously indicate most H atom sites but will also reveal subtler effects, such as free electron pairs or σ-orbital electrons at midpoints of covalent bonds (Fig. 23.3).

For methodological correctness, accurate electron density maps should be contoured in absolute e/Å^3 units, but the values from Fourier summation are

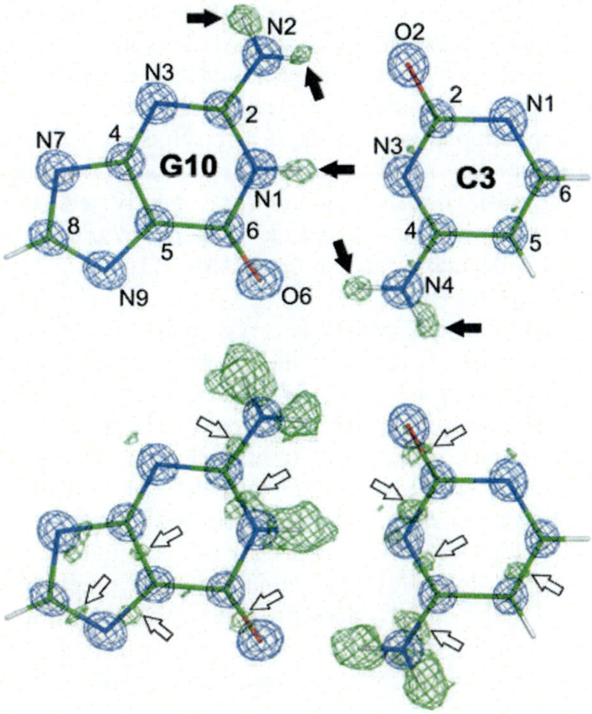

Fig. 23.3 A fragment of the Z-DNA structure refined at 0.55 Å [10]. The experimental F_o map, reflecting the covalent structure of a GC base pair, is contoured at 3σ. The F_o–F_c map, calculated without H-atom contribution, is contoured at 3.5σ in the *upper panel*, where it clearly shows H-atom positions (*full arrows*), and at 2σ in the *lower panel*, where it also reveals bonding electrons between atoms (*empty arrows*)

not absolute for macromolecular structures where the astronomically strong low-order reflections are usually missing and the F(000) term is not known altogether. Consequently, maps continue to be contoured in σ units. It should be noted that because of the low level of noise in accurate maps, the σ unit is usually low and any meaningful features are represented by high-level contours, higher than in medium-resolution crystallography.

23.12 Multipolar Refinement and Deformation Density Studies *vs* "Interatomic Scatterers"

At ultimate resolution, higher than 0.7 Å, one may contemplate charge (or deformation) density studies and multipolar refinement. Deformation density studies aim at mapping deviations of atomic electrons from the classical (but incorrect in covalent molecules) spherical independent-atom models (IAM). Such studies require data of very high resolution and are special even in small-molecule crystallography. In multipolar expansion, atomic electrons are partitioned into core and valence shells, and the latter ones are described by multipolar functions [17]. Refinement of a multipolar atom requires 3 (for H) to 27 (for heavy elements) extra parameters (depending on the level of multipolar expansion), in addition to the usual 3 coordinates and 6 anisotropic ADPs. Usually, the first round of refinement uses reflections from the high-resolution shell alone, to distill thermal motion parameters of non-H atoms from the electron distribution functions. Subsequent cycles refine the multipolar parameters of a substructure with excellent order and low thermal motion.

Even if ultimate resolution is not fully achieved, deformation density studies are still possible through the principle of transferability of experimental library multipolar atom models (ELMAM; [43]) or of libraries obtained by theoretical calculations [22].

Experimental charge-density studies of macromolecules are extremely rare and are limited at present to the enzyme human aldose reductase, which is a TIM-barrel protein comprised of 316 residues, analyzed at 0.66 Å resolution [16], and to the 46-residue crambin, analyzed at the record-setting resolution of 0.48 Å [29]. There is no charge-density analysis or multipolar refinement of a nucleic acid structure reported in the literature, although work has already started in this direction (M. Kubicki, personal communication).

Rigorous deformation density studies require sophisticated refinement software, such as MoPro [20]. As an alternative, a simple-minded approximation has been proposed to refine "pseudo-atom" scatterers at midpoints of covalent bonds that would take care of bonding electrons [3]. This simplistic approach is not quite on a par with the meticulous and accurate character of high-resolution studies.

23.13 Solvent Structure

As a rule of thumb, one is allowed to model about (3-d_{min}/Å) water molecules per residue in protein structures [41]. For ultrahigh resolution structures, for which the Matthews [24] volume is usually low, it is often possible to locate nearly all water molecules. The situation is, of course, not crystal clear because solvent molecules also (or particularly) in high resolution structures show a high degree of disorder, populating many sites with partial occupancy. In fact, modeling the outer hydration shell in high resolution macromolecular structures is usually the most frustrating step, well justifying the opinion that "a macromolecular refinement against high-resolution data is never finished, only abandoned" [32]. Despite the near-complete atomic interpretation of the solvent region at high resolution, it is a common (and not very harmful) practice to include in the refinement a bulk-solvent correction, for instance based on Babinet's principle (Fourier transforms of a mask and its complement have the same amplitudes, but opposite phases), which holds only at very low resolution ($d > 15$ Å).

The site occupation factors of (even all) water molecules could be refined together with their ADPs but a more prudent approach (at least at modestly high resolution) is to fix them after manual or automatic adjustment. In a pragmatic approach, after a round of occupancy (occ) refinement, one would (i) eliminate phantom molecules (occ <0.2), (ii) fix those refined to occ >0.9 at 1.0, (iii) couple the occupancies of alternate sites (O . . . O distance <2 Å), and (iv) let the remaining occupancies refine freely. A water molecule that is retained in the model should have, after refinement, clear $2F_o - F_c$ electron density at the 1σ level, should form at least one reasonable hydrogen bond (2.3–3.2 Å), and should not have prohibitively short contacts, e.g. with C atoms; however, the possibility of forming C–H. . . O hydrogen bonds (which are usually long, with C. . . O ∼3 Å) should be taken into account.

Water molecules should not be confused with metal cations. Although such species can be isoelectronic (e.g. $H_2O/Na^+/Mg^{2+}$), metal cations are likely to form shorter bonds (e.g. Mg. . . O ∼2 Å), do not have typical proton donors (such as amide N–H) in their coordination sphere, and will often have more than four ligands, e.g. six in the case of octahedral Mg coordination.

23.14 Benefits of Atomic Resolution

The benefits of atomic-resolution macromolecular structures have been discussed in several excellent reviews. They are certainly worth the considerably higher effort that must be invested in the experiment, computations and interpretation of the results. By improving the d/p ratio, high-resolution data help to remove model bias, which often blights crystallographic structures, especially solved by molecular replacement. More reflections and better resolving power allow accurate interpretation of multiple conformations, thus yielding more realistic models and better agreement with experiment. Unusual stereochemical features are best confirmed at atomic resolution. Conversely, the possibility to refine macromolecular models with relaxed or eliminated stereochemical restraints is the best road to scientific discovery of phenomena that could be blurred by paucity of data and/or prejudiced ideas about the result. Restraint-free refinement can ultimately produce accurate dictionaries of macromolecular stereochemistry, for use as restraints at lower resolution. From a methodological point of view, restraint-free refinement with sufficiently high d/p ratio allows the application of full-matrix least-squares, from which standard uncertainties of the geometrical parameters can be estimated. Calculation of both, the parameters and their error estimates, places the discussion of macromolecular geometry on an entirely new, statistically significant level. Although H atoms have only minimal contribution to X-ray scattering and are normally omitted from models of macromolecular structure, they are often of key importance for understanding the functioning of macromolecules, e.g. in enzyme catalysis or fine-tuned intermolecular recognition. Any sensible experimental interpretation of H atoms requires very high resolution X-ray diffraction data. Indeed, there is evidence suggesting that careful ultrahigh resolution X-ray analysis could be superior in this respect to macromolecular neutron diffraction, which requires prohibitively large crystals (∼1 mm) and is normally limited to only medium resolution. Even if H atoms are not visualized in electron density maps, their placement (e.g. in carboxylic groups) can be often unambiguously predicted in atomic-resolution structures from the pattern of bond distances between the heavier atoms, which changes on protonation [40]. Also, solvent molecules, which are often disordered and not very amenable to accurate modeling at lower resolution, can get sensible interpretation (even if involving circuits with extensive disorder) at atomic resolution. Finally, when ultimately high resolution data are available, it becomes possible to interpret the macromolecular structure at a level of detail that goes far beyond the localization of atoms. Such charge density studies, which involve refinement of multipolar parameters describing atomic orbitals, are still very rare but they are beginning to unveil a fascinating inner world of macromolecules, at the level of electrons in atoms, in interatomic bonds and in intermolecular interactions.

23.15 Survey of Highest-Resolution Structures in the PDB

The first protein structures at atomic resolution appeared in the PDB in mid 1980s (although the current definition of "atomic resolution" was coined much later). The very first case was the structure of BPTI determined at 1.0 Å resolution by joint refinement using X-ray and neutron diffraction data [39]. Even by today's standards, it is an outstanding achievement. Currently there are over 1,700 atomic resolution structures in the PDB, i.e. only 2 % of all entries. A surprising observation from Table 23.2 is that the relative rate of accumulation of atomic resolution structures is today not much faster than nearly 30 years ago. This leads to a rather pessimistic conclusion that the tremendous methodological progress in macromolecular crystallography is mainly fueling quantity (more and bigger structures) but not necessarily quality. Ultrahigh resolution macromolecular studies are still very rare, although the technical possibilities are available (primarily high-brilliance synchrotron sources) to expect more. There are only a few structures refined at the ultimate resolution of at least 0.66 Å and most of them are for rather small biomolecules. In Table 23.3, only lysozyme and aldose reductase have a molecular weight >10 kDa. In this list, the case of human aldose reductase has to be singled out as an exceptionally large protein (316 residues) for this type of study. It is to be particularly admired as it has been also studied (without a PDB deposit) by multipolar refinement, the only other such case being the small protein crambin. Crambin sets currently the limit of macromolecular crystal diffraction at 0.48 Å and it is not likely that this

Table 23.2 Growth of atomic- and ultrahigh-resolution entries in the Protein Data Bank

Resolution (Å) at least	All holdings	1.2	0.8
1981	98	0	0
1991	1,103	15	0
2001	17,694	284	6
2011	78,191	1,609	34
July 2012	83,000	1,723	36

Table 23.3 Eight highest-resolution structures in the Protein Data Bank (March 2012)

Code	Macromolecule	Resolution (Å)	R	Year	Authors	Remarks
3NIR	Crambin	0.48	**0.127**	2011	Schmidt et al.	Charge density; **R!**
1EJG	Crambin	0.54	0.090	2000	Jelsch et al.	Charge density
3P4J	Z-DNA	0.55	*0.078*	2011	Brzezinski et al.	Lowest R
1I0T	Z-DNA	0.60	**0.160**	2001	Tereshko et al.	Suboptimal; **R!**
1J8G	RNA tetraplex	0.61	0.103	2001	Deng et al.	
1UCS	Antifreeze protein	0.62	**0.137**	2003	Ko et al.	Suboptimal; **R!**
2VB1	Lysozyme	0.65	0.085	2007	Wang et al.	
1US0	Aldose reductase	0.66	0.094	2004	Howard et al.	IAM refinement[a]

Convergence with unexpectedly high R-factor is highlighted in *bold*. The lowest R-factor is shown in *italics*

[a]Followed by a charge-density study (model not deposited) via substructure ELMAM multipolar refinement with R = 0.087 [16]

limit will be pushed very much farther very soon. Although there are protein and nucleic acid (DNA and RNA) structures in Table 23.3, there is no charge density study of a nucleic acid structure with multipolar refinement. This situation should be rectified because electron density distribution in nucleic acids is as important, or possibly more important, than in proteins. It is noticeable that some of the structures in Table 23.3 were refined with surprisingly high R-factors. At this resolution one should expect an R factor of 0.10 or less, as illustrated by the refinement of the Z-DNA structure 3P4J at 0.55 Å resolution with R = 0.078 [10].

References

1. Adams PD, Grosse-Kunstleve RW, Hung L-W, Ioerger TR, McCoy AJ, Moriarty NW, Read RJ, Sacchettini JC, Sauter NK, Terwilliger TC (2002) PHENIX: building new software for automated crystallographic structure determination. Acta Crystallogr D 58:1948–1954
2. Addlagatta A, Czapinska H, Krzywda S, Otlewski J, Jaskolski M (2001) Ultrahigh-resolution structure of a BPTI mutant. Acta Crystallogr D 57:649–663
3. Afonine PV, Grosse-Kunstleve RW, Adams PD, Lunin VY, Urzhumtsev AG (2007) On macromolecular refinement at subatomic resolution with interatomic scatterers. Acta Crystallogr D 63:1194–1197
4. Allen FH (1986) A systematic pairwise comparison of geometric parameters obtained by X-ray and neutron diffraction. Acta Crystallogr B 42:515–522
5. Allen FH (2002) The Cambridge Structural Database: a quarter of a million crystal structures and rising. Acta Crystallogr B 58:380–388
6. Berkholz DS, Shapovalov MV, Dunbrack RLJ, Karplus PA (2009) Conformation dependence of backbone geometry in proteins. Structure 17:1316–1325
7. Berman HM, Westbrook J, Feng Z, Gilliland G, Bhat TN, Weissig H, Shindyalov IN, Bourne PE (2000) The protein data bank. Nucleic Acids Res 28:235–242
8. Bricogne G, Morris RJ (2003) Sheldrick's 1.2 Å rule and beyond. Acta Crystallogr D 59:615–617
9. Brünger AT (1992) Free R-value: a novel statistical quantity for assessing the accuracy of the crystal structures. Nature 335:472–475

10. Brzezinski K, Brzuszkiewicz A, Dauter M, Kubicki M, Jaskolski M, Dauter Z (2011) High regularity of Z-DNA revealed by ultra high-resolution crystal structure at 0.55 Å. Nucleic Acids Res 39:6238–6248
11. Czapinska H, Otlewski J, Krzywda S, Sheldrick GM, Jaskolski M (2000) High resolution structure of bovine pancreatic trypsin inhibitor with altered binding loop sequence. J Mol Biol 295:1237–1249
12. Dauter Z, Lamzin V, Wilson K (1997) The benefits of atomic resolution. Curr Opin Struct Biol 7:681–688
13. Diederichs K, Karplus PA (1997) Improved R-factors for diffraction data analysis in macromolecular crystallography. Nature Struct Biol 4:269–274
14. Engh RA, Huber R (1991) Accurate bond and angle parameters for X-ray protein structure refinement. Acta Crystallogr A 47:392–400
15. Evans P (2012) Resolving some old problems in protein crystallography. Science 336:986–987
16. Guillot B, Jelsch C, Podjarny A, Lecomte C (2008) Charge-density analysis of a protein structure at subatomic resolution: the human aldose reductase case. Acta Crystallogr D 64:567–588
17. Hansen NK, Coppens P (1978) Testing aspherical atom refinements on small-molecule data sets. Acta Crystallogr A 34:909–921
18. Howard ER, Sanishvili R, Cachau RE, Mitschler A, Chevrier B, Barth P, Lamour V, Van Zandt M, Sibley E, Bon C, Moras D, Schneider TR, Joachimiak A, Podjarny A (2004) Ultrahigh resolution drug design I: details of interactions in human aldose reductase-inhibitor complex at 0.66 Å. Proteins 55:792–804
19. Jaskolski M, Gilski M, Dauter Z, Wlodawer A (2007) Stereochemical restraints revisited: how accurate are refinement targets and how much should protein structures be allowed to deviate from them? Acta Crystallogr D 63:611–620
20. Jelsch C, Guillot B, Lagoutte A, Lecomte C (2005) Advances in protein and small-molecule charge-density refinement methods using MoPro. J Appl Crystallogr 38:38–54
21. Karplus PA, Diederichs K (2012) Linking crystallographic model and data quality. Science 336:1030–1033
22. Koritsanszky T, Volkov A, Coppens P (2002) Aspherical-atom scattering factors from molecular wave functions. 1. Transferability and conformation dependence of atomic electron densities of peptides within the nultipole formalism. Acta Crystallogr A 58:464–472
23. Lamzin VS, Wilson KS (1997) Automated refinement for protein crystallography. Methods Enzymol 277:269–305
24. Matthews BW (1968) Solvent content of protein crystals. J Mol Biol 33:491–497
25. Mooers BHM, Matthews BW (2004) Use of an ion-binding site to bypass the 1000-atom limit to structure determination by direct methods. Acta Crystallogr D 60:1726–1737
26. Murshudov GN, Vagin AA, Dodson EJ (1997) Refinement of macromolecular structures by the maximum-likelihood method. Acta Crystallogr D 53:240–255
27. Parkinson G, Vojtechovsky J, Clowney L, Brünger AT, Berman HM (1996) New parameters for the refinement of nucleic acid-containing structures. Acta Crystallogr D 52:57–64
28. Read RJ (1997) Model phases: probabilities and bias. Methods Enzymol 277:110–128
29. Schmidt A, Teeter M, Weckert E, Lamzin VS (2011) Crystal structure of small protein crambin at 0.48 Å resolution. Acta Crystallogr F 67:424–428
30. Schröder GF, Levitt M, Brünger AT (2010) Super-resolution biomolecular crystallography with low-resolution data. Nature 417:806–807
31. Sheldrick GM (1990) Phase annealing in SHELX-90: direct methods for larger structures. Acta Crystallogr A 46:467–473
32. Sheldrick GM (2008) A short history of SHELX. Acta Crystallogr A 64:112–122
33. Thaimattam R, Jaskolski M (2004) Synchrotron radiation in atomic-resolution studies of protein structure. J Alloys Comp 362:12–20
34. Thaimattam R, Tykarska E, Bierzynski A, Sheldrick GM, Jaskolski M (2002) Atomic resolution structure of squash trypsin inhibitor: unexpected metal coordination. Acta Crystallogr D 58:1448–1461

35. Tronrund DE, Karplus PA (2011) A conformation-dependent stereochemical library improves crystallographic refinement even at atomic resolution. Acta Crystallogr D 67:699–706
36. Vaguine AA, Richelle J, Wodak SJ (1999) SFCHECK: a unified set of procedures for evaluating the quality of macromolecular structure-factor data and their agreement with the atomic model. Acta Crystallogr D 55:191–205
37. Weiss M (2001) Global indicators of X-ray data quality. J Appl Crystallogr 34:130–135
38. Weiss M, Hilgenfeld R (1997) On the use of the merging R factor as a quality indicator for X-ray data. J Appl Crystallogr 30:203–205
39. Wlodawer A, Walter J, Huber R, Sjolin L (1984) Structure of bovine pancreatic trypsin inhibitor. Results of joint neutron and X-ray refinement of crystal form II. J Mol Biol 180:301–329
40. Wlodawer A, Li M, Gustchina A, Dauter Z, Uchida K, Oyama H, Goldfarb NE, Dunn BM, Oda K (2001) Inhibitor complexes of the *Pseudomonas* serine-carboxyl proteinase. Biochemistry 40:15602–15611
41. Wlodawer A, Minor W, Dauter Z, Jaskolski M (2008) Protein crystallography for non-crystallographers, or how to get the best (but not more) from published macromolecular structures. FEBS J 275:1–21
42. Wlodawer A, Lubkowski J, Minor W, Jaskolski M (2010) Is too ‘creative’ language acceptable in crystallography? Acta Crystallogr D 66:1041–1042
43. Zarychta B, Pichon-Pesme V, Guillot B, Lecomte C, Jelsch C (2007) On the application of an experimental multipolar pseudo-atom library for accurate refinement of small-molecule and protein crystal structures. Acta Crystellagr A 63:108–125

Chapter 24
Publishing in Proteopedia: The Guide

Jaime Prilusky, Wayne Decatur, and Eric Martz

Abstract Proteopedia (http://proteopedia.org) is an interactive web resource with 3D rotating models that react and change following user interaction. The main goals of Proteopedia are to collect, organize and disseminate structural and functional knowledge about protein, RNA, DNA, and other macromolecules, and their assemblies and interactions with small molecules, in a manner that is relevant and broadly accessible to students and scientists. This Guide provides instructions for the potential author of pages in Proteopedia, and for educators and teachers in incorporating Proteopedia when teaching proteins, and protein-functions. Of course you can use Proteopedia as a reference resource without authoring content.

Keywords Teaching protein structure function • Molecular biology • Computer aided instruction • 3D protein models • Interactive media • Education • Wiki • Online • Tutorial • Molecular visualization • Jmol

24.1 Introduction

This guide for authoring in Proteopedia [5, 8] has three sections. Section 24.1 is oriented to the prospective Proteopedia author, providing step by step directions on how to get started with Proteopedia and how to achieve professional

J. Prilusky (✉)
Bioinformatics Unit, Weizmann Institute of Science, Rehovot, Israel
e-mail: jaime.prilusky@weizmann.ac.il

W. Decatur
Department of Molecular, Cellular and Biomedical Sciences,
University of New Hampshire, Durham, NH, USA

E. Martz
University of Massachusetts, Amherst, MA, USA

R. Read et al. (eds.), *Advancing Methods for Biomolecular Crystallography*,
NATO Science for Peace and Security Series A: Chemistry and Biology,
DOI 10.1007/978-94-007-6232-9_24,

looking pages. Section 24.2 can be used as handout material for teaching, providing an easy and short path to creating a real Proteopedia page. Section 24.3 describes advanced Proteopedia authoring and its use in teaching. If you have questions that are not answered here, please email the Proteopedia Staff at contact@proteopedia.org

24.2 Section A. Help: Getting Started in Proteopedia

The online version for this part is available at http://proteopedia.org/w/Proteopedia:Getting_Started

This article, *Getting Started*, is intended to help orient beginners who wish to author content in Proteopedia. A good way to get oriented to using Proteopedia is to watch some of the Proteopedia:Video Guide (http://proteopedia.org/w/Proteopedia:Video_Guide).

24.2.1 Login

Everything below assumes that you have applied for, and been given, a login account. Applications are usually approved within 24 h. Your account name will be your real, full name, as it would appear when you are an author on a scientific publication. Indeed, your name will appear at the bottom of every Proteopedia page to which you contribute. Look at the bottom of any page and you will see the names of members who have contributed to it. (The name "OCA" refers to automated software.)

Please login now. You can tell when you are logged in because your name appears at the very top of the page.

24.2.2 Your User Page

When you are logged in, your name appears at the top, above every page in Proteopedia. Click on your name. Now you should see your User: page, with the title User:Your_Name. This is a good opportunity to expand or correct the biographical information in your User page, if you wish. Your User page is also a great place to keep an organized set of links to articles in Proteopedia to which you have contributed – see below for more about this.

24.2.3 *Editing in Proteopedia*

The process of editing a page in Proteopedia is explained in one of the videos in the Proteopedia:Video Guide (http://proteopedia.org/w/Proteopedia:Video_Guide). This would be a good time to watch that video. Most of the points in this video will not be repeated below.

Click on the tab *edit this page* at the top of your User page. A box will appear containing the wikitext for your User page. The **wikitext** of a Proteopedia article is the editable text and wiki markup that appears in the box after you click the tab *edit this page*.

You can type plain text in this box, change text, or delete text. The **history** tab of every page keeps a record of all changes made to that page. If necessary, using the history tab, you can undo a change by reverting the page to an earlier step in its history.

24.2.4 *Formatting and Styling*

24.2.4.1 Appearance

An overview of wikitext formatting and styling markup is given in the Cheatsheet (http://en.wikipedia.org/wiki/Wikipedia:Cheatsheet). Some is covered in the Proteopedia:Video Guide (http://proteopedia.org/w/Proteopedia:Video_Guide). More details are available at Help:Editing (http://proteopedia.org/w/Help:Editing). Below we cover some basics to get you started.

Paragraphs: Separate paragraphs with a blank line. If you want greater separation, use several blank lines.

Boldface, *italics*, subtitles, etc. have convenience buttons at the top of the wikitext box.

If you need to push a paragraph below a graphic object (image, table, Jmol), insert {{Clear}}.

You can always see how a page is formatted by clicking on *edit this page* to see the wikitext, then clicking **Cancel** instead of making any changes.

24.2.4.2 Links

Internal links, to other pages within Proteopedia, are made by enclosing the exact title of the page within double square brackets. For example, [[Help:Contents]] generates the link Help:Contents (http://proteopedia.org/w/Help:Contents), while [[Help contents]] produces the red link Help Contents because the colon was omitted. When a link is red, it means that the page does not exist.

If you forget the double square brackets, [[. . .]], you can simply type (or paste) the title, block it, and then click the convenience button **Ab** at the top of the wikitext box. If you want to change the label of the link, put a vertical bar "|" after the page title followed by the new label. For example, you could link to Help:Contents but label the link "list of help pages": [[Help:Contents|list of help pages]] appears as list of help pages.

External links to pages outside Proteopedia are made by enclosing the http address within single square brackets. However, if the topic concerns structural biology, the link should be to a page within Proteopedia (which you may need to create – see How To Create A New Page [http://proteopedia.org/w/Help:Editing#How_To_Create_A_New_Page]). That page within Proteopedia is the correct place to put a link to an external source, such as an article in Wikipedia (see Proteopedia:Policy). For example, if you wish to link to the Wikipedia page on HIV, you should link to the Proteopedia page Human Immunodeficiency Virus, which in turn contains a link to HIV in Wikipedia (http://proteopedia.org/w/Human_Immunodeficiency_Virus).

If the topic does not concern structural biology, and needs no page in Proteopedia, you can link within the text of an article, or in a footnote (see below). For example, [http://en.wikipedia.org/wiki/Wikipedia:Cheatsheet Wiki Markup Cheatsheet] makes the link Wiki Markup Cheatsheet (http://en.wikipedia.org/wiki/Wikipedia:Cheatsheet).

The label in an external link is placed within the single square brackets, after a space. For example, [http://proteopedia.org/w/Human_Immunodeficiency_Virus HIV in Wikipedia] makes the link HIV in Wikipedia (http://proteopedia.org/w/Human_Immunodeficiency_Virus).

24.2.4.3 References and Footnotes

References to published literature are easy to make – please see Citing Literature References (http://proteopedia.org/w/Help:Editing#Citing_Literature_References). Footnotes are also straightforward – please see http://proteopedia.org/w/Help:Editing#Footnotes.

24.2.4.4 Advanced

HTML (if you happen to know some) can sometimes be used. Only a very limited subset of HTML is allowed, but it often gives you more control than wiki markup. You can use unordered list or ordered list tags, or make tables in HTML.

CSS styling can also be used to a limited extent. For example, to make a font 150 % of the base size, <span style="font-size:150 %">Larger Font</span>.

Not covered here? Please ask: contact@proteopedia.org.

24.2.5 Make A Personal Sandbox

A good place to get started is to create a page *Sandbox 1* in your User space. Here are step by step instructions. A *Sandbox* page is a page reserved for temporary, practice work – see http://proteopedia.org/w/Help:Sandboxes.

Note that by placing the page in your user space (http://proteopedia.org/w/User:Your_Name/Sandbox_1) it is protected. Nobody else can edit it. However, anyone can read it. Please see http://proteopedia.org/w/Help:Protected_Pages. It is also possible to create hidden, protected pages – please see http://proteopedia.org/w/Proteopedia:Workbench.

There are also numerous Sandbox pages that can be edited by anyone – many are, however, reserved for use by university classes. See Teaching Strategies Using Proteopedia.

24.2.6 Link Your Work to Your User Page

Notice that by using a link on your User page to create *Sandbox 1* (see previous section), whenever you login, you are just 2 clicks away from that page (click on your name, then on the link to *Sandbox 1*).

A good way to make it easy to find pages you are working on, or have worked on, is to put links to those pages in your User page. You can always click *my contributions* (at the top of the page), but that lists individual edits. By making your own links, you can organize them, and have only a single link per page.

24.2.7 Best Practices

Guides are available to help you produce articles of good quality:

http://proteopedia.org/w/Proteopedia:How_to_Make_a_Page and http://proteopedia.org/w/Proteopedia:Guidelines_for_Ethical_Writing

24.2.8 Molecular Scenes

The present article concerns primarily the text content of an article, but of course the most exciting part of Proteopedia is the ease with which you can create *green links* that show customized interactive molecular scenes. Plenty of help is available to get you started with these:

Several of the videos in the Proteopedia:Video Guide (http://proteopedia.org/w/Proteopedia:Video_Guide) are the best place to start. The page http://proteopedia.org/w/Proteopedia:Primer describes creation of a molecular scene, step by step (see also Sect. 24.2 below).

24.2.9 *How Do I Display My Molecule?*

24.2.9.1 Published Model

First, you should search pdb.org (http://www.pdb.org) (the Protein Data Bank) to see if the molecule you wish to display has a published empirically-determined structure. The best way to search is by protein sequence (Advanced Search, Sequence). If you find a suitable model, be sure to write down the 4-character PDB code that uniquely identifies that model. The PDB code can simply be entered into Proteopedia's Molecular Scene Authoring Tools. Proteopedia will get the model automatically, directly from the PDB. Watch the relevant videos in the Video Guide (http://proteopedia.org/w/Proteopedia:Video_Guide) to see how easy this is, once you have a PDB code in hand.

24.2.9.2 Biological Unit

Often, the model published in the Protein Data Bank is not the "biological unit" or specific oligomeric form the molecule is believed to adopt when functional. However, it is easy to generate the biological unit and show it in Proteopedia: see Biological Unit: Showing (http://proteopedia.org/w/Biological_Unit:_Showing).

24.2.9.3 Hiding Parts of the Molecule

It is easy to hide parts of a molecule. In the Scene Authoring Tools, select the part(s) you wish to display. Then, under *selections*, click the button *invert current selection*. Then under *representations*, click *hide selection*. Now, invert the selection again, so you can color and display the visible part as you wish.

24.2.9.4 Uploading

If you have a custom model, not available (at least in the form you wish to use) from the Protein Data Bank, you can upload it into Proteopedia, and then use it in a molecular scene. Instructions: Help:Uploading molecules (http://proteopedia.org/w/Help:Uploading_molecules). Please note that if the model in question is available from the Protein Data Bank, there is no need to upload it. Simply request it using its PDB code when creating a molecular scene.

24.2.9.5 Homology Models

If there is no empirically-determined 3D structure for the protein sequence of interest, you may be able to create (and then upload) a homology model. For more about this, please see Homology modeling (http://proteopedia.org/w/Homology_modeling). An example of an article based on a homology model is Structure of E. coli DnaC helicase loader (http://proteopedia.org/w/Structure_of_E._coli_DnaC_helicase_loader).

24.2.10 Publishing Your Sandbox

Even while you are working on an article in a *Sandbox* page, any visitor to Proteopedia who searches for terms in your article will find your *Sandbox* page, and be able to read your work in progress.

Once you feel your article is reasonably complete, you can move it to a permanently titled page (see instructions, http://proteopedia.org/w/Help:Sandboxes#Publishing_Your_Completed_Article). By doing so, you are announcing that the page is reasonably complete, and inviting others to contribute additional information or to improve the portions you have written. Proteopedia, like all wikis, is intended to foster such collaborations. For example, see the many contributors to the article Avian Influenza Neuraminidase, Tamiflu and Relenza (http://proteopedia.org/w/Avian_Influenza_Neuraminidase%2C_Tamiflu_and_Relenza).

24.2.11 Other Help

Please consult http://proteopedia.org/w/Help:Contents for a list of help articles in Proteopedia. If you don't find an answer easily, don't hesitate to ask: contact@proteopedia.org

24.3 Section B. Proteopedia:Primer

Note for teachers: For teaching purposes, you may copy, adapt and distribute this document. Please remember to replace '_YOURSCHOOL' by some unique identifier for your group, so Sandboxes from other courses will not interfere. Download this Primer as a Word document at http://proteopedia.org/support/ProteopediaPrimer.doc

24.3.1 A Proteopedia Worksheet

24.3.1.1 Getting to Know Each Other

Use a web browser to access http://proteopedia.org. Proteopedia's Main Page should greet you. Take a few minutes to familiarize yourself with the Proteopedia layout.

Proteopedia has a top banner, a left hand side bar and a central area. The left hand side bar has three sections, from top to bottom: navigation, search, toolbox.

From the navigation area, remember the 'Help' link. It will become handy.

From the toolbox, the 'Export this page' link allows you to save a self-contained version of a page, complete with interactive 3D molecules and working green links. Excellent for exporting pages for display or lessons, even when you don't have internet access.

The search area has two text input boxes. The top one uses Proteopedia's own searching engine and also allows for jumping directly to any page, as long as you enter it's full name. Let's call this input area 'Proteopedia's search'. The bottom text input area uses Google's powerful search engine to find pages in Proteopedia related to the word or words you enter. Let's call this bottom slot 'Google's search'.

24.3.1.2 Logging into Proteopedia

Proteopedia allows free anonymous access, but editing and creation of pages is allowed only to registered users. To identify yourself to Proteopedia, click on the "Log in/request account" link on the top right hand side of the top banner.

Enter your Username, Password and click 'Log in'. For the purpose of this exercise, your instructor will provide you with a temporary Username and Password. You may also request your own Username by clicking on the "request one" link above the Username input field.

24.3.1.3 Creating Your First Proteopedia Page

Type into the Proteopedia's search slot the name of the page you want to create. For the purpose of this exercise, enter Sandbox_YOURSCHOOL## (where ## is a unique number you will get from the instructor) and click 'Go'. From now own we'll use the name 'Sandbox_YOURSCHOOL##', but, of course, you will see the name you composed yourself.

After you click 'Go', as expected, you will get a message reporting that 'There is no page with the exact title . . . ' and, a little further, a red link 'You can create a page titled Sandbox_YOURSCHOOL##'. Click the red link and Proteopedia will enter into editing mode. You should get a page entitled 'Editing Sandbox_YOURSCHOOL##' and a large central text input area, called the wikitext box. This is the area where you will enter the text for the page.

A little further down, you see the access link to SAT, the 'Scene authoring tools'. We will use them later on.

Then you will find three buttons: 'Save page', 'Show preview' and 'Show changes', that do exactly what they are named for. You can use 'Show preview' to evaluate the effect of the editing you are working on, and 'Save page' to permanently save the page, in its current state, into Proteopedia's database.

24.3.1.4 Let's Start with Some Text on a Clean Page

For this tutorial, we want to have a clean and fresh page, so please select anything you find in the text input area and delete it, before starting.

Now, type *enter* two or three times into the text input area, making several empty lines, and then 'Hello World, this is my Proteopedia page'. Now, click the button 'Show preview' at the bottom, and you will see, under the 'Editing...' line a preview of your page, with a clear label 'Preview', still displaying the large text input area down below.

We will now apply some boldface and italics. Scroll down the browser window, back to the wikitext box. Use the mouse to select (highlight) the word Proteopedia and click the button with the bold 'B'. Now, select (highlight) the words *Hello World*, and click the button with the *slanted I*. Click 'Save page' and behold: your first Proteopedia page is ready.

24.3.1.5 Next, a Rotating 3D Protein Structure

Click the tab [edit this page] located on the top banner. Click in the text input area/wikitext box, and position the cursor one or two lines below the text you entered before.

Note: please be careful when copying words in examples – file names, load = and scene = parameters, and PDB ids are case sensitive.

Click on the button labeled 3D. This button inserts a long line, but don't worry about understanding everything in it for now. We want to replace the selected text 'Insert PDB code or filename here' with the PDB id of a structure we'll use for testing, **1acj**. Once replaced, the line should start like this '<applet load='1acj 'size = ...'

Click 'Save page' and observe the result of your work. Your page should now contain text and a fully functional Jmol applet, displaying a rotating 3D representation of the protein structure of the Protein Data Bank entry 1acj. (For background information on 1acj, go to http://proteopedia.org/w/1acj, or simply open a new browser tab and enter 1acj into the search slot.).

Optional: You may edit the page again and play a little with the parameters for the < applet line. The value following size = sets the size in pixels for the applet; the text in quotes following caption = (*if any*) *will be displayed as caption of the applet, under the rotating structure*; *the align* = parameter can have values left, center, right, and will locate the applet on the page accordingly. Most of the time, align = 'right' is the best option.

24.3.1.6 Now a Static Image

Click the tab [edit this page] located on the top banner, to go back into editing mode. As before, click in the text input area/wikitext box, and position the cursor one or two lines below the text you entered. Now click the button with the picture, the sixth from the left. You should get this line [[Image:Example.jpg]], with the word *Example.jpg* highlighted. You will replace this *Example.jpg* with the actual name of a sample Image already in Proteopedia: MW_Folding_Simulations.gif

Alternatively, you may copy and paste the following line: [[Image:MW_Folding_Simulations.gif]]

Click 'Save page' and you will now have text, image and a 3D representation of 1acj.

24.3.1.7 Looking into a Structure More Closely

We will now start working with the 3D structure. But let's use 1pgb (http://proteopedia.org/w/1pgb), a simpler structure than 1acj (http://proteopedia.org/w/1acj).

Edit the page and replace the PDB id '1acj' with '1pgb'. You should have <applet load = '1pgb' ... and click 'Save page'. This loads and displays a smaller structure file. You may also want to replace the caption text with 1pgb (caption = '1pgb').

Click on 'Save page' and examine the structure. You may drag the structure with the mouse to rotate it. There are several accepted ways of representing and coloring a 3D structure to highlight different aspects related to functionality, structure, elements, etc. The current and default representation is called 'cartoon', where alpha helices are represented as helical ribbons and beta-strands as relatively straight ribbons. Each secondary structure element (helix or strand) has an arrowhead at one end, pointing from the N to the C terminus.

24.3.1.8 Using Color to Bring the Molecule to Life

Let's create a *green link* in your page to color the 3D structure with colors that emphasizes the N terminus to C terminus sequence of the whole structure.

(a) Enter the edit mode with the tab [edit this page].
(b) Below the text input box/wikitext box, click on [show] at *Scene Authoring Tools*.
(c) Click the 'load molecule' tab and type **1pgb** into the PDB code slot, and click the 'load' button closest to the PDB slot.
(d) Click the 'select all' button below the molecule.
(e) Click the 'colors' tab and there, the button 'N→C rainbow (named chain)'.

(f) Click the 'save scene' tab, and type 'N to C rainbow' in the 'Scene name' slot.
(g) Use the mouse to drag the molecule into a pleasing orientation. The green link will start in whatever orientation you have when you save the scene.
(h) Click the button 'save current scene'.
(i) Copy the <scene ...> ... </scene> that appears in the Wikitext box.
(j) Scroll up, and paste it into the main text input/wikitext box above.
(k) Above the Scene Authoring Tools, click [hide].
(l) Click 'Save page'.

Now you should see a green link *TextToBeDisplayed*. When you click this green link, the N→C rainbow color scheme will be applied to the molecule.

Let's insert a color key to explain this color scheme.

(a) Open a new browser tab and go to Proteopedia.Org.
(b) Type 'color keys' in the Search slot at the left side of the page and Enter.
(c) One of the pages found is Help:Color_Keys. Click on this.
(d) At the Color Keys page, click on 'N to C rainbows'. There, click on the link to 'DRuMS'.
(e) At DRuMS, click on Rainbows. Select one of the wikitext templates and copy it, such as {{Template:ColorKey_Amino2CarboxyRainbow}}
(f) Go back to your Sandbox page, and paste the color key template into the text input/wikitext box. Make sure you have double curly brackets at each end {{...}}.
(g) In your green link, change 'TextToBeDisplayed' to something that describes the scene, such as 'N to C sequence'.
(h) Save your page.

Now you have a green link that colors the protein ribbon trace with the N to C rainbow color scheme, and a color key that explains that scheme.

In a similar manner, you could follow the steps above to create a green link for a different color scheme, such as 'Secondary Structure'. Also you can paste in a template with a color key for the secondary structure color scheme.

By clicking on such a new green link, you can distinguish clearly the two main types of secondary structure in this model: alpha helix and beta strand. This scheme uses four different colors to distinguish four types of protein secondary structures (helices, beta strands and sheets, turns, and loops) and DNA vs. RNA.

24.3.1.9 Adding Some Explanations to the Page

Proteopedia's green links are much like the standard links on a HTML page, where you frame a word with a start/end tag to make it 'hot' and responsive to a mouse click. Try to enter some explanation around the last link you created. Enter the edit mode, and type some text before and after the green link and click 'Save page'. Here's a suggestion:

Let us color the two main forms of regular **secondary structure** in this protein.

24.3.1.10 Quiz Anyone?

Now, we will create a simple quiz for self-evaluation. Enter the edit mode, click in the text input area, to have the cursor located one or two lines below the text you entered and copy these lines:

<quiz display=simple>
{How many alpha helices are in this structure?
|type="[]"}
- None.
+ One.
- Four.
</quiz>

Save the page and have fun by testing it. More possibilities are explained at Help:Quiz (http://proteopedia.org/w/Help:Quiz)

24.3.1.11 This Is the END

This marks the end of this exercise. With it, you should be able to create a Proteopedia page with text, static images and 3D models of protein structures. You should be able to interact with the model and create quizzes for student evaluation.

A more complete description is available on Proteopedia.org, by clicking the 'Help' link on the top left hand-side of the screen.

24.3.1.12 ... and Beyond, the Real Fun

How to create your own scenes: The Scene Authoring Tool (SAT) is a unique and powerful Proteopedia feature for easily creating 3D scenes. To create additional and more complicated scenes, go to Proteopedia.org in a new web browser tab, click on 'Help' (top left) and then click on 'Proteopedia:DIY:Scenes' (http://proteopedia.org/w/Proteopedia:DIY:Scenes).

On your page, enter the edit mode with the tab [edit this page], click in the text input area, position the cursor located one or two lines below the text you entered, click on the '3D' button and type 1acj where the selected text states 'Insert PDB code or filename here'.

Click on [show] Scene authoring tools (SAT), and follow the step-by-step instructions on the Tacrine section of the Proteopedia:DIY:Scenes page.

An annotated applet: The line '<Structure... ' created when you click on the button 3D provides a rotating 3D structure that can be dragged and zoomed with the mouse, or popped up into a separate resizable window with the 'popup' button below the molecule.

You may also want to display an applet with complete Functional, Evolutionary and Structural information of a PDB file. The following line shows how.

{{STRUCTURE_1acj | PDB=1acj}}

Enter this line and replace '1acj' with the PDB id you want to display. Automatic data mining processes refresh the added information every week.

24.4 Section C. Advanced Proteopedia Authoring and Its Use in Teaching

The online version of this part is available at http://proteopedia.org/w/Proteopedia:Practical_Guide_to_Advanced_Proteopedia_Authoring_and_Its_Use_in_Teaching

24.4.1 Advanced Authoring and Uses

There are myriad ways to further your skills in authoring Proteopedia content; however, none is probably more helpful than just further exploring the site. Here we will briefly list here a few routes we hope you'll consider.

You can incorporate scrollable sections to your pages, for an example see http://proteopedia.org/w/Glutamate_receptor_(GluA2). Having earlier in this guide learned to cite references from PubMed using the PMID shortcut, you could learn to efficiently handle using the same citation multiple times on your page, see for example http://proteopedia.org/w/Help:Editing#Repeat_citations. Develop a page for presenting a structural paper in journal club, class presentation, or a lab group meeting. Using a restricted access workbench for development might be suitable for such efforts because collaborative development is allowed via this mechanism, which is not possible when authoring in one's own user space, see http://proteopedia.org/w/Workbench. As you can develop in private and then make the page public, workbenches are also useful for developing interactive 3D content in Proteopedia that complements published research work or review articles, see for examples http://proteopedia.org/w/Interactive_3D_Complement_in_Proteopedia.

Advanced efforts in authoring are also possible via the Scene Authoring Tools. You could alter the transition between your scene views, see http://proteopedia.org/w/Scene_authoring_tools. You could add custom interfaces for controlling the Jmol structure scene window using templates and Jmol commands, see for example http://proteopedia.org/w/Template:Button_Toggle_AnimationOnPause used on http://proteopedia.org/w/Citrate_Synthase. Related methods allow you to develop very complex animations that can be triggered via green scene link, see a list of morphs at http://proteopedia.org/w/Jmol#Complex_Animations. An advanced scene that can be developed is a morph. These show as the transition between two structural states as rotatable, interactive 3D structures and are particularly good for illustrating conformational changes, such as those involved in ligand binding; see http://proteopedia.org/w/Morphs#Morphing_Methods.

24.4.2 *Practical Guide for Teachers and Educators: Publishing Pages That Your Students Will Use to Explore Structural Biology*

24.4.2.1 Consider Protected Namespaces for Publishing Class Related Content

As a registered user you have a protected namespace for publishing within as was covered earlier in this primer when discussing personal sandboxes. Only the user owning that space can edit pages there while any user can view the page. Most likely anything you want to be available for a class should be authored in your protected user space. Otherwise you may find your page edited to a form not suitable for your purposes if you choose to leave it solely in the public space. Ultimately this would not be too problematic because your content as you wrote it would still exist in the page's history; however, placing it in your protected user space from the outset may save you from having to ponder which version as you stand before a audience of students and makes it easier to direct students to the page when outside of class. Much of the earlier parts of this guide, particularly the first section, will be applicable for steps in making such a protected page. Instead of making a personal sandbox though, you'd title the page as you see fit with a descriptive title that is preceded by 'User:' followed by your username. See Eric Martz's Nucleosomes page for example (http://proteopedia.org/w/User:Eric_Martz/Nucleosomes).

Keep in mind that you can author quizzes like those described earlier in this guide on your page.

24.4.2.2 Publish a Version of Your Class-Related Content to the Public Space as Well

Unless much of the content is already present in Proteopedia in some form, any content you produce may be useful to others outside of your class as well. Please consider publishing any content to the public user space in addition. Most often these pages are more readily found relevant in searches. Additionally, you will be attributed with your contribution at the bottom of the page and the page can serve as resource for others for learning or for further development. And others can edit the public version and improve the content. The version in your protected user space will still remain untouched though.

Copying a protected page into a public page is done by copying the entire wikitext of the protected page (from the wikitext box in edit mode), and then pasting it into the wikitext box of a newly-created public page. For example, the above page on nucleosomes, protected for class use, was copied into http://proteopedia.org/w/Nucleosomes. Subsequently many other people contributed to this page.

24.4.3 Practical Guide for Teachers and Educators: Choosing the Right Pages for Your Students to Work on in Proteopedia

24.4.3.1 Options

Beyond a way to convey structural biology to your students in a visually informative way, Proteopedia offers options for involving students at a wide-range of levels actively in educational ways. For example very advanced projects have emerged at both the University (see http://proteopedia.org/w/CBI_Molecules) and high school level (see http://proteopedia.org/w/Group:SMART:Teams). These are the exceptions rather than the rule. Most teachers use Proteopedia as a resource and engagement tool with the majority of content being produced meant to be temporary starting point if the students wish to take their efforts farther. Some have done so with tremendous outcomes, see http://proteopedia.org/w/Student_Projects and http://proteopedia.org/w/Proteopedia:News#Adoptions_in_University_Classes.

Considering the scope of this guide, we will be discussing using Proteopedia as a resource and engagement tool with an emphasis on students authoring and publishing pages. We will only be able to cover the basics here. We suggest Proteopedia's Teaching Strategies page http://proteopedia.org/w/Teaching_Strategies_Using_Proteopedia as a resource for additional information.

Going beyond being passive users of Proteopedia can empower students to be more knowledgeable about structural biology and also be better scientific communicators. Regardless of whether ultimately students produce anything of reasonable substance and quality that would add to Proteopedia, attempts at producing any sort of content, no matter how ephemeral, within Proteopedia often engages students in a manner that makes them think differently about how to consider structural data, extract meaningful insights, and share structural information productively with others. As with any class, the goals for an individual module of a course have to be factored against the overall pedagogical goals for the course and considered against the major limiting factors. Generally time of both the instructor and students is the major limiting factor. We hope that by outlining a few of the basic considerations here we will obviate many of the technical hurdles an educator might come across when using Proteopedia for teaching. Account type and namespace for your students to use within Proteopedia are some of the primary issues faced. As these two actually turn out to be linked by the permissions system built in to Proteopedia we will first touch upon accounts and then outline a number of page types suitable for places to have students author content within Proteopedia. We suggest exploring Proteopedia in order to familiarize yourself with them. To aid in this endeavor, after we discuss types of accounts we outline many of the types of pages in order to hopefully aid considering which of the possibilities best match with your pedagogical goals.

24.4.3.2 Individual Users or a Shared Student Account?

In order to publish, ultimately users need an individual account, and encouraging students to register as Proteopedia users sooner, rather than later, is in the best interest overall. However, as part of the efforts to control quality and prevent abuse, registrations are not handled in an automated manner; each registration is considered by a Proteopedia staff member and upon approval accounts are set up, which takes time. Typically, accounts are approved within 24 h of application. In a class this can often hamper efforts to smoothly transition from being a passive user of Proteopedia to being an author. If possible, encouraging your students to register well in advance of any class in which authoring may be involved and contacting the staff of Proteopedia ahead of such a class so they are expecting registrations can help mitigate such minor hiccups. Keep in mind though Proteopedia is an international, voluntary effort and sometimes delays cannot be avoided.

Importantly, educators who plan accordingly can contact Proteopedia staff (contact@proteopedia.org) to request special shared student accounts that can be used in classes to get students quickly authoring without having to wait for the registration process. A shared, student access account is not as fully privileged as an individual user account and only allows editing in the student userspace or sandboxes, either reserved or basic. In fact, if teaching using Proteopedia it is best to at least obtain the credentials for shared account access, as it may become useful to keep class moving productively. If using a shared account, be sure to keep in mind its limitations and you'd be well advised to use it ahead of the class to perform any planned exercises.

Teachers should point out that those students that are registered users are best served by signing in and working in their own account even if the shared account is available. As touched upon earlier in this guide, Proteopedia has mechanisms for tracking contributions and efforts by which the student will benefit by being logged into their own account.

24.4.3.3 Basic Sandboxes

Basic personal sandboxes were covered earlier in this guide from the point of view of someone beginning authoring in Proteopedia. While this type of page would be suitable for students work, a basic sandbox requires the users be logged in (either in a shared or a personal account) in order to edit pages. Additionally, for any assessment of the student's work the educator is required to learn the name of the particular page the student edited.

24.4.3.4 Reserved Sandboxes

A solution that avoids the issues associated with basic personal sandboxes is for the educator to set up a number of Reserved Sandboxes for students' work.

In conjunction with reserving these sandboxes, an educator can contact Proteopedia staff and request a student login account that can be shared with the class as described earlier. There is a page on Proteopedia (http://proteopedia.org/w/Special:SandboxReservation) where Reserved Sandboxes can be set up and populated with text at the time of set up. Because of this latter issue, it is best for the educator to have the content with which to seed the pages chosen at the time of request. This will save the educator from possibly needing to edit a lot of pages. It is suggested to look at examples of Reserved Sandboxes set up by other educators, preferably those not subsequently altered by students, when designing the seed content. It is suggested links to the individual pages in the block of reserved sandboxes be placed on a page where students are directed. As reserved sandboxes are numbered, a student or group of students can be assigned a specific page number by the Educator. It is suggested students be directed to add their name into the page proper as part of the editing process for the class, or else the specific assignments need to be recorded. Integrating the student's name into the text of the sandbox is especially important if using the option of the shared student account; unlike normal registered users, users logged into the shared student account do not have their individual names added to the contributors list at the bottom of every page in Proteopedia. Furthermore, it is suggested educators caution students to be careful to edit their own page; however, remind them that if issues should arise in the course of editing that most can be sorted out easily given the page history component of the Wiki. For example, if one student accidentally saves their page over another student's assigned page, there remains a record of the first student's page among the history of the page and the student or instructor can restore the content by copying and pasting if needed.

At the time the sandboxes are reserved, an expiration date is set and the pages may be cleaned out some time after that. If you or your students are planning to publish the content elsewhere (see below), they should at least copy the content to somewhere else in a reasonable time frame, prior to the agreed-upon expiration date.

24.4.3.5 Workbenches

Workbenches are pages that can be developed while hidden from the public. Only the author, and specified other account holders, can read Workbench pages. They are intended for development of pages that complement journal articles.

Eventually they can be converted to the public namespace, typically when the journal article is published, see http://proteopedia.org/w/Workbench. Because their set-up borders on being an advanced authoring task and relies on your students allowing you access for grading, Workbench pages are not suitable for teaching and are mainly included here to illustrate the logical progression between the concepts of Sandboxes and Studios.

24.4.3.6 Studios for Supervised, Private Group Collaboration and Page Development

A modified form of a Workbench, a Studio is a restricted access namespace where authorized users can work. In the current implementation, there's a landlord that creates the Studio page, for example Studio:G5SecL04 and grants access to a group of tenants, who can be students, TA's, and/or a mix of other registered Proteopedia users. Users that are either landlords or tenants. Each tenant can be given read only, or read and write privileges.

Studio shares with a Workbench the privacy properties, however, Studio tenants cannot grant access to the page to other users. The credits area at the bottom area of a Studio page will list all those tenants that actually did some editing, like a regular page. This is a nice complement to the workbench environment. And similar to the workbench environment, the studio can be converted to the public namespace at a later point and maintain the list of contributors to the page.

24.4.3.7 Where to Go Next?

Finally, you may wish that your students publish the content they have authored in Proteopedia. This can either be a requirement for a component of the grade or an option suggested as follow-up to particularly motivated students. In the latter case you may wish to direct the students to the earlier parts of this primer within Proteopedia to start thinking about best practices and publishing. If publishing is a requirement, we suggest significant involvement of the instructor at each step of the work to insure production of content meeting expectations, especially if this is to be published to the public user space. Excitingly, student users who have developed pages have even had articles published describing their efforts in a peer-reviewed journal, see for example http://proteopedia.org/w/Citrate_Synthase. This possibility is related to the joint effort between the journal *Biochemistry and Molecular Biology Education* (BAMBED) and Proteopedia to publish high-quality Proteopedia articles in BAMBED as features in a section titled ‘Multimedia in Biochemistry and Molecular Biology Education’, see http://proteopedia.org/w/Proteopedia:BAMBED [1–4, 6, 7, 9].

For additional ideas for where to proceed next, we also suggest exploring more within Proteopedia in order to maximize your use of Proteopedia as a component of teaching in the future. Moreover, we suggest Proteopedia's Teaching Strategies page http://proteopedia.org/w/Teaching_Strategies_Using_Proteopedia as a next step for additional information.

References

1. Canner D (2011) Proteopedia entry: HMG-CoA reductase. Biochem Mol Biol Educ 39(1):64. PMID: 21433257 doi:10.1002/bmb.20481
2. Decatur WA (2010), Proteopedia entry: the large ribosomal subunit of *Haloarcula marismortui*. Biochem Mol Biol Educ 38(5):343. PMID: 21567858 doi:10.1002/bmb.20444
3. Decatur WA, Eddelman DB (2011) Proteopedia entry: citrate synthase. Biochem Mol Biol Educ 39(3):229. PMID: 21618391 doi:10.1002/bmb.20519
4. Dornfeld CL, Hoelzer M, Forst S (2012) Proteopedia entry: beta-prime subunit of bacterial RNA polymerase. Biochem Mol Biol Educ 40(4):284. PMID: 22807435 doi:10.1002/bmb.20630
5. Hodis E, Prilusky J, Martz E, Silman I, Moult J, Sussman JL (2008) Proteopedia – a scientific 'wiki' bridging the rift between three-dimensional structure and function of biomacromolecules. Genome Biol 9(8):R121. Epub 2008 Aug 3. PMID: 18673581 doi:10.1186/gb-2008-9-8-r121
6. Oberholser K (2010) Proteopedia entry: Ramachandran plots. Biochem Mol Biol Educ 38(6):430. doi:10.1002/bmb.20457. PMID: 21567876 doi:10.1002/bmb.20457
7. Prilusky J, Hodis E (2012) Proteopedia entry: "tutorial: how we get the oxygen we breathe". Biochem Mol Biol Educ 40(5):339. PMID: 22987558 doi:10.1002/bmb.20646
8. Prilusky J, Hodis E, Canner D, Decatur W, Oberholser K, Martz E, Berchanski A, Harel M, Sussman JL (2011) Proteopedia: a status report on the collaborative, 3D web-encyclopedia of proteins and other biomolecules. J Struct Biol 23. PMID: 21536137 doi:10.1016/j.jsb.2011.04.011
9. Sagar A, Oberholser K (2012) Proteopedia entry: deoxyribonucleic acid (DNA). Biochem Mol Biol Educ 40:74. doi:10.1002/bmb.20566

Chapter 25
Proteolysis, Complex Formation and Conformational Changes Drive the Complement Pathways

Piet Gros and Federico Forneris

Abstract The complement system is an important part of the mammalian immune defense in blood and interstitial fluids. This set of ~30 plasma proteins and receptors enables the host to recognize and clear invading pathogens and altered host cells, while protecting healthy host cells and tissues. Over the last 7 years, we have resolved the structural details of the central components of this system, which is referred to as the Alternative Pathway of complement activation, and deduced the molecular mechanisms that underlie the amplification and regulation of this protein network. In short, we revealed that large domain-domain rearrangements of these multi-domain proteins, upon proteolysis and complex formation, determine the specificity that provides a local and brief burst to mark targets cells for immune clearance. Most recently, we and others have revealed structural details of the Terminal Pathway that leads to pore formation by Membrane-Attack-Complexes in cell membranes yielding lysis.

Keywords Immunology • Complement system • Proteolytic casade • Conformational changes • Large macromolecular complexes

25.1 Introduction

The complement system can be initiated through three main routes, either through the Classical Pathway (that may be mediated by antibodies forming immune complexes), the Lectin-mediated Pathway or through a 'tick-over' mechanism due

---

This text is based on the research highlights of the Gros lab, see www.crystal.chem.uu.nl/gros/researchhighlights.htm

P. Gros (✉) • F. Forneris
Crystal and Structural Chemistry, Bijvoet Center, Utrecht University, Utrecht, The Netherlands
e-mail: p.gros@uu.nl

R. Read et al. (eds.), *Advancing Methods for Biomolecular Crystallography*, NATO Science for Peace and Security Series A: Chemistry and Biology, DOI 10.1007/978-94-007-6232-9_25, © Springer Science+Business Media Dordrecht 2013

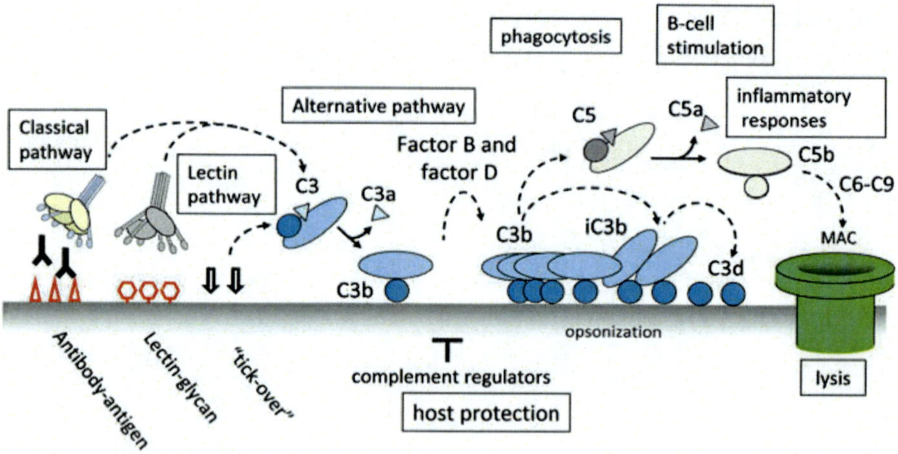

Fig. 25.1 Complement scheme

to a low level of hydrolysis of complement component C3 (for reviews see e.g. [4, 25]). The initiation routes converge in the proteolytic activation of the central protein C3; see Fig. 25.1. C3 is cleaved into C3a and C3b. C3b covalently binds to target surfaces through its reactive thioester moiety. As part of the Alternative Pathway, this signal is amplified by the generation of C3 convertases. This C3 convertase is formed by binding of pro-protease factor B to surface-bound C3b. Next, protease factor D cleaves factor B. This results in release of the pro-peptide fragment Ba yielding the C3b-Bb (or C3bBb) complex, which is the active C3 convertase of the Alternative Pathway that cleaves C3 into C3a and C3b, amplifying C3b attachment to the targeted surface. Surfaces covered by C3b, and subsequent proteolytic fragments iC3b and C3dg, are marked for destruction and clearance either through initiation of the Terminal Pathway of complement leading to cell lysis, binding to macrophages for phagocytosis or binding to B-cells stimulating the adaptive immune response.

The covalent binding of C3b is indiscriminate of self and foreign surfaces. Therefore, the host requires complement regulators to protect healthy host cells and surfaces [30]. This is a critical step as witnessed by the many mutations in regulators causing dysfunctional regulation and hence disease. Typical examples of disease are: age-related macular degeneration (AMD), atypical hemolytic uremic syndrome (aHUS) and membranoproliferative glomerulonephritis type II (MPGN-II) or dense-deposit disease (DDD) [3, 9, 25]. Complement regulators consist of strings of complement-control-protein (CCP) domains. They are either expressed on host cell surfaces (e.g., the membrane cofactor protein CD46/MCP) or are present in plasma and bind host cells and matrix specifically, e.g. the soluble, abundant regulator factor H (FH).

Activation of the Terminal Pathway of complement results in formation of membrane-attack-complexes (MAC) that form large (100 Å wide) pores in the target membrane [22]. These complexes are formed when C5 is cleaved into C5b by the C5 convertase. The C5 convertase is formed by association of one (or possibly more) C3b molecule to the C3 convertase. This somehow shifts the substrate specificity from C3 to its homologue C5. C5b binds subsequently C6, C7, hetero-trimeric C8αβγ and multiple copies of C9. C6 to C9 are homologous proteins build-up from a central MAC-perforin (MACPF) domain and several N- and C-terminal regulatory domains. Together these multi-domain proteins change from soluble proteins into a pore-forming complex that perforates the membrane.

25.2 Activation States of the Central Component C3

Human C3 is a large (1,641 amino-acid residues) protein with a remarkable structural arrangement of 13 domains formed by two protein-chains β and α [10]; see Fig. 25.2. The internal domain homologies indicate that this type of protein molecule evolved from a core of eight homologous "macroglobulin" (MG) domains, marking the beginning of a generic host defense mechanism more than 1,300 million years ago; well before the emergence of antibodies. Five other domains are attached to these eight MG domains: a linker and an anaphylatoxin (ANA/C3a) domain are inserted into MG6, a CUB and thioester domain (TED/C3d) are located in between MG7 and MG8, and a netrin-like C345C domain is attached via an anchor region to the C-terminus of domain MG8. In native C3 the reactive thioester is protected in two ways. First, the thioester is sequestered from water to prevent hydrolysis and second, domain-domain interactions prevent transformation of the reactive thioester into a highly reactive thiolate and acyl-imidazole intermediate [17].

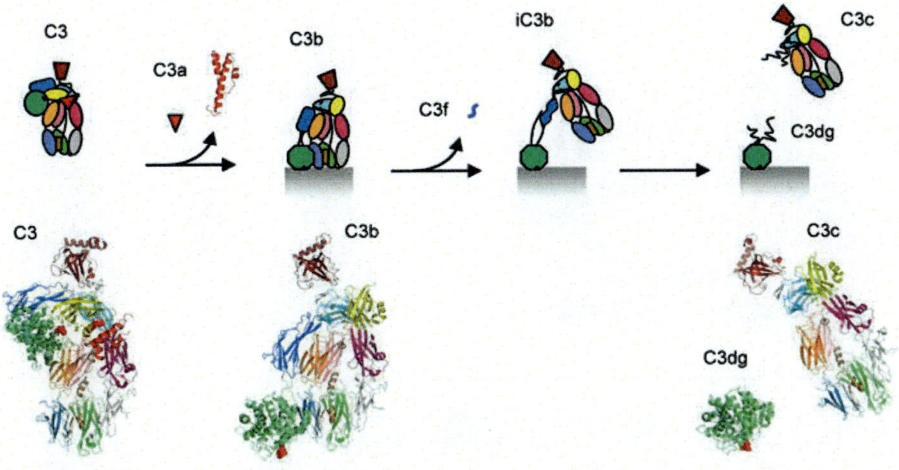

Fig. 25.2 Structures of C3 and its proteolytic fragments

Proteolytic activation of C3 produces anaphylatoxin C3a and opsonin C3b. The major fragment C3b changes its conformation markedly compared to that of native C3 [11, 35]; see Fig. 25.2. The thioester region moves over 85 Å, becomes fully exposed and is activated into the acylimidazole intermediate for reaction with hydroxyls on the targeted surfaces. Moreover, the rearranged structure now exposes previously hidden binding sites for pro-protease factor B and a variety of regulator proteins that function to protect healthy host cells and tissue.

25.3 Native Pro-enzyme Factor B

Factor B is the pro-protease that after assembly and proteolysis provides the serine-protease activity of the central C3 convertase complex (C3bBb) responsible for amplification of the complement response.

The structure of the pro-protease factor B [21] consists of three N-terminal complement-control-protein (CCP) domains, a linker helix αL (which together with the CCP domains forms the pro-peptide segment Ba), a central Von Willebrand A-type (VWA) domain and C-terminal serine protease (SP) domain (the two latter forming together the protease segment Bb); see Fig. 25.3. The VWA domain is homologous to the regulatory "inserted" (I) domains of integrins [32]. Similar to integrin I-domains, VWA has a metal-ion (Mg^{2+}) dependent adhesion site (MIDAS) critical for ligand (C3b) binding and a C-terminal helix α7 that may be involved in determining ligand-binding affinity at the MIDAS and transmitting conformational changes upon ligand binding [20]. In the pro-enzyme factor B, however, we observe a novel arrangement of these elements in the VWA domain. Helix α7 is displaced from its canonical groove by helix αL of the pro-peptide and the MIDAS, which is critical for C3b binding, is disrupted. Furthermore, the P1 arginine residue of

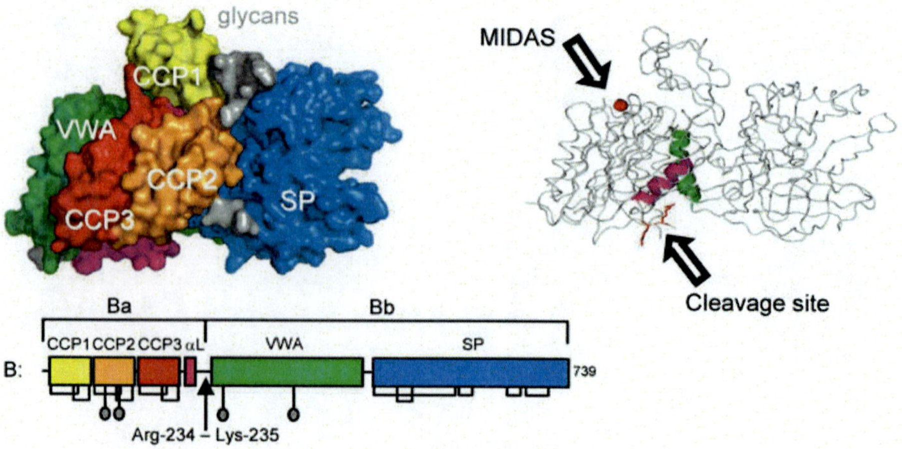

Fig. 25.3 Structure of factor B

the scissile bond (cleaved by factor D in activation of the pro-convertase C3bB) is located in a shallow cleft formed by the helices αL and α7 and forms salt bridges with both helices. These arrangements suggest a "locked" state for native factor B and that binding of C3b likely induces large conformational rearrangements of the helices αL and α7 that expose the scissile bond and make factor B susceptible to cleavage by factor D.

25.4 Convertase Formation

The C3 convertase of the Alternative Pathway is formed in two steps. First, pro-protease factor B binds in an Mg^{2+}-dependent manner to C3b yielding the pro-convertase C3bB. Second, protease factor D cleaves C3bB releasing the pro-peptide fragment Ba resulting in the active C3 convertase, C3bBb.

We first studied the pro-convertase formed by factor B and cobra-venom factor (CVF), which is a potent homologue of C3b [13]. The isolated structure of CVF [14] and the complex CVF-B showed that CVF is structurally homologous to C3b. Factor B binds CVF through two interfaces: one formed by the CCP domains of the pro-peptide segment Ba and one through the VWA domain of the protease segment Bb. The interactions with the Ba segment are apparently critical to "load" factor B onto CVF or C3b, because the Bb fragment alone does not bind to CVF or C3b. The Bb segment interacts with CVF through its MIDAS present in the VWA domain, where the carboxy terminus of CVF chelates the Mg^{2+} ion. Surprisingly, no overall domain rearrangement in factor B is observed upon binding to CVF. Notably, the scissile bond remains occluded as in the structure of native factor B. Negative-stain EM data indicated, however, conformational changes as factor B binds to C3b [33]. We hypothesized that CVF-B may represent a "loading" state whereas the EM data of C3bB represents an "activation" state of the pro-convertase, which can be cleaved by factor D. Subsequently, this hypothesis was supported by EM data revealing the co-existence of a closed and open state of the pro-convertase [34].

A crystal structure of the C3bB complex revealed a large conformational change in factor B [5]; see Fig. 25.4. Whereas the crystal structure of CVF-B correlated well with the EM data of the closed (loading) form [34], the structure of C3bB correlated well with the open (activation) form [33]. The conformational change involved an unexpected, marked, reorientation of the catalytic serine protease (SP) domain of factor B. This rotation partially extends the VWA-SP linker and unwinds the C-terminal α7 helix of the VWA domain. In conjunction, helix αL that precedes the scissile bond in factor B extends by two turns positioning hydrophobic residues in pockets that are vacated by the scissile loop. Together, these changes destabilize the binding of the P1 arginine (Arg234) to the αL and α7 helices, resulting in exposure of the whole scissile loop.

Factor D circulates in blood in an inactive conformation, in which the catalytic Ser-His-Asp triad is distorted and the P1-binding pocket is blocked by an arginine [23]. Factor D binds factor B specifically in the open/activation state of C3bB.

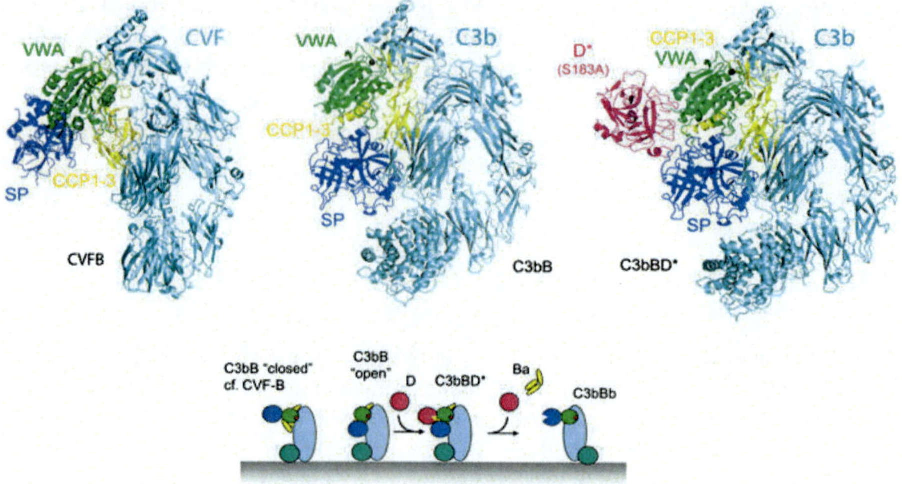

Fig. 25.4 Convertase formation

The interface formed between the "exosite" of factor D and both the VWA and SP domains of factor B explains the high affinity (Kd = 9 nM). Binding of factor D to C3bB activates factor D. Arg202, which blocked the P1-binding pocket, swings out and the self-inhibitory loop rearranges allowing the catalytic Ser-His-Asp triad to be restored [5]. What causes these conformational changes is not fully clear, though. Interaction between Glu230 of the scissile loop in factor B and Arg202 of factor D contributes to the substrate specificity and possibly provides a trigger for the conformational changes that activate factor D. Finally, factor D cleaves factor B and liberates the Ba fragment. The proteolytic fragment Bb, formed by the VWA and SP domains, remains bound to C3b yielding the active C3bBb protease complex.

25.5 C3 Convertase Activity and Specificity

C3 convertases are unstable complexes that dissociate irreversibly (with a half life time of ~90 s for C3bBb) and thus provide a local and brief burst of complement amplification.

The crystallographic studies of the labile C3bBb protease complex were facilitated by an immune evasion protein, called staphylococcal complement inhibitor (SCIN), which is secreted by *S. aureus*, that inhibits the C3 convertase, while stabilizing it [26, 27]. The crystal structure of the triple complex (C3b-Bb-SCIN) revealed a dimeric arrangement of convertases (C3bBb) stabilized by two bridging SCIN molecules [6, 27]; see Fig. 25.5. Two C3b molecules form the center of the dimer. The protease fragment Bb is bound to the C-terminal C345C domain of C3b

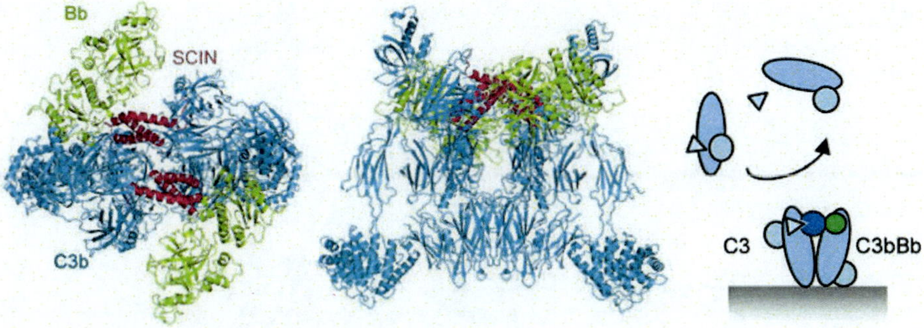

Fig. 25.5 The C3bBb-SCIN dimer

through its VWA domain, whereas the SP domain is oriented side-ways without making contacts to C3b. SCIN interacts with both C3b and Bb stabilizing the loose arrangement. The observed C3b:C3b contacts suggests a putative substrate:enzyme complex, consistent with earlier considerations based on inhibitor binding sites that block substrate binding [12]. Replacement of one C3b by native C3 positions the scissile bond of C3 in front of the catalytic site of the SP domain in Bb. A swing of the Bb fragment towards substrate would result in a productive orientation of the scissile loop in the active site, while the inhibitor SCIN prevents such a movement in the inhibited complex. Such a putative dimerization of the substrate C3 with C3b of the C3 convertase (C3bBb) explains the high specificity and activity of the central C3 convertases [27].

The C3b:C3 dimerization model for the enzyme:substrate (C3bBb:C3) complex is now confirmed. A dense-deposit disease mutant $C3_{923\Delta DG}$ cannot be cleaved by the C3 convertase; however, when activated to C3b by proteases this C3 mutant forms active convertases [19]. This is consistent with the position of the loop containing the deletion in C3 vs. C3b in the putative C3bBb:C3 complex. In addition, a crystal structure of CVF in complex with C3-homologue C5 reveals a CVF:C5 binding consistent with the C3bBb:C3 dimerization model [16].

25.6 Host Protection

Factor H (FH) is an abundant soluble regulator that protects tissues and cells with limited surface regulators. It consists of 20 CCP domains. The first 4 CCP domains are essential and sufficient for the functional activity, whereas the other domains CCP5-20 are involved in distinguishing self from foreign [24]. We have determined the structure of FH domains 1–4 with its target C3b [36]; see Fig. 25.6. FH(1–4) binds C3b in an extended arrangement forming a 100-Å long binding site. Comparison of this structure with that of C3bBb shows that domains CCP1-2

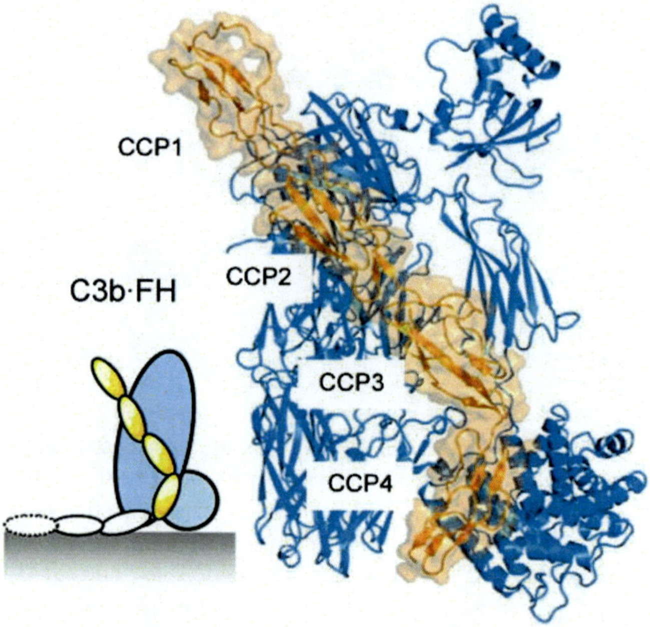

Fig. 25.6 The structure of C3b-FH(1–4)

of FH are directly involved in displacing Bb from C3b, which explains the "decay-accelerating activity" that breaks down C3 convertases and thereby stops the amplification and production of C3b. In a second mechanism, referred to as "cofactor activity", FH serves as a cofactor to bind the protease factor I (FI). FI cleaves C3b in the CUB domain yielding inactive iC3b that cannot bind factor B to form convertases and thus blocks complement amplification. Mutational data (in particular those of the homologue from variola virus done by Sahu and co-workers) suggest a role for CCP2-3 in binding FI. Domains CCP2-3 lie adjacent to the CUB domain of C3b and, thus, provide an appropriate binding site for FI to cleave C3b. A recent crystal structure of FI reveals an arrangement of its five domains that would putatively be consistent with binding of FI in between CCP2-3 of FH and C345C of C3b with the catalytic SP domain oriented towards the scissile bonds in the CUB domain of C3b [29].

25.7 Membrane-Attack Complex

The structure of the membrane-insertion (MACPF) domain of complement component C8α revealed a surprising structural resemblance to the bacterial cholesterol-dependent cytolysins [7, 28]. This suggests a common membrane perforation

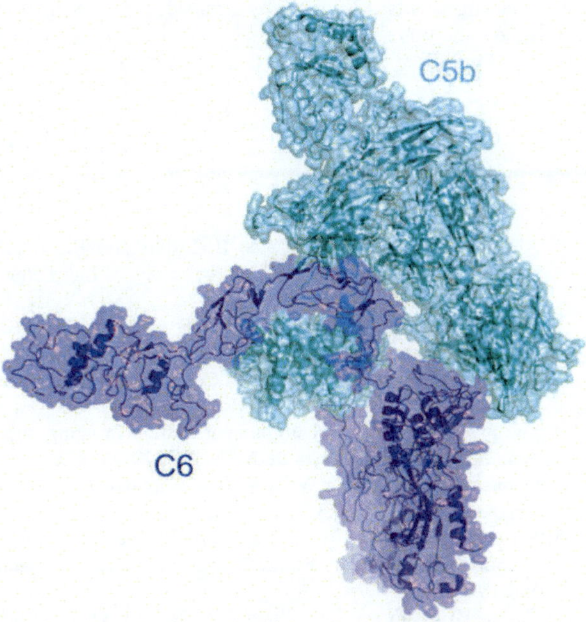

Fig. 25.7 Structure of the C5b6 complex

mechanism for MAC and perforin of the mammalian immune system and these bacterial pore forming proteins. Sodetz and co-workers resolved additional structures, C8α-MACPF-γ and that of the hetero-trimer C8αβγ, revealing both the interactions with the N- and C-terminal ancillary domains and those within the α-β MAC dimer of C8 [18, 31].

Next, we solved the structure of the C5b-C6 (C5b6) complex [1, 8]; see Fig. 25.7. This structure reveals that, when C5 is cleaved into C5b, large structural changes occur [15]. In part, these changes are similar to those of the C3 to C3b conversion. However, in the case of C5 to C5b the TED/C5d domain ends up in a position halfway the MG ring. This conformation of C5b is captured by C6. C6 consists of a core of domains similar to C8 (with an additional TSP domain at the N-terminus) and a C-terminal extension of two CCP domains and 2 FIMACs [2]. The core of C6 binds the bottom part of the MG ring of C5b. The C-terminal linker and CCP domains of C6 wrap around the TED of C5b. As a consequence the C6 MACPF domain with its pore-forming segments is positioned below C5b. In collaboration with the labs of Susan Lea (Oxford) and Oscar Llorca (Madrid) we placed the C5b6 crystal structure into the cryo-EM reconstruction map of the soluble MAC (sMAC or sC5b6-9) [8]. These data indicate that the arrangement of C5b-C6-C7-C8β-C8αγ-C9 yields an arc of the MAC proteins with a protrusion at the beginning formed by C5b. Below the arc large blobs of density indicate the likely presence of clusterin and vitronectin that enwrap the pore-forming segments, thereby providing protection to host cells to bystander damage.

Acknowledgements We gratefully acknowledge the help of many lab members and collaborators throughout the years and financial support from NWO-CW, NIH and ERC. Special thanks to Bert Janssen, Fin Milder, Jin Wu and Michael Hadders.

References

1. Aleshin AE, Discipio RG, Stec B, Liddington RC (2012) Crystal structure of c5b-6 suggests structural basis for priming assembly of the membrane attack complex. J Biol Chem 287:19642–19652
2. Aleshin AE, Schraufstatter IU, Stec B, Bankston LA, Liddington RC, Discipio RG (2012) Structure of complement C6 suggests a mechanism for initiation and unidirectional, sequential assembly of the Membrane Attack Complex (MAC). J Biol Chem 287:10210–10222
3. Anderson DH, Radeke MJ, Gallo NB, Chapin EA, Johnson PT, Curletti CR et al (2010) The pivotal role of the complement system in aging and age-related macular degeneration: hypothesis re-visited. Prog Retin Eye Res 29:95–112
4. Dunkelberger JR, Song WC (2010) Complement and its role in innate and adaptive immune responses. Cell Res 20:34–35
5. Forneris F, Ricklin D, Wu J, Tzekou A, Wallace RS, Lambris JD et al (2010) Structures of C3b in complex with factors B and D give insight into complement convertase formation. Science 330:1816–1820
6. Garcia BL, Ramyar KX, Tzekou A, Ricklin D, McWhorter WJ, Lambris JD et al (2010) Molecular basis for complement recognition and inhibition determined by crystallographic studies of the staphylococcal complement inhibitor (SCIN) bound to C3c and C3b. J Mol Biol 402:17–29
7. Hadders MA, Beringer DX, Gros P (2007) Structure of C8alpha-MACPF reveals mechanism of membrane attack in complement immune defense. Science 317:1552–1554
8. Hadders MA, Bubeck D, Roversi P, Hakobyan S, Forneris F, Morgan BP et al (2012) Assembly and regulation of the membrane attack complex based on structures of C5b6 and sC5b9. Cell Rep 1:200–207
9. Holers VM (2008) The spectrum of complement alternative pathway-mediated diseases. Immunol Rev 223:300–316
10. Janssen BJ, Huizinga EG, Raaijmakers HC, Roos A, Daha MR, Nilsson-Ekdahl K et al (2005) Structures of complement component C3 provide insights into the function and evolution of immunity. Nature 437:505–511
11. Janssen BJ, Christodoulidou A, McCarthy A, Lambris JD, Gros P (2006) Structure of C3b reveals conformational changes that underlie complement activity. Nature 444:213–216
12. Janssen BJ, Halff EF, Lambris JD, Gros P (2007) Structure of compstatin in complex with complement component C3c reveals a new mechanism of complement inhibition. J Biol Chem 282:29241–29247
13. Janssen BJ, Gomes L, Koning RI, Svergun DI, Koster AJ, Fritzinger DC et al (2009) Insights into complement convertase formation based on the structure of the factor B-cobra venom factor complex. EMBO J 28:2469–2478
14. Krishnan V, Ponnuraj K, Xu Y, Macon K, Volanakis JE, Narayana SV (2009) The crystal structure of cobra venom factor, a cofactor for C3- and C5-convertase CVFBb. Structure 17:611–619
15. Laursen NS, Gordon N, Hermans S, Lorenz N, Jackson N, Wines B et al (2010) Structural basis for inhibition of complement C5 by the SSL7 protein from *Staphylococcus aureus*. Proc Natl Acad Sci U S A 107:3681–3686
16. Laursen NS, Andersen KR, Braren I, Spillner E, Sottrup-Jensen L, Andersen GR (2011) Substrate recognition by complement convertases revealed in the C5-cobra venom factor complex. EMBO J 30:606–616

17. Law SK, Dodds AW (1997) The internal thioester and the covalent binding properties of the complement proteins C3 and C4. Protein Sci 6:263–274
18. Lovelace LL, Cooper CL, Sodetz JM, Lebioda L (2011) Structure of human C8 protein provides mechanistic insight into membrane pore formation by complement. J Biol Chem 286:17585–17592
19. Martinez-Barricarte R, Heurich M, Valdes-Canedo F, Vazquez-Martul E, Torreira E, Montes T et al (2010) Human C3 mutation reveals a mechanism of dense deposit disease pathogenesis and provides insights into complement activation and regulation. J Clin Invest 120:3702–3712
20. Milder FJ, Raaijmakers HC, Vandeputte MD, Schouten A, Huizinga EG, Romijn RA et al (2006) Structure of complement component C2A: implications for convertase formation and substrate binding. Structure 14:1587–1597
21. Milder FJ, Gomes L, Schouten A, Janssen BJ, Huizinga EG, Romijn RA et al (2007) Factor B structure provides insights into activation of the central protease of the complement system. Nat Struct Mol Biol 14:224–228
22. Muller-Eberhard HJ (1986) The membrane attack complex of complement. Annu Rev Immunol 4:503–528
23. Narayana SV, Carson M, el-Kabbani O, Kilpatrick JM, Moore D, Chen X et al (1994) Structure of human factor D. A complement system protein at 2.0 A resolution. J Mol Biol 235:695–708
24. Pangburn MK (2000) Host recognition and target differentiation by factor H, a regulator of the alternative pathway of complement. Immunopharmacology 49:149–157
25. Ricklin D, Hajishengallis G, Yang K, Lambris JD (2010) Complement: a key system for immune surveillance and homeostasis. Nat Immunol 11:785–797
26. Rooijakkers SH, Milder FJ, Bardoel BW, Ruyken M, van Strijp JA, Gros P (2007) Staphylococcal complement inhibitor: structure and active sites. J Immunol 179:2989–2998
27. Rooijakkers SH, Wu J, Ruyken M, van Domselaar R, Planken KL, Tzekou A et al (2009) Structural and functional implications of the alternative complement pathway C3 convertase stabilized by a staphylococcal inhibitor. Nat Immunol 10:721–727
28. Rosado CJ, Buckle AM, Law RH, Butcher RE, Kan WT, Bird CH et al (2007) A common fold mediates vertebrate defense and bacterial attack. Science 317:1548–1551
29. Roversi P, Johnson S, Caesar JJ, McLean F, Leath KJ, Tsiftsoglou SA et al (2011) Structural basis for complement factor I control and its disease-associated sequence polymorphisms. Proc Natl Acad Sci U S A 108:12839–12844
30. Sjoberg AP, Trouw LA, Blom AM (2009) Complement activation and inhibition: a delicate balance. Trends Immunol 30:83–90
31. Slade DJ, Lovelace LL, Chruszcz M, Minor W, Lebioda L, Sodetz JM (2008) Crystal structure of the MACPF domain of human complement protein C8 alpha in complex with the C8 gamma subunit. J Mol Biol 379:331–342
32. Springer TA (2006) Complement and the multifaceted functions of VWA and integrin I domains. Structure 14:1611–1616
33. Torreira E, Tortajada A, Montes T, Rodriguez de Cordoba S, Llorca O (2009) 3D structure of the C3bB complex provides insights into the activation and regulation of the complement alternative pathway convertase. Proc Natl Acad Sci U S A 106:882–887
34. Torreira E, Tortajada A, Montes T, Rodriguez de Cordoba S, Llorca O (2009) Coexistence of closed and open conformations of complement factor B in the alternative pathway C3bB(Mg2+) proconvertase. J Immunol 183:7347–7351
35. Wiesmann C, Katschke KJ, Yin J, Helmy KY, Steffek M, Fairbrother WJ et al (2006) Structure of C3b in complex with CRIg gives insights into regulation of complement activation. Nature 444:217–220
36. Wu J, Wu YQ, Ricklin D, Janssen BJ, Lambris JD, Gros P (2009) Structure of complement fragment C3b-factor H and implications for host protection by complement regulators. Nat Immunol 10:728–733

Chapter 26
Monoamine Oxidase Inhibitors: Diverse and Surprising Chemistry with Expanding Pharmacological Potential

Claudia Binda, Dale E. Edmondson, and Andrea Mattevi

Abstract The prominence of monoamine oxidases (MAO's) in pharmacology arose from initial findings in the 1960s that arylalkyl hydrazines, originally used to treat tuberculosis patients, exhibited a mood-elevating response by the irreversible inhibition of MAO. This finding sparked considerable investigations into various MAO inhibitors by both academic and pharmaceutical laboratories to develop drugs that could be used as anti-depressants and active research in this field continues. This chapter provides an account of the contribution given by structural studies in this field.

Keywords Neurotransmitter • Drug-design • Cofactor • FDA

26.1 Introduction

Monoamine oxidases (MAOs; [7]) are membrane-bound mitochondrial enzymes that are responsible for the metabolism of various amine neurotransmitters. Forty years of intense research (>20,000 papers in Pubmed) have produced seven FDA-approved antiMAO drugs that are used for the treatment of depression and Parkinson's disease. Most of the MAO inhibitors were developed before thorough chemical and structural biology was possible. So, their mechanism of function remained unclear despite their widespread clinical usage. In the past years, we have extensively investigated MAO inhibition including the elucidation of some 50 enzyme-inhibitor/drug complexes.

C. Binda • A. Mattevi (✉)
Department of Biology and Biotechnology, University of Pavia, Pavia, Italy
e-mail: andrea.mattevi@unipv.it

D.E. Edmondson
Departments of Chemistry and Biochemistry, Emory University, Atlanta, USA

R. Read et al. (eds.), *Advancing Methods for Biomolecular Crystallography*, NATO Science for Peace and Security Series A: Chemistry and Biology, DOI 10.1007/978-94-007-6232-9_26,

26.2 Currently Known Drugs

Such "*a posteriori*" biochemical and structural analysis has revealed that most of the MAO inhibitors and all FDA-approved antiMAO drugs function through formation of a covalent bond with the flavin cofactor (i.e. they are covalent inhibitors). Remarkably, these drugs are chemically diverse and form different types of covalent adducts with the flavin. To illustrate this point, we refer to three well-known classes of MAO inhibitors that we have extensively investigated in structural and biochemical terms in our laboratories.

1. The hydrazines (initially developed as antitubercular drugs) act as suicide substrates: they are oxidized the flavin and subsequently form a covalent adduct. However, the reaction is not fully coupled because the covalent adduct does not form in every catalytic cycle. Importantly, the mechanism of covalent inhibition, despite a 40-year long history for these compounds, is far from being fully defined [2].
2. The propargylamines are probably the most successful MAO inhibitors; a member of this class, rasagiline, has been approved by FDA as an anti-Parkinson's drug in 2005. These inhibitors form a covalent adduct with the flavin in a reaction that, unlike that for the hydrazines, is fully coupled – it apparently results from the direct covalent reaction of the inhibitor with the flavin.
3. Tranylcypromine (parnate) has been one the first antidepressive drugs ever used. This inhibitor features a cyclopropyl ring that forms a covalent adduct with the MAO flavin. The adduct, however, is different from that formed by the propargylamines and hydrazines forms through opening of the cyclopropyl ring [6] (Fig. 26.1).

Tranylcypromine features a further outstanding property: it is able to enhance the binding of a non-covalent group of MAO B inhibitors, the imidazolines, well known

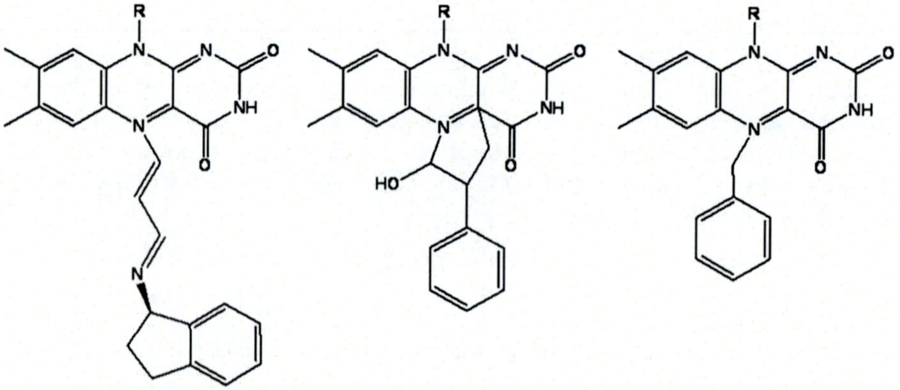

Fig. 26.1 Flavin adducts: rasagiline in MAO B, propargylamine in LSD1, and phenylhydrazine in MAO B

as antihypertensive agents. A sub-set of imidazoline class of compounds target MAOs and their binding is strongly (>100-fold) enhanced by tranylcypromine. The structural and biochemical analysis shows that tranylcypromine to MAO B causes some small conformational changes that favour binding of the imidazoline molecule in a cavity (so-called *entrance cavity*) that is adjacent but distinct from the substrate-binding site. This is a rare case of synergistic binding between two mechanistically and structurally distinct inhibitors and our studies provide one of the few cases in which simultaneous binding of two inhibitors to the same enzyme has been visualized.

26.3 Future Avenues in Drug Discovery

Most interestingly and relevant in pharmacologically terms, it is now becoming evident that the most promising and selective non-covalent MAO B inhibitors bind to the enzyme by occupying both the entrance and substrate cavities as highlighted by the tranylcypromine-imidazoline synergistic binding. An example of this notion is safinamide, a highly selective, high-affinity non-covalent MAO B inhibitor in advanced clinical trials as anti-Parkinson's drug [1]. Another intriguing example is pioglitazione, a very widely used anti-diabetes drug that has been found to exert a neuroprotective effect. We have found that pioglitazone is a selective non-covalent inhibitor of MAO B that exploits the entrance and substrate cavities for binding to the enzyme [5]. In this regard, pioglitazione provides an excellent example for the notion that known drugs can be re-purposed for different usages.

The knowledge on MAO inhibition mechanisms is acquiring much broader relevance with the discovery that MAO inhibitors or their analogues function also against a more recently discovered flavoenzyme, the histone demethylase LSD1 [8–10]. This enzyme was the first discovered histone demethylase and it is attracting considerable attention as possible target for epigenetic therapies against drugs. LSD1 has weak homology with MAOs but substantially different structure (including the active site) and biology. Many researchers have found that a few antiMAO drugs do inhibit also LSD1. These findings have at least three important implications (1) current antiMAOs have far-from optimal selectivity ("dirty drugs"); (2) current antiMAOs apparently exploit intrinsic chemical properties of protein-bound flavins rather than highly specific features in the binding sites; and (3) the chemical knowledge gained from MAO studies can be applied to other flavoenzymes. Indeed, several papers (also from our laboratory; [3, 4]) have been published in the last years that present various inhibitors targeting LSD1 (and now also LSD2, a newly found histone demethylase) which were developed based on the antiMAOs described in the previous paragraph. In our laboratories, a few compounds derived from tranylcypromine with partial selectivity for LSD1 were identified. The biological activity of one of these new inhibitors was evaluated with a cellular model of acute promyelocytic leukemia chosen since its pathogenesis includes aberrant activities of several chromatin modifiers. Marked effects on cell differentiation and an unprecedented synergistic activity with antileukemia drugs were observed. These data demonstrate that these LSD1/2 inhibitors are of potential relevance for the treatment of promyelocytic leukemia and, more generally, as tools to alter chromatin state with promise of a block of tumor progression. A confirmation of this notion from in vivo studies have been recently published. Thus, after more of 40 years of history, drug design and development studies on the "old" MAO enzymes continue to illuminate fundamental aspects of flavoenzyme biochemistry and are finding an explanding number of potential applications also by targeting MAO-related enzymes.

References

1. Binda C, Wang J, Pisani L, Caccia C, Carotti A, Salvati P, Edmondson DE, Mattevi A (2007) Structures of human monoamine oxidase B complexes with selective noncovalent inhibitors: safinamide and coumarin analogs. J Med Chem 50:5848–5852
2. Binda C, Wang J, Li M, Hubalek F, Mattevi A, Edmondson DE (2008) Structural and mechanistic studies of arylalkylhydrazine inhibition of human monoamine oxidases A and B. Biochemistry 47:5616–5625
3. Binda C, Aldeco M, Mattevi A, Edmondson DE (2010) Interactions of monoamine oxidases with the antiepileptic drug zonisamide: specificity of inhibition and structure of the human monoamine oxidase B complex. J Med Chem 54:909–912
4. Binda C, Valente S, Romanenghi M, Pilotto S, Cirilli R, Karytinos A, Ciossani G, Botrugno OA, Forneris F, Tardugno M, Edmondson DE, Minucci S, Mattevi A, Mai A (2010) Biochemical, structural, and biological evaluation of tranylcypromine derivatives as inhibitors of histone demethylases LSD1 and LSD2. J Am Chem Soc 132:6827–6833
5. Binda C, Aldeco M, Geldenhuys WJ, Tortorici M, Mattevi A, Edmondson DE (2012) Molecular insights into human monoamine oxidase B inhibition by the glitazone antidiabetes drugs. ACS Med Chem Lett 3:39–42
6. Bonivento D, Milczek EM, McDonald GR, Binda C, Holt A, Edmondson DE, Mattevi A (2010) Potentiation of ligand binding through cooperative effects in monoamine oxidase B. J Biol Chem 285:36849–53686
7. Edmondson DE, Binda C, Wang J, Upadhyay AK, Mattevi A (2009) Molecular and mechanistic properties of the membrane-bound mitochondrial monoamine oxidases. Biochemistry 48:4220–4230
8. Forneris F, Binda C, Battaglioli E, Mattevi A (2008) LSD1: oxidative chemistry for multifaceted functions in chromatin regulation. Trends Biochem Sci 33:181–189
9. Schenk T, Chen WC, Göllner S, Howell L, Jin L, Hebestreit K, Klein HU, Popescu AC, Burnett A, Mills K, Casero RA Jr, Marton L, Woster P, Minden MD, Dugas M, Wang JC, Dick JE, Müller-Tidow C, Petrie K, Zelent A (2012) Inhibition of the LSD1 (KDM1A) demethylase reactivates the all-trans-retinoic acid differentiation pathway in acute myeloid leukemia. Nat Med 18:605–611
10. Shi Y, Lan F, Matson C, Mulligan P, Whetstine JR, Cole PA, Casero RA, Shi Y (2004) Histone demethylation mediated by the nuclear amine oxidase homolog LSD1. Cell 119:941–953

Chapter 27
Structure of the Eukaryotic Ribosome: Tips and Tricks

Sergey Melnikov

Abstract Ribosomes are biological assemblies consisting of more than 50 different proteins and several thousands of RNA bases. Recently the structure of the eukaryotic 80S ribosome was determined by methods of X-ray crystallography. This chapter highlights several steps and ideas of the experimental approach that can be applied for a broader range of biological complexes. The biochemistry for the 80S ribosome project (purification and crystal treatments) was derived from procedures and principles developed for structural studies of labile plant complexes and resulted in elaboration of a rapid and gentle purification protocol. The structure determination of this giant enzyme implied model building of 3.3MDa of RNA and protein moieties: remodeling of ~2.3MDa residues of the conserved core, known from the structures of prokaryotic ribosomes, and *de novo* building of ~1MDa of eukaryote-specific components.

Keywords Eukaryotic ribosome • X-ray crystallography • Postcrystalization treatments • Radiation damage • Ribosomal proteins • Ribosomal RNA

27.1 Introduction

Here we discuss the principal steps of experimental procedure and several curious cases of model building underlying the recent structure determination of the eukaryotic ribosome – the 80S ribosome from yeast *S. cerevisiae* [5, 6].

S. Melnikov (✉)
Department of Integrated Structural Biology, Institut de Génétique et de Biologie Moléculaire et Cellulaire, Illkirch F-67400, France
e-mail: sergey@igbmc.fr

R. Read et al. (eds.), *Advancing Methods for Biomolecular Crystallography*, NATO Science for Peace and Security Series A: Chemistry and Biology,
DOI 10.1007/978-94-007-6232-9_27,

27.2 Sample Preparation

Crystallization of biological molecules (or complexes) is the first and most frequently crucial step of the structure determination. It largely relies on three major qualities of a biological sample: (1) amount, sufficient for crystallization trials, (2) purity from other biological molecules that may affect crystal growth and (3) conformational homogeneity, that may also interfere with crystallization and reduce diffraction quality of crystals. As ribosomes are one of the most abundant molecules in a living cell [18] and, thus, could be purified in a sufficient amount, the major issue of their isolation was:

27.2.1 How to Make Ribosomes Pure and Homogenous?

Purification procedure can be simplified by physiological treatments of living cells that cause homogenization of ribosomes *in vivo*. We used glucose starvation as a treatment that arrests protein biosynthesis and induces accumulation of a single ribosome population in *S. cerevisiae* cells (Fig. 27.1a) [2]. This approach allowed us to prevent association of ribosomes with various ligands (as tRNAs, mRNAs, translation factors). Surprisingly, we found that glucose starvation induces stoichiometric association of the 80S ribosomes with a stress-related protein Stm1 that occupies the mRNA- and tRNAs-binding sites on the ribosomes, thus hibernating ribosomes' activity under starvation conditions (Fig. 27.1b).

27.2.2 How to Minimize Contaminations and Degradation of Ribosomes After Cell Lysis?

Rapid and mild purification procedure aimed to decrease contamination and degradation is highly favourable for the quality of purified biological complexes. As eukaryotic cells contain 70S and 80S ribosomes (in mitochondria and the cytoplasm respectively), cell breaking procedure was used to keep mitochondria intact [14]. Furthermore, two modifications were introduced in the *S. cerevisiae* strain used for ribosomes purification. Firstly, the gene encoding for vacuolar proteinase A was deleted to prevent potential damage to ribosomal proteins when the protease is released upon cell breaking. Secondly, the L-A virus, commonly present in budding yeasts, was eliminated to prevent ribosomes contamination by viral particles known to have a similar sedimentation coefficient as the 80S ribosomes.

Another important step of the purification procedure was based on use of polyethylene glycol (PEG) [3, 4]. The Fig. 27.2 illustrates how PEG 20,000 can be used to rapidly concentrate the 80S ribosomes and to separate them from the total cell extract.

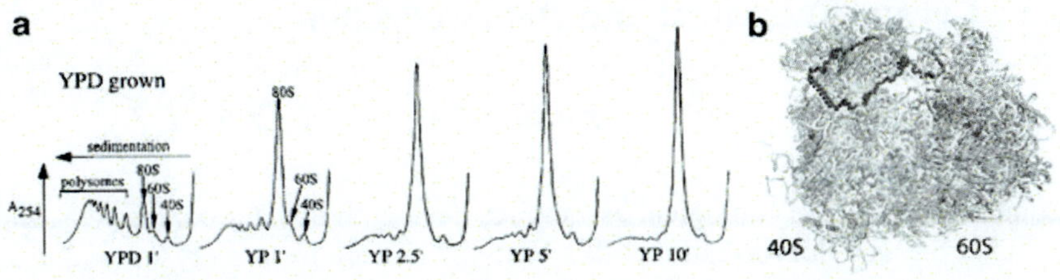

Fig. 27.1 (**a**) Time-lapse diagrams showing how transfer of yeast cells from normal growth media (*YPD*) to glucose-lacking media (*YP*) causes translation inhibition: in 10 min heterogeneous translating ribosomes (polysomes) turn into a single population of the 80S ribosomes (Adopted from Ref. [2]). (**b**) The 80S ribosomes purified from starved cells are associated with stress-related protein Stm1 (colored in *black*) that blocks ribosome access to mRNAs and tRNAs (Adopted from Ref. [6])

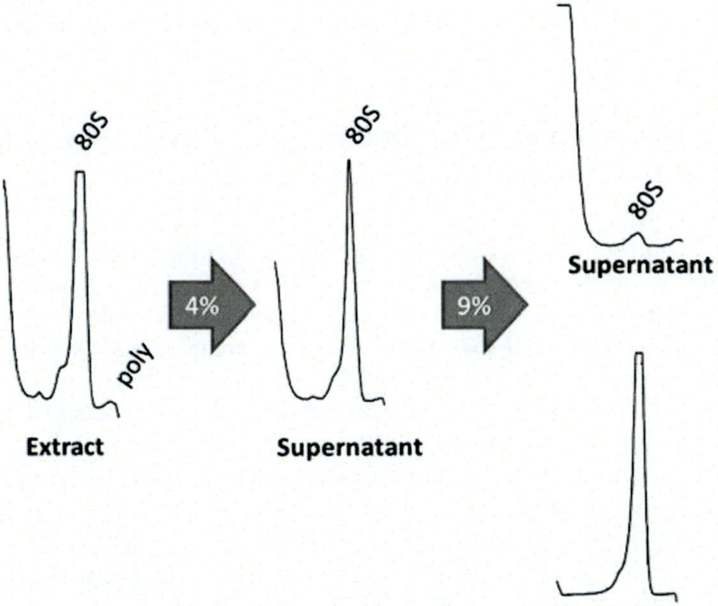

Fig. 27.2 Analytical sedimentation profiles (sucrose gradient centrifugation) illustrating principle of PEG-based purification (**a**) Total cell extract of starved *S. cerevisiae*. (**b**) Addition of PEG 20,000 up to 4 % (w/v) and subsequent centrifugation cause precipitation of polysomes and other components of a cell extract. (**c**) The further increase of PEG 20,000 concentration up to 9 % and centrifugation result in the 80S ribosomes precipitation, while other components of the extract are mainly remain in the supernatant. (**d**) The pellet of the precipitated 80S ribosomes can be dissolved and purified by sucrose gradient centrifugation. The peak corresponds to a highly homogenous pool of the 80S ribosomes

27.3 Crystal Treatment and Data Collection

27.3.1 *How to Improve Diffraction Quality of Crystals?*

Frequently, crystals of biological assemblies have a high water content that allows local disorder and conformational flexibility of molecules in a crystal and usually weakens diffraction qualities of a crystal. One of the most potent ways to deal with weak diffraction qualities of a crystal is dehydration – post-crystallization treatment, based on crystal's air drying or introducing chemical agents that reduce the water content [8, 12].

Crystals of the 80S ribosome were initially soft to the touch and poorly diffracting (fuzzy reflections maximum to ~20 Å resolution). However, step-wise transfer of these crystals in a solution, containing high amounts of glycerol (cryoprotectant) and PEG 6,000 (dehydrating agent) (Fig. 27.3) substantially improved their mechanical properties and diffraction quality (reflections up to 3 Å resolution).

27.3.2 *Use of Metal Ions Can Contribute Both to Phasing and to Improvement of Diffraction Quality*

Ribosomes possess a distinguishable feature that allows rapid ion exchange between their globules and a solution: ribosomal RNA folds in such a way that many cavities between RNA-helices are formed and filled with water molecules and ions. Numerous studies showed that one of the most common RNA-associated ions, fully hydrated magnesium ion $Mg(H_2O)_6^{2+}$, can be replaced by ions of osmium hexamine $Os(NH_3)_6^{3+}$ [7, 11] and increases diffraction properties of crystals if used for soaking or cocrystallization [9]. Introduction of $Os(NH_3)_6^{3+}$ ions – later used as an extra source of phase information, -at the last step of dehydration of

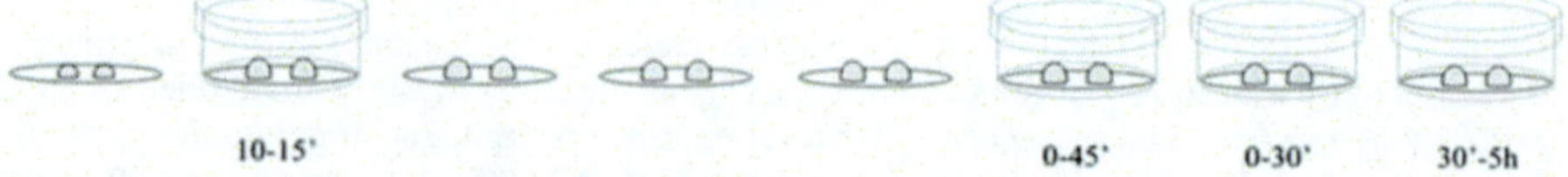

Fig. 27.3 Cover slips with crystallization drops were transferred into Petri dishes and iteratively mixed with dehydration solutions containing increasing amounts of PEG 6,000. *Numbers* in *brackets* (like 20 + 0) correspond to percentage of glycerol (*first number*, v/v) and PEG 6,000 (*second number*, w/v) in solution. "M" stands for 2 mM $Os(NH_3)_6Cl_3$. At the last step Petri dishes were closed with parafilm and kept at +4 °C before measurements

the 80S ribosome crystals provoked a conformational change of one of ribosomes in the asymmetric unit, shirked the unit cell parameters, and resulted in a further improvement of diffraction quality (reflections up to 2.5 Å resolution).

27.3.3 How to Optimize Data Collection?

Optimal data collection strategy relies on a good balance between accurate measurements of reflections' intensity and minimal radiation damage to crystals. To collect the final data set for the 80S ribosome structure determination we used the benefits of a new generation of detectors – single-photon counting detectors PILATUS 6M [13]. The frames of data were collected according to a strategy developed at the SLS that exploits the fast readout time and the absence of a readout noise of PILATUS 6M [15]: instead of maximum exposure time and maximum beam intensity we used highly attenuated beam (~7.5 % of the maximum intensity), shorter exposure time (0.5 compared to 1–2 s used for CCD-detectors) and reduced oscillation range (0.10° which is roughly ½ of a mosaicity value compared to 0.25–0.5°). It allowed us to collect up to 500 frames from a single spot of a crystal without significant radiation damage and scale these data with those from multiple crystals and multiple spots per each crystal. The final data set used to determine the 80S ribosome structure was represented by a highly redundant pool of data from 13 crystals: ~20,000 frames were collected giving the multiplicity equal to ~38 if Bijvoet pairs were merged. Position of ~1,360 anomalous scatters per asymmetric unit was defined and used as a source of experimental phases (Fig. 27.4).

27.4 Model Building

27.4.1 How to Improve Quality of the Maps?

Anomalous signal from $Os(NH_3)_6^{3+}$ ions allowed us to calculate occupancy values for each binding site with *Phaser* [16]. However, the resulting ($F_o - F_c$) electron density maps contained a lot of peaks overlapping with position of $Os(NH_3)_6^{3+}$ ions surrounded by a strong noise. This is due to the fact that some sites were occupied partially by $Os(NH_3)_6^{3+}$ and partially by $Mg(H_2O)_6^{2+}$ ions. As $Mg(H_2O)_6^{2+}$ ions, practically lacking anomalous signal and not modelled in the structure, contributed to electron density and, thus, to the positive difference in ($F_o - F_c$) maps. To take into account $Mg(H_2O)_6^{2+}$ contribution we refined occupancies of $Os(NH_3)_6^{3+}$ sites with *Phenix* [1] that resulted in accurate ($F_o - F_c$) maps and simplified model building of neighbouring RNA and protein moieties.

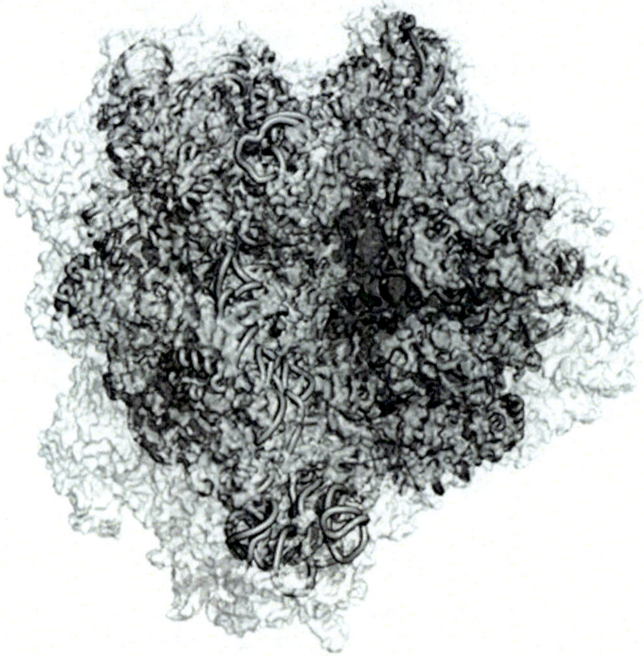

Fig. 27.4 The starting point for the 80S modeling: structures of the 30S and 50S (shown as *ribbons*) used for molecular replacement in the middle of the full-size 80S ribosome (semitransparent surface). The 80S ribosome contained a layer of previously unmodeled RNA and protein moieties

27.4.2 What Features of the 80S Structure Made Model Building More Complicated?

Eukaryote-specific ribosomal proteins and proteins' extensions frequently form intricate networks of interactions where they are involved in multiple contacts with numerous neighbours (Fig. 27.5). It makes modelling especially difficult if proteins contain flexible loops and tails so that the exact protein identity and registry could not be deduced from continuity of the electron density and relies on a solution of the whole puzzle.

Furthermore, there are several examples of ribosomal proteins that are located in unexpected areas of the 80S ribosome. For example, protein rpL24e, component of the large ribosomal subunit, contains C-terminal domain that is associated with the small subunit (Fig. 27.6). Two domains of rpL24e are connected with a flexible linker so that the identity of rpL24e C-terminus was deduced only when the structure of the 40S subunit was completely built and the extra density corresponded to the size and secondary structure predictions for rpL24e while the density contained a few distinguishable blobs of bulky amino acids.

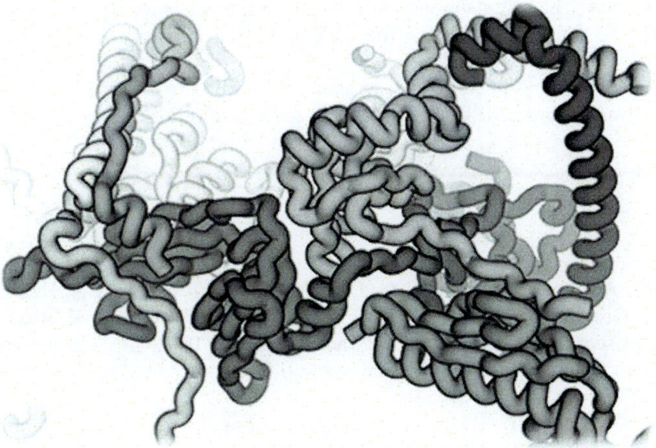

Fig. 27.5 Intricate networks of protein-protein interactions make the process of protein assignment puzzling. Seven proteins are shown at the central protuberance of the 60S subunit of the 80S ribosome

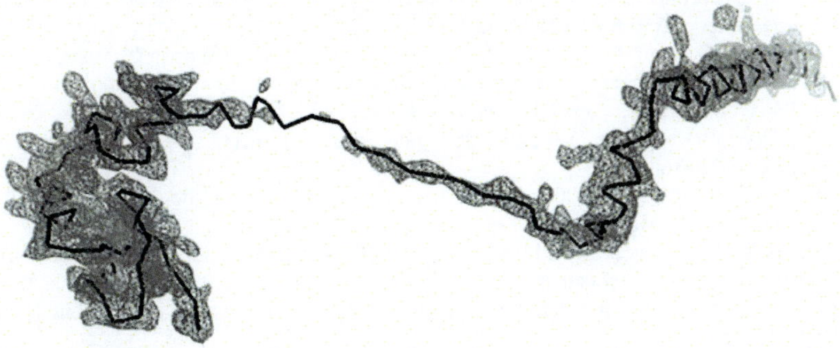

Fig. 27.6 Unexpected location of proteins within the 80S ribosome structure exemplified by protein rpL24e. *Black ribbon* shows rpL24e backbone path. The ($2F_o - F_c$) electron density map around the ribbon shows that two domains of rpL24e are connected by a linker barely resolved in the X-ray structure that makes it difficult to assign the part on the *right side* of the figure as a C-terminus of rpL24e

27.4.3 *How to Use Specific Protein or RNA Features to Simplify Model Building?*

We used several sources of data to properly assign a particular region of the electron density map to a particular protein or RNA. Among them were tools for secondary [10] and tertiary structure predictions [17], Christmas tree-like structure of alpha-helices and Zn-finger domains present in a few ribosomal proteins.

Acknowledgments The original protocol of the 80S ribosome purification and crystal treatments was elaborated by Adam Ben-Shem. The data collection strategy was suggested by Clemens Schulze-Briese and Marcus Mueller, beam scientists at the SLS, x06sa. The work on the 80S structure determination has been greatly assisted by the authors of original publications and was supported by awards from EMBO, HFSP, CNRS, ANR and SPINE2.

References

1. Adams PD, Afonine PV, Bunkoczi G, Chen VB, Davis IW, Echols N, Headd JJ, Hung LW, Kapral GJ, Grosse-Kunstleve RW, McCoy AJ, Moriarty NW, Oeffner R, Read RJ, Richardson DC, Richardson JS, Terwilliger TC, Zwart PH (2010) PHENIX: a comprehensive Python-based system for macromolecular structure solution. Acta Crystallogr D Biol Crystallogr 66(Pt 2):213–221
2. Ashe MP, De Long SK, Sachs AB (2000) Glucose depletion rapidly inhibits translation initiation in yeast. Mol Biol Cell 11(3):833–848
3. Ben-Shem A, Frolow F, Nelson N (2003) Crystal structure of plant photosystem I. Nature 426(6967):630–635
4. Ben-Shem A, Nelson N, Frolow F (2003) Crystallization and initial X-ray diffraction studies of higher plant photosystem I. Acta Crystallogr D Biol Crystallogr 59(Pt 10):1824–1827
5. Ben-Shem A, Jenner L, Yusupova G, Yusupov M (2010) Crystal structure of the eukaryotic ribosome. Science 330(6008):1203–1209
6. Ben-Shem A, Garreau de Loubresse N, Melnikov S, Jenner L, Yusupova G, Yusupov M (2011) The structure of the eukaryotic ribosome at 3.0 Å resolution. Science 334(6062):1524–1529
7. Cate JH, Gooding AR, Podell E, Zhou K, Golden BL, Kundrot CE, Cech TR, Doudna JA (1996) Crystal structure of a group I ribozyme domain: principles of RNA packing. Science 273(5282):1678–1685
8. Chayen NE, Saridakis E (2008) Protein crystallization: from purified protein to diffraction-quality crystal. Nat Methods 5(2):147–153
9. Clemons WM Jr, Brodersen DE, McCutcheon JP, May JL, Carter AP, Morgan-Warren RJ, Wimberly BT, Ramakrishnan V (2001) Crystal structure of the 30S ribosomal subunit from Thermus thermophilus: purification, crystallization and structure determination. J Mol Biol 310(4):827–843
10. Cole C, Barber JD, Barton GJ (2008) The Jpred 3 secondary structure prediction server. Nucleic Acids Res 36(Web Server issue):W197–W201
11. Doudna JA (2005) Chemical biology at the crossroads of molecular structure and mechanism. Nat Chem Biol 1(6):300–303
12. Heras B, Martin JL (2005) Post-crystallization treatments for improving diffraction quality of protein crystals. Acta Crystallogr D Biol Crystallogr 61(Pt 9):1173–1180
13. Kraft P, Bergamaschi A, Broennimann C, Dinapoli R, Eikenberry EF, Henrich B, Johnson I, Mozzanica A, Schleputz CM, Willmott PR, Schmitt B (2009) Performance of single-photon-counting PILATUS detector modules. J Synchrotron Radiat 16(Pt 3):368–375
14. Lang B, Burger G, Doxiadis I, Thomas DY, Bandlow W, Kaudewitz F (1977) A simple method for the large-scale preparation of mitochondria from microorganisms. Anal Biochem 77(1):110–121
15. Mueller M, Wang M, Schulze-Briese C (2012) Optimal fine phi-slicing for single-photon-counting pixel detectors. Acta Crystallogr D Biol Crystallogr 68(Pt 1):42–56
16. Read RJ, McCoy AJ (2011) Using SAD data in Phaser. Acta Crystallogr D Biol Crystallogr 67(Pt 4):338–344
17. Soding J, Biegert A, Lupas AN (2005) The HHpred interactive server for protein homology detection and structure prediction. Nucleic Acids Res 33(Web Server issue):W244–W248
18. Warner JR (1999) The economics of ribosome biosynthesis in yeast. Trends Biochem Sci 24(11):437–440

Chapter 28
Neutron Protein Crystallography. How to Proceed the Experiments to Obtain the Structural Information of Hydrogen, Protons and Hydration in Bio-macromolecules

Nobuo Niimura

Abstract NPC takes the next step beyond the folding structure and gives information about the hydrogen bonding, protonation states, and hydration configuration of the biomolecule—all of which are critical features for understanding how it actually functions. This unique information makes NPC a valuable tool for structural biology. The basic principles for analysis of the structure of the biomolecules are that the calculated structure factor, *Fc*, is subtracted from the observed structure, *Fo*, the result should include only the positions of the missing hydrogen atoms. One of the most difficult problems in NPC is to obtain a single crystal that is large enough to obtain diffraction results with the low flux of a neutron beam. A large single crystal can be grown in the metastable region in the crystallization phase diagram. The NPC of RNase A has been given as a walk-through example.

Keywords Neutron protein crystallography • Hydrogen bonds • Protonation states • Hydration • Crystallization phase diagram • Metastable region • H/D exchange

28.1 Why Neutron Protein Crystallography?

The application of neutron diffraction to the study of bio-macromolecules is known as neutron protein crystallography (NPC), in keeping with the term X-ray protein crystallography (XPC) for the technique that uses X-ray diffraction. There are two key differences between these techniques. The first difference is that neutrons are

N. Niimura (✉)
Frontier Research Center for Applied Atomic Sciences, Ibarak Quantum Beam Research Center, Ibaraki University, Sirakata 162-1, Tokai-mura, Ibaraki Prefecture 319-1106, Japan
e-mail: niimura@mx.ibaraki.ac.jp

R. Read et al. (eds.), *Advancing Methods for Biomolecular Crystallography*, NATO Science for Peace and Security Series A: Chemistry and Biology,
DOI 10.1007/978-94-007-6232-9_28, © Springer Science+Business Media Dordrecht 2013

much more difficult to produce than are X-rays, in terms of both cost and particle flux. Indeed, there are presently only about five locations in the world that can conduct NPC experiments, although more facilities are being constructed. Even so, the fluxes from these facilities are about 10 orders of magnitude smaller than the fluxes from X-ray sources, a fact that forces NPC experiments to take much more time (7–100 days) than XPC experiments take (30–60 min). Therefore, NPC experiments should be performed only when they can provide unique information.

The statistics of the Protein Data Bank (PDB) bear out the consequences of this difference in cost and ease of use. Since the beginning of the PDB, the number of folding structures registered in the PDB has increased exponentially year by year; by January 2012, it was close to 80,000. Most of these structures have been determined by XPC or by nuclear magnetic resonance (NMR). In contrast, the number of structures determined by NPC is very small (less than 50 as of January 2012). Thus, it could be said that NPC has not substantially contributed to the determination of the folding structures of biological macromolecules.

Although NPC is not used for determining folding structures, the second difference between NPC and XPC is the main reason for using NPC: Neutrons can be strongly diffracted by hydrogen nuclei (i.e., protons). X-rays are not diffracted by atomic nuclei but by atomic electron density, and since hydrogen has only one electron, X-rays are not strongly diffracted by hydrogen or its isotopes, deuterium and tritium. Therefore, neutrons can provide unique information about hydrogen atoms, protonation states, and hydrogen orientations in water molecules.

In general, about half of the constituent atoms of a protein are hydrogen, and in principle, all the hydrogen atoms (protonation states) can be identified by neutron diffraction. So, the purpose of NPC is not to determine the folding structure of biomolecules but rather the structure that is "beyond the folding structure." In other words, by identifying the hydrogen atoms in and around the biomolecule, NPC takes the next step beyond the folding structure and gives information about the hydrogen bonding, protonation states, and hydration configuration of the biomolecule—all of which are critical features for understanding how it actually functions. This unique information makes NPC a valuable tool for structural biology.

The general subject of NPC has been reviewed by several authors and here only recent ones were given. [2, 4, 5].

28.2 The Specific Features of the Interactions Between Neutrons And Bio-Macromolecules

The relationship between energy and wavelength is important for the radiation damage of protein crystals. Because a neutron is a baryon particle with a significant mass (1.0087 amu) but an X-ray is a photon particle (an electromagnetic wave) with no mass, the relationship between the energy and the wavelength is very different for each particle. The energies of a 1 Å X-ray and a 1 Å neutron are about 12.4 keV and

Table 28.1 Neutron scattering lengths and X-ray atomic scattering factors

Neutron scattering lengths and X-ray atomic scattering factors		
Atom	Neutron b_{coh} (10^{-12} cm)	X-ray $f_{x\text{-}ray}$ (10^{-12} cm)
D+	0.67	0
D	0.67	0.28
H	−0.37	0.28
H(−+) (anti-parallel neutron and proton spins)	−4.7420	0.28
H(++) (parallel neutron and proton spins)	1.0817	0.28
C	0.67	1.69
N	0.94	1.97
O	0.58	2.25
P	0.513	4.20
S	0.29	4.48

82 meV, respectively. Since the energy of a 1 Å neutron is less than the energies of chemical bonds (about several eV), neutrons do not cause radiation damage to proteins. Therefore, it is not necessary in NPC experiments to freeze crystals to liquid-helium or liquid-nitrogen temperatures to avoid radiation-induced damage. NPC can determine the protein structure at ambient temperatures, such as 20 °C or so. The same crystal which has been used for a neutron diffraction experiment can be measured later in an X-ray diffraction experiment to provide the initial model for NPC.

Table 28.1 shows the neutron scattering lengths and X-ray atomic scattering factors of some elements which constitute proteins. Distinctive features of neutrons are summarized as follows:

One technical issue that must be pointed out is that hydrogen atoms have very large incoherent-scattering cross sections (80 barns). The nature of the incoherent scattering causes a high background in the diffraction pattern, and in order to avoid it, it is recommended that hydrogen atoms should be replaced by deuterium, because the incoherent-scattering cross sections of deuterium atoms are very small (2 barns). In order to perform this replacement, which can be partial or complete, hydrogenated protein crystals can be soaked in heavy water (D_2O; *D* stands for deuterium) solutions. In addition, we can obtain the H/D exchange ratio of exchangeable hydrogen atoms in proteins. Otherwise, cells can be grown in D_2O to obtain fully deuterated proteins and crystals can be grown in D_2O solutions.

The b_H of hydrogen atom is negative. At medium or low resolution (>2.0 Å), the Fourier map of a methylene group ($-CH_2-$) nearly vanishes the carbon and two hydrogen atoms are merged in such a resolution because the sum ($b_C + 2{*}b_H$) becomes close to zero.

The b_N of a nitrogen atom is comparatively large. The hydrogen atoms of the $C(NH_2)_3$ group of arginine, the NH_2 group of lysine, and the NH group of amide are often replaced by deuterium atoms and become $C(ND_2)_3$, ND_2, and ND when the crystal is soaked in D_2O. They are easily visible in Fourier map.

The b_S of a sulfur atom is comparatively small; therefore, the –S–S– bonds are rather difficult to be observed. Number of different variations on the starting template, the overall success rate can be increased significantly.

28.3 The Basic Principles for Analysis of the Structure of Single Crystals

The analysis of crystal structures using neutron-diffraction data is the same as the analysis using X-ray-diffraction data, except that the scattering element is the nucleus in neutron diffraction, whereas the scattering element is the electron in X-ray diffraction.

Several methods have been developed to solve the phase problem for the case of X-ray diffraction. For the case of NPC, the folding structure of the protein is usually already known from XPC analysis, and the task is simply to locate the hydrogen atoms. Thus, at the start of the NPC analysis, the positions of most of the atoms in the protein molecule are known, including all of the carbon, nitrogen, oxygen, and heavier atoms and also some of the non-exchangeable hydrogen atoms. This knowledge enables the calculation of a first approximation of $F(hkl)$. In Fig. 28.1, the variable Fc refers to this calculated structure factor. Fo refers to the observed

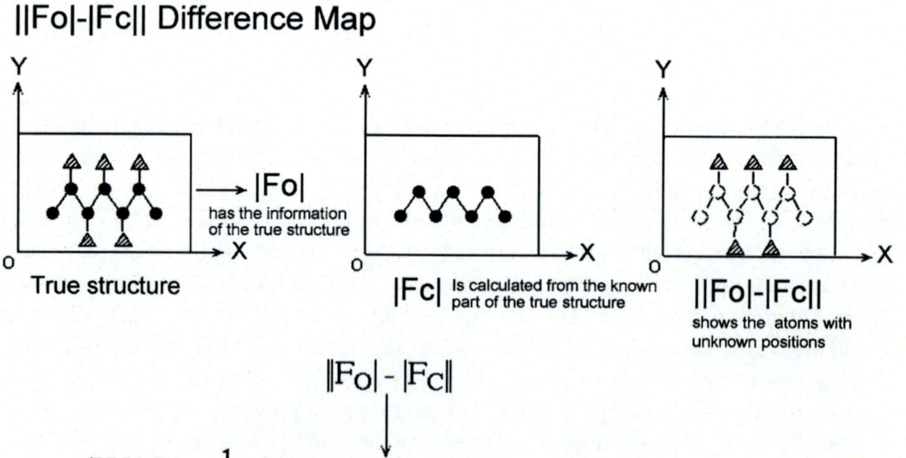

$$\rho(XYZ) = \frac{1}{V}\sum_h\sum_k\sum_l |Fo(hkl)|\cos 2\pi\{(hX + kY + lZ) - \alpha(hkl)\}$$

$$\alpha(hkl) = \frac{B(hkl)}{A(hkl)}$$

Fig. 28.1 Scheme of the use of difference maps to obtain the position of hydrogen atoms structure factor, which includes the information of all hydrogen atoms. If the calculated structure factor, *Fc*, is subtracted from the observed structure, *Fo*, the result should include only the positions of the missing hydrogen atoms. This result is called an omit map or a (*Fo* – *Fc*) difference Fourier map. Figure 28.1 illustrates this calculation. In practice, some hydrogen atoms do not appear in the first trial. Therefore, the newly found hydrogen atoms are incorporated into the model, a new calculated structure factor (*Fc*) is generated, and another omit map is made, which then reveals more hydrogen atoms. This iterative process is continued, and gradually the number of determined hydrogen atoms increases until most or all of them are found.

The refinement program of neutron protein crystallography is modified on the basis of X-ray protein crystallography of CNS1.1. The most important modifications are in the topology and parameter files because hydrogen atoms are included there. The general refinement proceeded as follows:

1. A positional minimization and B-factor refinement for all the non-hydrogen atoms of the initial model were performed.
2. After the initial refinement, putative non-exchangeable hydrogen atoms were added to the initial model.
3. An omit map was calculated for each amino acid residue; and exchangeable D atoms were then added to the model if corresponding nuclear densities were seen.
4. An occupancy refinement was then carried out only for the peptide amide H/D atoms to determine the H/D exchange ratios.
5. The model was fit to (2*Fo* – *Fc*) and (*Fo* – *Fc*) maps using the program *XtalView*.
6. D_2O molecules were added to the structure if the nuclear density could accommodate them. For small densities, only water oxygen atoms were located.
7. The protonation states of side chains, orientations of methyl groups and water positions were confirmed by calculating a (2*Fo* – *Fc*) map and an omit map for each amino acid residue and water molecule.

These refinement cycles were repeated to obtain the final structure.

28.4 Crystallization

One of the most difficult problems in NPC is to obtain a single crystal that is large enough to obtain diffraction results with the low flux of a neutron beam. Often, a crystal with a volume of several cubic millimeters (mm^3) is required. To grow such a large single crystal of a protein, a seed crystal is placed into a metastable growth environment, in which new crystals do not nucleate but protein molecules still crystallize on the seed crystal. The problems, then, are to determine this metastable region in the crystallization phase diagram (CPD) and to maintain the environment within those parameters. Figure 28.2 shows a typical, protein Crystallization Phase Diagram (CPD). The CPD consists of two regions, the undersaturated and supersaturated regions, which are separated by a solubility

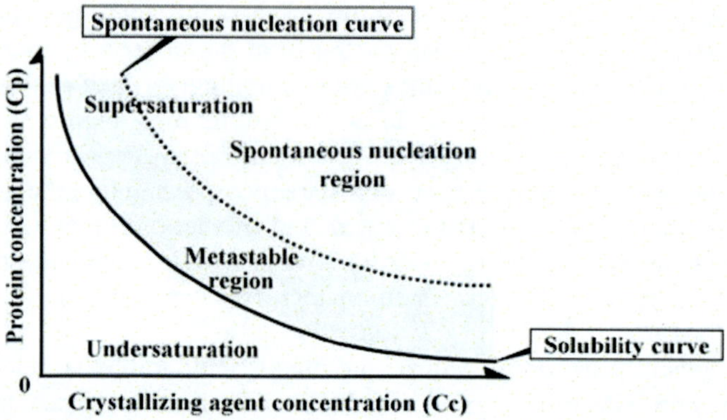

Fig. 28.2 A typical protein crystallization phase diagram (*CPD*)

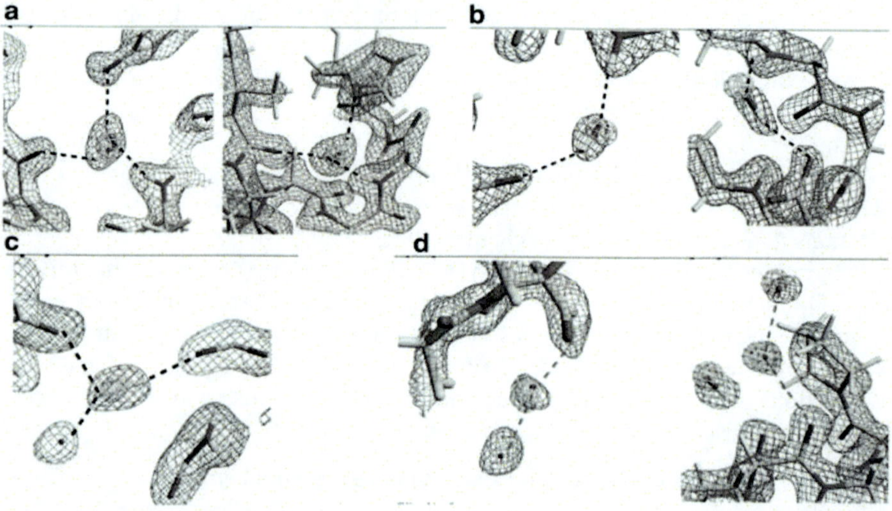

Fig. 28.3 Typical examples of the shapes of a water molecule: (**a**) triangular, (**b**) short ellipsoidal, (**c**) long ellipsoidal, (**d**) spherical

curve (SC). Furthermore, the supersaturated region is divided by the spontaneous nucleation curve (SNC), the dotted line in Fig. 28.3 into the metastable region and the spontaneous nucleation region. The SNC is also known as the supersolubility curve and as the nucleation border line. Consequently, the CPD consists of three regions, as follows: (i) The undersaturated region is a solution phase, and crystals dissolve in this region. (ii) In the metastable region, nucleation of new crystals does not occur, but a seed crystal will grow if it is placed there. (iii) In the spontaneous nucleation region, both crystal growth and nucleation occur.

A large single crystal can be grown in the metastable region. In the supersaturation region beyond the SNC, many nucleation seeds appear, and the protein molecules present in the solution are consumed by the growth of many tiny crystals. On the other hand, when a seed crystal is put in the metastable region, the protein molecules are concentrated onto this crystal, because new nucleation does not occur in the metastable region. Thus, the determination of the CPD is necessary to know the location and boundaries of the metastable region, so that large single crystals can be grown.

28.5 Walk-Through Example: RNaseA

28.5.1 Data Collection and Processing

The protein RNase A crystallizes in the monoclinic space group $P2_1$ has a unit cell with parameters $a = 30.38$ Å, $b = 38.56$ Å, $c = 53.40$ Å, and $\beta = 105.78°$ [6]. The neutron diffraction experiment was carried out at room temperature (*1) using the BIX-4 single-crystal diffractometer installed at the JRR-3 reactor of the Japan Atomic Energy Agency. A step scanning method with an interval of 0.3° was used to collect data, and the exposure time (*2) was 30 min per frame. After 608 frames were collected, the crystal rotation axis was changed by about 90° to a different rotation axis, and 603 more frames were collected. The intensities of reflections were integrated and scaled using the programs *DENZO* and

Table 28.2 Statistics for the data processing and structure refinement of RNase A

Crystallographic data		Refinement	
Wavelength (Å)	2.6	Initial model (PDB ID)	1KF4
Exposure time/frame (min.)	30	Resolution (Å)	80–1.4 (1.49–1.40)
Space group	$P2_1$	R_{cryst} (%)[a]	19.3 (33.0)
Unit cell dimensions (Å, °)	$a = 30.38$, $b = 38.56$ $c = 53.40$ $\beta = 105.78$	R_{free} (%)[b]	23.6 (36.1)
Temperature (K)	293	No. non-hydrogen atoms	1,043
Resolution (most outer shell) (Å)	80–1.4 (1.49–1.40)	No. H atoms	783
No. total frames	1,211	No. D atoms	385
No. observed reflections	31,649	No. DOD molecules	84
No. unique reflections	15,039	No. water molecules (O form)	8
Redundancy	2.1		
Completeness (%)	63.9 (23.7)		
R_{merge} (%)	7.1(20.5)		

[a] $R_{merge} = \Sigma_h \Sigma_i |I_{hi} - <I_h>| / \Sigma_h \Sigma_i |I_{hi}|$
[b] $R_{cryst} = \Sigma |F_o - F_c| / \Sigma |F_o|$

SCALEPACK. The statistics for the data processing and the structure refinement are summarized in Table 28.2. Although the outermost shell (1.45–1.40 Å) has only 20 % completeness because of the geometrical restrictions of the detector and limited machine time, it has significant diffraction, with an average $I/\sigma(I)$ of 3.07. Therefore, all the diffraction data up to 1.4 Å resolution were included in the refinement, to increase the data/parameter ratio. The effective resolution was estimated to be 1.7 Å, by analyzing the observed structure factors. A total of 15,039 independent reflections were obtained with an overall R_{merge} of 8.6 % from 31,649 observed reflections.

28.5.2 Structure Analysis and Refinement

Randomly selected 5 % reflections were assigned as a test set for cross-validation. The refinement was carried out using the program *CNS*1.1, which was modified for neutron diffraction studies. The refinement procedure was as follows: First, a positional-minimization, B-factor refinement for all the atoms was performed; occupancy refinement was then carried out only for the peptide amide hydrogen atoms, to determine the H/D exchange ratios. The model was fit to $(2Fo - Fc)$ and $(Fo - Fc)$ maps using the program *XtalView*. The refinement cycles were repeated to obtain the final structure. The X-ray structure of RNase A that was solved at 1.1 Å (PDB ID:1KF4, [1]) was used as the initial model. After initial refinement, putative, non-exchangeable H atoms were put onto the initial model. An omit map was calculated for each amino acid residue; then exchangeable D atoms were added to the model if there was scattering length density. The peptide amide hydrogen atoms were introduced to the structure as alternate conformations; then the occupancy factors were refined. (*3) Afterward, D_2O molecules were added to the structure, if the scattering length density could accommodate them. For weak densities, only the oxygen atoms of the water molecules were located. At the end of the refinement, 92 water molecules (D_2O: 84, O: 8) were included in the model. (*4) The protonation states of side chains, the directions of methyl groups, and the positions of water molecules were confirmed by calculating a $(2Fo - Fc)$ map and an omit map for each amino acid residue and water molecule. The final values of R_{cryst} and R_{free} were 19.5 and 23.8 %, respectively, for 15,029 unique reflections up to a resolution of 1.7 Å. The coordinates and structure factors have been deposited asPDB ID :3A1R.

(*1) To hold the protein crystal within the neutron beam and to keep it from drying out during the experiment, a capillary tube made of quartz glass is usually used. The crystal is placed inside the capillary tube, along with a sufficient amount of liquid (either deuterated crystallization solution or heavy water, according to the crystal stability) to keep the crystal from drying out. The tube is then tightly sealed. Unlike the capillary tubes used with X-ray-diffraction experiments, the ones used with neutron-diffraction experiments should be made out of only quartz (SiO_2) glass, rather than ordinary glass. This distinction is necessary because ordinary glass sometimes contains neutron-absorbing elements, such as Li and B, whereas quartz glass is completely transparent to the neutron beam.

(*2) Because access to the neutron source is usually limited in duration for a given experiment, it is important to estimate beforehand the necessary and sufficient neutron irradiation time to collect all the Bragg reflections for a crystal.

(*3) If the normalized nuclear scattering length density value at a particular peptide group's hydrogen site in the Fourier map is denoted by *Ex*, then the H/D exchange ratio, *X*, for that peptide group is related to *Ex*. By

$$X = \frac{Ex - b_H}{b_D - b_H}$$

Here, b_H and b_D are the neutron scattering lengths of hydrogen atom and deuterium atom, which are -0.375×10^{-12} and 0.667×10^{-12} cm, respectively.

(*4) The soaking ordinary protein crystals in D_2O will quickly replace all of the H_2O in the crystal with D_2O, regardless of whether or not any hydrogen in the protein is replaced. Thus, the water molecules are D_2O. With this fact in mind, three types of water molecules can be identified in nuclear scattering length density maps [3] (Fig. 28.3): (i) Boomerang (triangular) type water molecules. In this case, the water molecule is ordered, and a clear signal appears for the oxygen atom and both deuterium atoms. (ii) Short, ellipsoidal (stick) type water molecules. This type of density signal appears when the oxygen and one deuterium atom are ordered, but the second deuterium atom is disordered. (iii) Long, ellipsoidal (stick) type water molecules. This type of density signal appears when the two deuterium atoms are ordered, but the oxygen atom is disordered. (iv) Sphere-like water molecules. This type of density appears when only the oxygen atom or only one deuterium atom is ordered.

28.6 Summary

The NPC can provide unique information about hydrogen atoms, protonation states, and hydrogen orientations in water molecules. The analysis of NPC data is the same as the one of XPC except that the scattering element is the nucleus in neutron diffraction, whereas the scattering element is the electron cloud in X-ray diffraction. Therefore, those, who have carried out the XPC experiments, can do NPC after learning the differences between them. However, the NPC experiment is still a time-consuming process. The most important task in the NPC experiment to overcome the problem is to grow a large single crystal. Finally, we must never forget that NPC experiments should be performed only when they can provide unique information on hydrogen, protons, and hydration in Bio-macromolecules.

References

1. Berisio R et al (2002) Atomic resolution structures of ribonuclease A at six pH values. Acta Crystallogr D 58:441–450
2. Blakely M (2009) Neutron macromolecular crystallography. Cryst Rev 15(3):157–218
3. Chatake T et al (2003) Hydration in proteins observed by high-resolution neutron crystallography. Proteins 50:516–523
4. Niimura N, Bau R (2008) Neutron protein crystallography: beyond the folding structure of biological macromolecules. Acta Crystallogr A 64:12–22
5. Niimura N, Podjarny A (2011) Neutron protein crystallography. Hydrogen, protons, and hydration in bio-macromolecules, vol 25, IUcr monographs on crystallography. Oxford University Press, Oxford. ISBN 978-0-19-957886-3
6. Yagi D et al (2009) A neutron crystallographic analysis of phosphate-free ribonuclease A at 1.7 Å resolution. Acta Crystallogr D 65:892–899

Chapter 29
Coherent Diffraction and Holographic Imaging of Individual Biomolecules Using Low-Energy Electrons

Tatiana Latychevskaia, Jean-Nicolas Longchamp, Conrad Escher, and Hans-Werner Fink

Abstract Modern microscopy techniques are aimed at imaging an individual molecule at atomic resolution. Here we show that low-energy electrons with kinetic energies of 50–250 eV offer a possibility of overcome the problem of radiation damage, and obtaining images of individual biomolecules. Two experimental schemes for obtaining images of individual molecules – holography and coherent diffraction imaging – are discussed and compared. Images of individual molecules obtained by both techniques, using low-energy electrons, are shown.

Keywords Phase problem • Individual molecule • Coherent diffraction imaging • Holography • Low-energy electrons

29.1 Introduction

Investigating the structure of biomolecules at the atomic scale has always been of utmost importance for healthcare, medicine and life science in general, since the three-dimensional shape of proteins, for example, relates to their function. At the moment, these structural data are predominantly obtained by X-ray crystallography, cryo-electron microscopy and NMR. Despite there being an impressive database address (www.pdb.org) [35] obtained with these methods, they all require large quantities of a particular protein. This leads to averaging over fine conformational details in the recovered structure. The goal of modern imaging techniques is to visualize *an individual biomolecule* at atomic resolution.

T. Latychevskaia (✉) • J.-N. Longchamp • C. Escher • H.-W. Fink
Physics Institute, University of Zurich, Winterthurerstrasse 190, 8057 Zurich, Switzerland
e-mail: tatiana@physik.uzh.ch

R. Read et al. (eds.), *Advancing Methods for Biomolecular Crystallography*, NATO Science for Peace and Security Series A: Chemistry and Biology, DOI 10.1007/978-94-007-6232-9_29,

29.2 Imaging an Individual Molecule: Choice of Radiation

A direct visualization of an individual molecule at Ångstrom resolution can be achieved using electron or X-ray waves which have a wavelength of about 1 Å, see Fig. 29.1. Although both, X-rays and high-energy electrons possess sufficiently short wavelengths to resolve the individual atoms constituting a protein, the resolution achieved is mainly limited by radiation damage inherent to both types of radiation.

29.2.1 Imaging with High-Energy Electrons (80–200 keV)

In cryo-electron microscopy [1], cooling the sample to the temperature of liquid nitrogen allows a higher electron dose to be used for the same amount of radiation damage. Depending on the resolution required, typical electron exposures vary between 5 and 25 e/Å^2 [12]. Due to the very low signal-to-noise ratio in the images obtained, over 10,000 images of individual molecules typically need to be collected and averaged to arrive at the reconstruction of the structure [32].

29.2.2 Imaging with X-rays

Visualization of an individual molecule at atomic resolution by employing X-rays is planned at the X-ray Free Electron Lasers (XFELs) facilities which are being developed worldwide. Here the radiation damage problem [13] is circumvented by employing ultra-short X-ray pulses, which allow the diffraction pattern of an

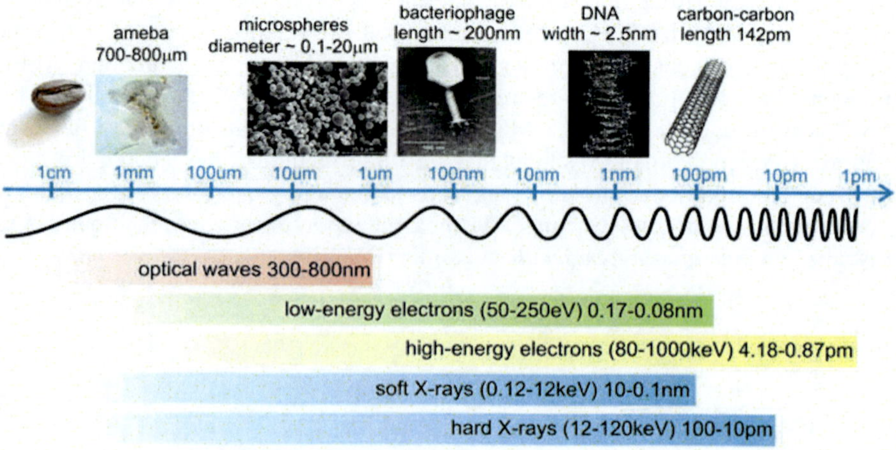

Fig. 29.1 Spectrum of radiation used for imaging. The bars show the range of the sizes of the objects which can be imaged with the assigned radiation

individual molecule to be recorded before it decomposes due to the strong inelastic scattering [2, 23]. High intensity X-ray laser pulses will provide the intensity in the diffraction pattern detected at the high scattering angles, which is required for further numerical recovery of the molecular structure at high resolution.

29.2.3 Imaging with Low-Energy Electrons (50–250 eV)

Low-energy electrons (with kinetic energies of 50–250 eV, corresponding to wavelengths of 0.78–1.73 Å) can be employed to visualize individual biomolecules directly. It has been shown [9] that individual DNA molecules can withstand low-energy electron radiation having energy of 60 eV (corresponding to a wavelength of 1.58 Å) for at least 70 min. This in total amounts to a radiation dose of 10^6 e/Å^2, which is at least six orders of magnitude larger than the permissible dose in high-energy electron microscopy or X-ray imaging.

29.3 Imaging an Individual Molecule: The Phase Problem

The principle of lensless imaging of an individual molecule is as follows: when a coherent wave is scattered by a molecule, it carries both, amplitude and phase information imposed by the scattering event. The phase distribution is especially important since it carries information about the position of the atoms constituting the molecule. However, detectors are not sensitive to the phase information; instead they just record the intensity which is the square of the wave amplitude. Hence, the recovery of the complex-valued scattered wave requires a solution to the so-called *phase problem*. Today there are two known solutions to the phase retrieval problem: holography and coherent diffraction imaging (CDI), both schematically shown in Fig. 29.2. Their proper implementation would ultimately allow the atomic mapping of an individual molecule in three dimensions.

In Fig. 29.3, the experimental set-ups for both, holographic and CDI recording with low-energy electrons designed and built in our laboratory [5, 30] are sketched.

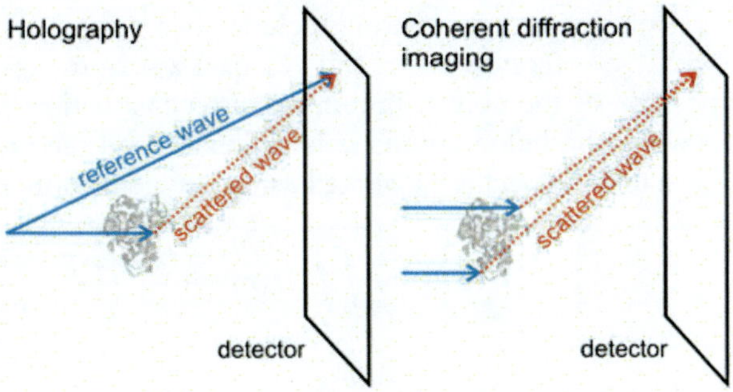

Fig. 29.2 Schematics of the lensless imaging of an individual molecule

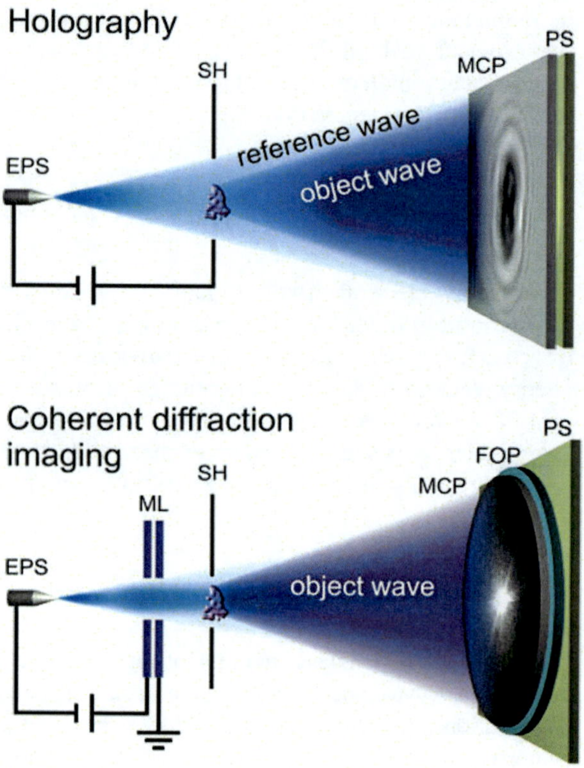

Fig. 29.3 Low-energy electron microscopes. Coherent low-energy electrons are extracted from the electron point source (*EPS*) by field emission. A biological molecule is fixed in the sample holder (*SH*). In the holographic microscope, the interference between the scattered (object) wave and unscattered (reference) wave is recorded by the detector unit (consisting of a micro-channel plate (*MCP*) followed by a phosphor screen (*PS*)). In the CDI microscope, the electron beam is collimated by a microlens (*ML*) and the detector unit consists of an MCP followed by a fibre optic plate (*FOP*) with a thin phosphorous layer (*PS*)

29.4 Holography

In holography the unknown wave that is scattered by an object is superimposed with a known reference wave. A hologram is the interference pattern formed by constructive and destructive interference between these two waves [7] and is illustrated in Fig. 29.4. The holography technique uniquely solves the phase problem in one step because of the presence of the reference wave. However, it lacks high resolution due to the higher-order scattered signal being buried in the experimental noise of the reference wave.

Hologram reconstruction includes two steps: (1) illumination of the hologram with the reference wave and (2) backward propagation of the wavefront to the position of the object. In numerical reconstruction, the complex-valued reference wave at the hologram plane is simulated and the propagation from the hologram back to the object is calculated using Huygens' principle and Fresnel formalism:

$$U(\boldsymbol{\rho}) = \frac{i}{\lambda}\iint H(\mathbf{r})\frac{\exp(ikr)}{r}\frac{\exp(-ik\,|\mathbf{r}-\boldsymbol{\rho}|)}{|\mathbf{r}-\boldsymbol{\rho}|}\mathrm{d}\sigma_{\mathrm{S}}, \qquad (29.1)$$

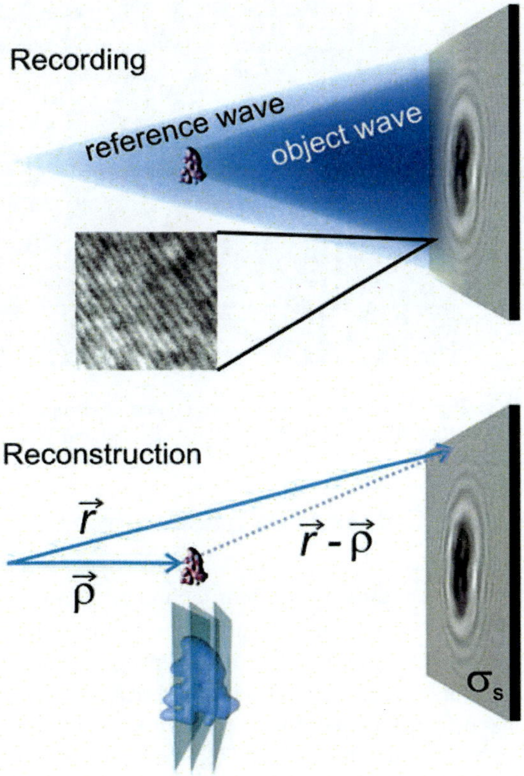

Fig. 29.4 Illustration of recording an inline hologram and reconstructing the object. Recording: the superposition of the reference wave and the wave scattered by the object is recorded; a magnified region of the hologram shows the fringes of the interference pattern. Reconstruction: back propagation of the recovered object wave from the hologram plane to the planes of the object's location (analogous to optical sectioning) results in a three-dimensional reconstruction

where $H(\boldsymbol{r})$ is the hologram's transmission function distribution, $\boldsymbol{r}$ and $\boldsymbol{\rho}$ are defined as illustrated in Fig. 29.4, and the integration is performed over the hologram's surface σ_S. The result of this integral transform is a complex-valued distribution of the object wavefront at any coordinate ρ, and, hence, a three-dimensional reconstruction.

An example of an inline hologram of an individual DNA molecule and its reconstruction is shown in Fig. 29.5. The successful trials during the last decade of imaging individual biological molecules by low-energy electron holography include the imaging of: DNA molecules [3, 6], phthalocyaninato polysiloxane molecules [11], the tobacco mosaic virus [33], a bacteriophage [31] and ferritin [16]. Despite a very short wavelength (1–2 Å) of the probing electron wave, the resolution in the reconstructed molecular structures remains in the order of a few nanometres. The reason is that the resolution in inline holography is limited by the detectability of the interference fringes at high diffraction angles [15, 29] (such as, for instance, the fringes shown in the magnified region in Fig. 29.4). The pattern of these fine fringes is very sensitive to the object's lateral movements and can be destroyed by the object shifting even by just the distance corresponding to the wavelength. In addition, these fine fringes are often buried in the experimental noise of the reference wave.

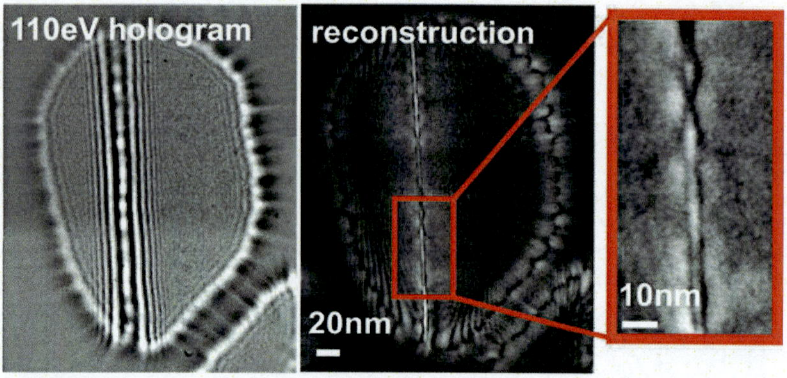

Fig. 29.5 Low-energy electron hologram of an individual ssDNA molecule stretched over a hole in a carbon film (sample courtesy by Michael and William Andregg, www.halcyonmolecular.com)

29.5 Coherent Diffraction Imaging

CDI is a relatively modern technique which combines the recording of a far-field diffraction pattern of a non-crystalline object and the numerical recovery of the object structure. In 1952, Sayre proposed that it was possible to recover the phase information associated with scattering off a non-crystalline specimen by sampling its diffraction pattern at a frequency higher than twice the Nyquist frequency (oversampling) [25]. In 1972, Gerchberg and Saxton proposed an iterative algorithm to recover the phase distribution from two amplitude measurements taken: at the object plane and at the far-field plane [8]. In 1998, Miao et al. combined these two ideas and successfully recovered an object from its oversampled diffraction pattern [18]. They demonstrated that the phase retrieval algorithm converges if the initial conditions are such that the surrounding of the molecule ("support") is known. The concept of knowing the support of the molecule is analogous to the solvent flattening technique in the phasing methods. The known surrounding of a molecule is usually mathematically described by zero-padding the object, which in turn leads to oversampling of the spectrum in the Fourier domain. Thus, reconstruction becomes possible if the diffraction pattern is recorded under the oversampling condition [17, 19, 21]; this is also illustrated in Fig. 29.6.

The basic iterative reconstruction loop [4] is shown in Fig. 29.7. It begins with the complex-valued wave distribution at the detector plane which is formed by the superposition of the square root of the measured intensity and a random phase distribution. In the object domain various constraints are applied. For instance, the electron density reconstructed from the X-ray diffraction images must be real and positive.

The resolution in CDI is defined by the outermost detected signal in the diffraction pattern, $R = \lambda/\sin\theta$, where θ is the scattering angle. The resolving power of the CDI technique has already been demonstrated by the reconstruction

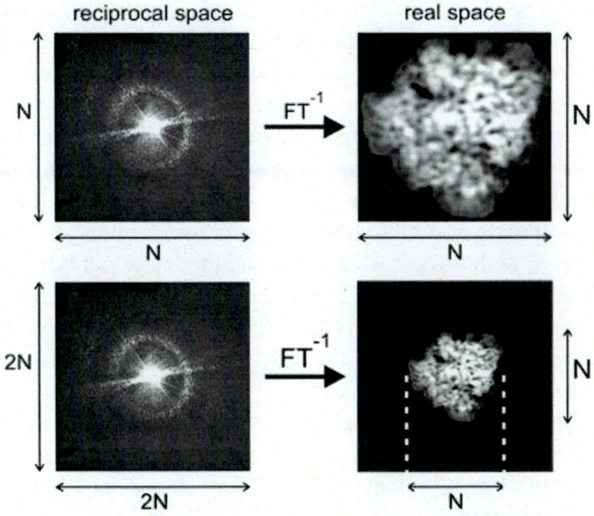

Fig. 29.6 Sampling at the Nyquist frequency (*upper row*) and twice the Nyquist frequency (*lower row*). The Fourier transform of the spectrum sampled at the Nyquist frequency results in the object distribution filling the entire reconstructed area. The Fourier transform of the spectrum sampled at twice the Nyquist frequency results in the zero-padded object distribution [17]

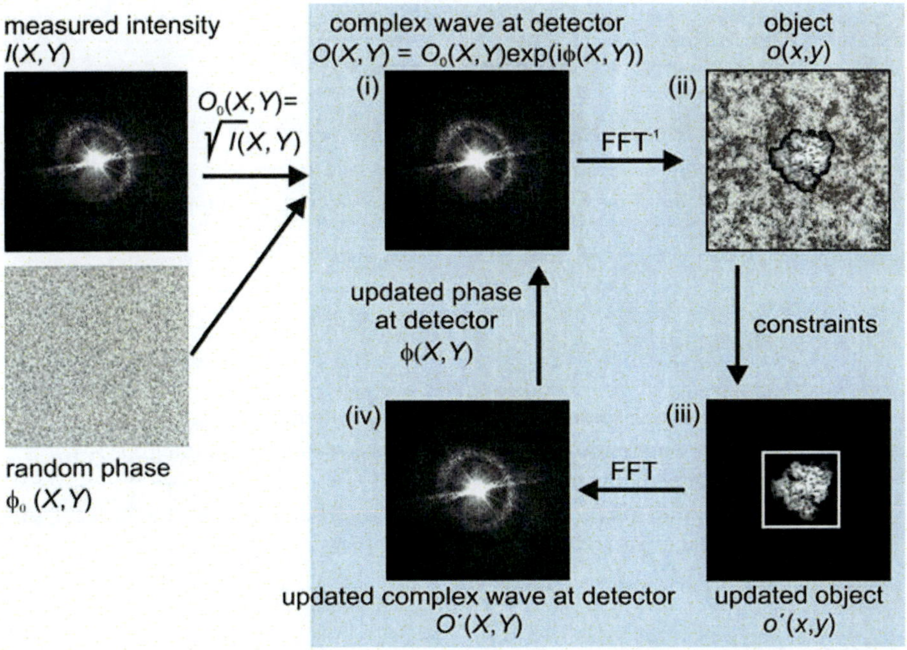

Fig. 29.7 Iterative reconstruction of a coherent diffraction pattern. *Left column*: amplitude and phase distributions at the detector plane initiating the iterative loop. *Right*: the steps (i)–(iv) showing the flow of the iterative loop

of a double-walled carbon nanotube at a resolution of 1 Å from a coherent diffraction pattern recorded using a 200 keV electron microscope exhibiting a nominal conventional TEM resolution of 2.2 Å [36].

The X-ray diffraction pattern of a crystal, unlike that of an individual molecule, displays a strong signal due to the periodicity of the crystal. Obtaining an X-ray diffraction pattern of *an individual molecule* in turn requires a much more intense X-ray beam. As a consequence, the resolution is limited by radiation damage and remains very moderate. A few biological specimens have been imaged by CDI using X-rays at a resolution of a few nanometres: E.coli bacteria [20], an unstained yeast cell [27], single herpes virions [28], malaria-infected red blood cells [34], a frozen hydrated yeast cell [14], human chromosomes [24], unstained and unsliced freeze-dried cells of the bacterium Deinococcus radiodurans by ptychography [10], and labelled yeast cells [22]. Ultra-short and extremely bright coherent X-ray pulses from XFEL allow the recording of a high-resolution diffraction pattern before the sample explodes [2, 23]. The first results from the first XFEL facility to be operational (the Linac Coherent Light Source) reported imaging an individual unstained mimivirus at 32 nm resolution [26]; in this experiment an X-ray pulse of 1.8 keV (6.9 Å) energy and 70 fs duration was focused to a spot 10 μm in diameter with 1.6×10^{10} photons per 1 μm^2. A sub-nanometre resolution could be achieved by employing shorter pulses and a higher photon flux [2, 26]; at present this is beyond the capabilities of the XFELs but might be realized with the next generation of XFELs.

29.6 Comparing Holography and CDI

Each of the two techniques has its pros (+) and cons (–), which are summarized below:

29.6.1 Holography

- Requires well-defined reference wave over entire detector area (–)
- Non-iterative reconstruction by calculating back-propagation integral (+)
- Three-dimensional reconstruction (+)
- Low resolution, due to high sensitivity of the interference pattern to object shifts and experimental noise in the reference wave (–)

29.6.2 CDI

- No reference wave is required (+)
- Reconstruction is done by an iterative procedure and does not always converge to a uniquely defined outcome (–)
- Reconstruction is not three-dimensional, it is limited to one plane (–)
- High resolution provided by stability of diffraction pattern being insensitive to shifts of the object (+)

29.7 HCDI: Combining Holography and CDI

Recently, we revealed the relationship between the hologram and the diffraction pattern of an object, which allows holography and CDI to be combined into a superior technique: holographic coherent diffraction imaging (HCDI). HCDI inherits fast and reliable reconstruction from holography and the highest possible resolution from CDI [15].

The Fourier transform of an inline hologram is proportional to the complex-valued object wave in the far-field, as illustrated for experimental images in Fig. 29.8. Thus, the phase distribution of the Fourier transform of the inline hologram provides the phase distribution of the object wave in the far-field and hence the solution to the "phase problem" in just one step. The diffraction pattern

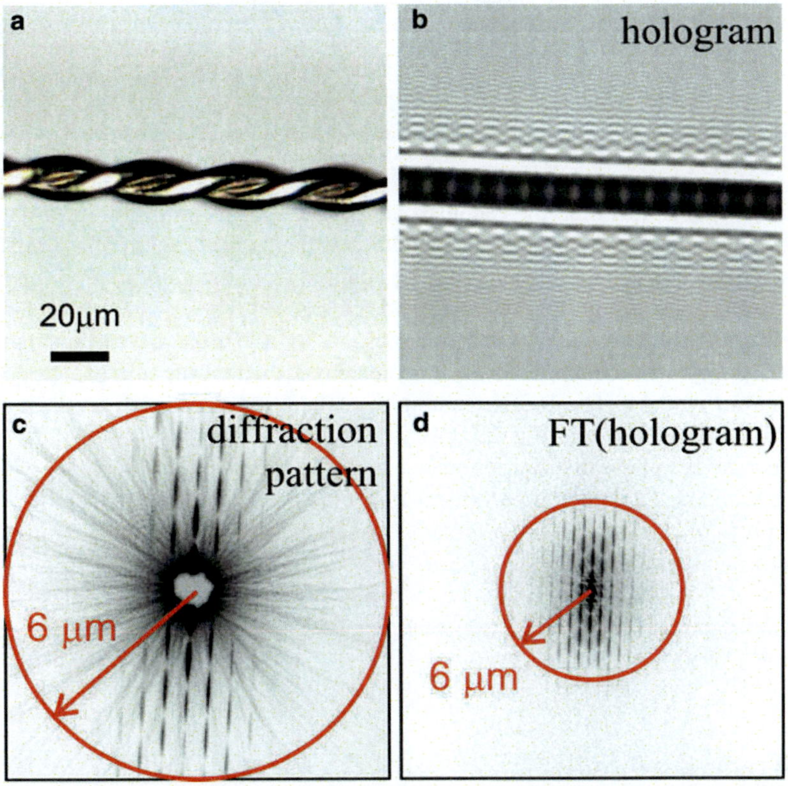

Fig. 29.8 Experimental verification of the relationship between a hologram and a diffraction pattern. (**a**) Reflected-light microscopy image of two twisted tungsten wires. (**b**) Inline hologram recorded with laser light. (**c**) Diffraction pattern. (**d**) The amplitude of the Fourier transform of the hologram is displayed using a logarithmic and inverted intensity scale. The diffraction pattern (**c**) provides the same resolution as the hologram – namely 6 μm – but it is recorded while fulfilling the oversampling condition

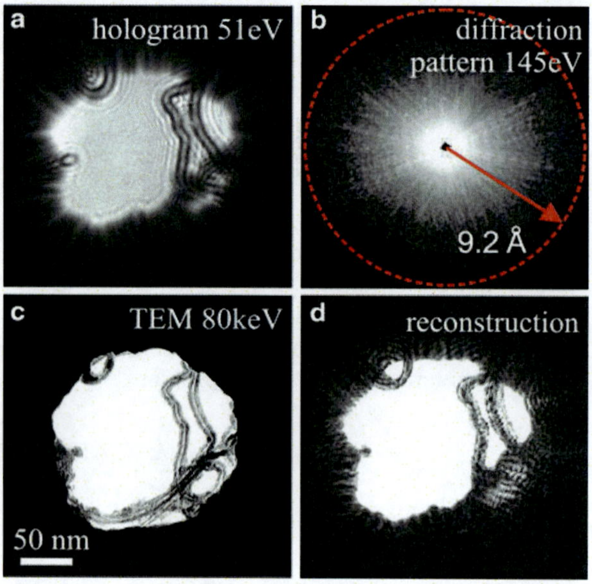

Fig. 29.9 (**a**) Hologram of carbon nanotubes acquired with 51 eV electrons. (**b**) Coherent diffraction pattern of the same nanotubes recorded with electrons of 145 eV kinetic energy. The *dashed circle* shows the highest order signal detected corresponding to 9.2 Å resolution. (**c**) TEM image of the very same area obtained with 80 keV electrons. (**d**) Reconstruction obtained by the HCDI method

is then required to refine the reconstruction of the high-resolution information by a conventional iterative procedure. In addition, the central region of the diffraction pattern, which is usually missing, can be adapted from the amplitude of the Fourier transform of the hologram; see Fig. 29.8d.

The hologram and the diffraction pattern of a bundle of carbon nanotubes recorded with the coherent low-energy electron diffraction microscope [30] are shown in Fig. 29.9. The HCDI technique was applied to reconstruct these images and the result is shown in Fig. 29.9d.

Because the phase distribution stored in a holographic image is uniquely defined and is associated with the three-dimensional object distribution, HCDI may offer the possibility of retrieving a three-dimensional object distribution from its diffraction pattern.

Acknowledgments We would like to thank the Swiss National Science Foundation for its financial support.

References

1. Adrian M, Dubochet J, Lepault J, McDowall AW (1984) Cryo-electron microscopy of viruses. Nature 308(5954):32–36
2. Bergh M, Huldt G, Timneanu N, Maia F, Hajdu J (2008) Feasibility of imaging living cells at subnanometer resolutions by ultrafast X-ray diffraction. Q Rev Biophys 41(3–4):181–204
3. Eisele A, Voelkel B, Grunze M, Golzhauser A (2008) Nanometer resolution holography with the low energy electron point source microscope. Z Phys Chem 222(5–6):779–787

4. Fienup JR (1982) Phase retrieval algorithms – a comparison. Appl Optics 21(15):2758–2769
5. Fink H-W, Stocker W, Schmid H (1990) Holography with low-energy electrons. Phys Rev Lett 65(10):1204–1206
6. Fink H-W, Schmid H, Ermantraut E, Schulz T (1997) Electron holography of individual DNA molecules. J Opt Soc Am A Opt Image Sci Vis 14(9):2168–2172
7. Gabor D (1949) Microscopy by reconstructed wave-fronts. Proc Phys Soc Lond Ser A 197(1051):454–487
8. Gerchberg RW, Saxton WO (1972) A practical algorithm for determination of phase from image and diffraction plane pictures. Optik 35(2):237–246
9. Germann M, Latychevskaia T, Escher C, Fink H-W (2010) Nondestructive imaging of individual biomolecules. Phys Rev Lett 104(9):095501
10. Giewekemeyer K, Thibault P, Kalbfleisch S, Beerlink A, Kewish CM, Dierolf M, Pfeiffer F, Salditt T (2010) Quantitative biological imaging by ptychographic x-ray diffraction microscopy. Proc Natl Acad Sci U S A 107(2):529–534
11. Golzhauser A, Volkel B, Jager B, Zharnikov M, Kreuzer HJ, Grunze M (1998) Holographic imaging of macromolecules. J Vac Sci Technol A Vac Surf Films 16(5):3025–3028
12. Henderson R (2004) Realizing the potential of electron cryo-microscopy. Q Rev Biophys 37(1):3–13
13. Howells MR, Beetz T, Chapman HN, Cui C, Holton JM, Jacobsen CJ, Kirz J, Lima E, Marchesini S, Miao H, Sayre D, Shapiro DA, Spence JCH, Starodub D (2009) An assessment of the resolution limitation due to radiation-damage in X-ray diffraction microscopy. J Electron Spectrosc 170(1–3):4–12
14. Huang X, Nelson J, Kirz J, Lima E, Marchesini S, Miao H, Neiman AM, Shapiro D, Steinbrener J, Stewart A, Turner JJ, Jacobsen C (2009) Soft X-ray diffraction microscopy of a frozen hydrated yeast cell. Phys Rev Lett 103(19):198101
15. Latychevskaia T, Longchamp J-N, Fink H-W (2012) When holography meets coherent diffraction imaging. Optics Express 20(27):28871–28892
16. Longchamp J-N, Latychevskaia T, Escher C, Fink H-W (2012) Non-destructive imaging of an individual protein. Appl Phys Lett 101(9):093701
17. Miao JW, Sayre D (2000) On possible extensions of X-ray crystallography through diffraction-pattern oversampling. Acta Crystallogr A 56:596–605
18. Miao JW, Sayre D, Chapman HN (1998) Phase retrieval from the magnitude of the Fourier transforms of nonperiodic objects. J Opt Soc Am A Opt Image Sci Vis 15(6):1662–1669
19. Miao JW, Charalambous P, Kirz J, Sayre D (1999) Extending the methodology of X-ray crystallography to allow imaging of micrometre-sized non-crystalline specimens. Nature 400(6742):342–344
20. Miao JW, Hodgson KO, Ishikawa T, Larabell CA, LeGros MA, Nishino Y (2003) Imaging whole Escherichia coli bacteria by using single-particle x-ray diffraction. Proc Natl Acad Sci U S A 100(1):110–112
21. Miao JW, Ishikawa T, Anderson EH, Hodgson KO (2003) Phase retrieval of diffraction patterns from noncrystalline samples using the oversampling method. Phys Rev B 67(17)
22. Nelson J, Huang XJ, Steinbrener J, Shapiro D, Kirz J, Marchesini S, Neiman AM, Turner JJ, Jacobsen C (2010) High-resolution x-ray diffraction microscopy of specifically labeled yeast cells. Proc Natl Acad Sci U S A 107(16):7235–7239
23. Neutze R, Wouts R, van der Spoel D, Weckert E, Hajdu J (2000) Potential for biomolecular imaging with femtosecond X-ray pulses. Nature 406(6797):752–757
24. Nishino Y, Takahashi Y, Imamoto N, Ishikawa T, Maeshima K (2009) Three-dimensional visualization of a human chromosome using coherent X-ray diffraction. Phys Rev Lett 102(1):018101
25. Sayre D (1952) Some implications of a theorem due to Shannon. Acta Crystallogr 5(6):843–843
26. Seibert MM, Ekeberg T, Maia FRNC, Svenda M, Andreasson J, Jonsson O, Odic D, Iwan B, Rocker A, Westphal D, Hantke M, DePonte DP, Barty A, Schulz J, Gumprecht L, Coppola N, Aquila A, Liang M, White TA, Martin A, Caleman C, Stern S, Abergel C, Seltzer V,

Claverie J-M, Bostedt C, Bozek JD, Boutet S, Miahnahri AA, Messerschmidt M, Krzywinski J, Williams G, Hodgson KO, Bogan MJ, Hampton CY, Sierra RG, Starodub D, Andersson I, Bajt S, Barthelmess M, Spence JCH, Fromme P, Weierstall U, Kirian R, Hunter M, Doak RB, Marchesini S, Hau-Riege SP, Frank M, Shoeman RL, Lomb L, Epp SW, Hartmann R, Rolles D, Rudenko A, Schmidt C, Foucar L, Kimmel N, Holl P, Rudek B, Erk B, Homke A, Reich C, Pietschner D, Weidenspointner G, Struder L, Hauser G, Gorke H, Ullrich J, Schlichting I, Herrmann S, Schaller G, Schopper F, Soltau H, Kuhnel K-U, Andritschke R, Schroter C-D, Krasniqi F, Bott M, Schorb S, Rupp D, Adolph M, Gorkhover T, Hirsemann H, Potdevin G, Graafsma H, Nilsson B, Chapman HN, Hajdu J (2011) Single mimivirus particles intercepted and imaged with an X-ray laser. Nature 470(7332):78–81

27. Shapiro D, Thibault P, Beetz T, Elser V, Howells M, Jacobsen C, Kirz J, Lima E, Miao H, Neiman AM, Sayre D (2005) Biological imaging by soft x-ray diffraction microscopy. Proc Natl Acad Sci U S A 102(43):15343–15346
28. Song CY, Jiang HD, Mancuso A, Amirbekian B, Peng L, Sun R, Shah SS, Zhou ZH, Ishikawa T, Miao JW (2008) Quantitative imaging of single, unstained viruses with coherent X-rays. Phys Rev Lett 101(15):158101
29. Spence JCH, Qian W, Silverman MP (1994) Electron source brightness and degeneracy from Fresnel fringes in-field emission point projection microscopy. J Vac Sci Technol A Vac Surf Films 12(2):542–547
30. Steinwand E, Longchamp J-N, Fink H-W (2011) Coherent low-energy electron diffraction on individual nanometer sized objects. Ultramicroscopy 111(4):282–284
31. Stevens GB, Krüger M, Latychevskaia T, Lindner P, Plückthun A, Fink H-W (2011) Individual filamentous phage imaged by electron holography. Eur Biophys J 40:1197–1201
32. van Heel M, Gowen B, Matadeen R, Orlova EV, Finn R, Pape T, Cohen D, Stark H, Schmidt R, Schatz M, Patwardhan A (2000) Single-particle electron cryo-microscopy: towards atomic resolution. Q Rev Biophys 33(4):307–369
33. Weierstall U, Spence JCH, Stevens M, Downing KH (1999) Point-projection electron imaging of tobacco mosaic virus at 40 eV electron energy. Micron 30(4):335–338
34. Williams GJ, Hanssen E, Peele AG, Pfeifer MA, Clark J, Abbey B, Cadenazzi G, de Jonge MD, Vogt S, Tilley L, Nugent KA (2008) High-resolution X-ray imaging of plasmodium falciparum-infected red blood cells. Cytom Part A 73A(10):949–957
35. www.pdb.org Protein Data Bank
36. Zuo JM, Vartanyants I, Gao M, Zhang R, Nagahara LA (2003) Atomic resolution imaging of a carbon nanotube from diffraction intensities. Science 300(5624):1419–1421

Chapter 30
Structure Analysis of Biological Macromolecules by Small-Angle X-ray Scattering

Dmitri I. Svergun

Abstract Small-angle X-ray scattering (SAXS) is a low resolution (1–2 nm) structural method, which is applicable to macromolecules in solution providing information about the overall structure and structural transitions. The method covers an extremely broad range of sizes (from a few kDa to hundreds MDa) and experimental conditions (temperature, pH, salinity, ligand addition *etc.*). Recent progress in instrumentation and novel data analysis methods significantly enhanced resolution and reliability of structural models provided by the technique and made SAXS a useful complementary tool to high resolution methods. Modern SAXS allows for rapid validation of crystallographic or theoretically predicted models, identification of biologically active oligomers and visualization of missing fragments in high resolution structures. Quaternary structure of complexes can be analyzed by rigid body movements/rotations of high resolution models of the individual subunits of domains. Recent developments made it possible also to quantitatively characterize flexible macromolecular systems, including intrinsically unfolded proteins. The basics of SAXS will be presented and illustrated by advanced applications to macromolecular solutions.

Keywords Quaternary structure • Macromolecular envelope • Flexible proteins • Structural transitions • Solution scattering

30.1 Introduction

The structural genomics initiatives aiming at large-scale expression and purification for subsequent structure determination using X-ray crystallography and NMR spectroscopy [1, 2]; have already yielded unprecedented numbers of high resolution

D.I. Svergun (✉)
European Molecular Biology Laboratory, Hamburg Outstation,
Notkestraße 85, D-22603 Hamburg, Germany
e-mail: svergun@embl-hamburg.de

R. Read et al. (eds.), *Advancing Methods for Biomolecular Crystallography*,
NATO Science for Peace and Security Series A: Chemistry and Biology,
DOI 10.1007/978-94-007-6232-9_30,

models for isolated proteins and/or their domains. These numbers are expected to grow rapidly in the coming years [3]. The focus of modern structural biology has largely shifted towards the study of macromolecular machines accomplishing most important cellular functions, see e.g. [4]. The macromolecular complexes are usually too large for the structural NMR studies, and they often possess inherent structural flexibility making them difficult to crystallize.

The structural analysis approach to macromolecular complexes includes new crystallographic initiatives complemented by the use of methods yielding structural information in solution at lower resolution. In particular, Cryo-EM allows one to obtain excellent results in many cases [5], but it is usually limited to relatively large macromolecular aggregates (starting from about 200 kDa).

Small-angle X-ray scattering (SAXS) is a rapid method to characterize low resolution structures of individual macromolecules and complexes in solution and to analyse structural changes in response to variation of external conditions. For establishing the three-dimensional structural models this technique needs monodisperse solutions of purified macromolecules but does not require special sample treatment (growth of crystals, isotopic labelling, cryo-freezing *etc.*). SAXS is applicable to a broad range of conditions and sizes (from a few kDa to hundreds MDa). Unlike most other structural methods, SAXS is able to quantitatively characterize equilibrium and non-equilibrium mixtures and monitor kinetic processes such as (dis)assembly and (un)folding.

Recently, the power of SAXS has been boosted by the significant improvements in instrumentation (most notably, by the high brilliance synchrotron radiation sources) accompanied by the development of novel data analysis methods. These developments made it possible to significantly improve resolution and reliability of the structural models constructed from the SAXS data. Here, the main aspects of SAXS including data processing and interpretation procedures and some applications will be briefly reviewed.

30.2 Basics of a SAXS Experiment

This section will briefly describe the basic theoretical and experimental aspects of SAXS to understand the main principles of the technique as applied to solutions of biological macromolecules. The reader is referred to textbooks [6, 7] or reviews [8–11] for more detailed description.

Conceptually, a SAXS experiment is rather simple, as illustrated in Fig. 30.1. The samples are exposed to a collimated monochromatic X-ray or neutron beam with the wave vector $k = 2\pi/\lambda$ where λ is the radiation wavelength (Fig. 30.1). The isotropic scattered intensity I is recorded as a function of the momentum transfer $s = 4\pi \sin\theta/\lambda$, where 2θ is the angle between the incident and scattered beam. The scattering from the solvent is measured separately and subtracted to remove the solvent and parasitic background signals.

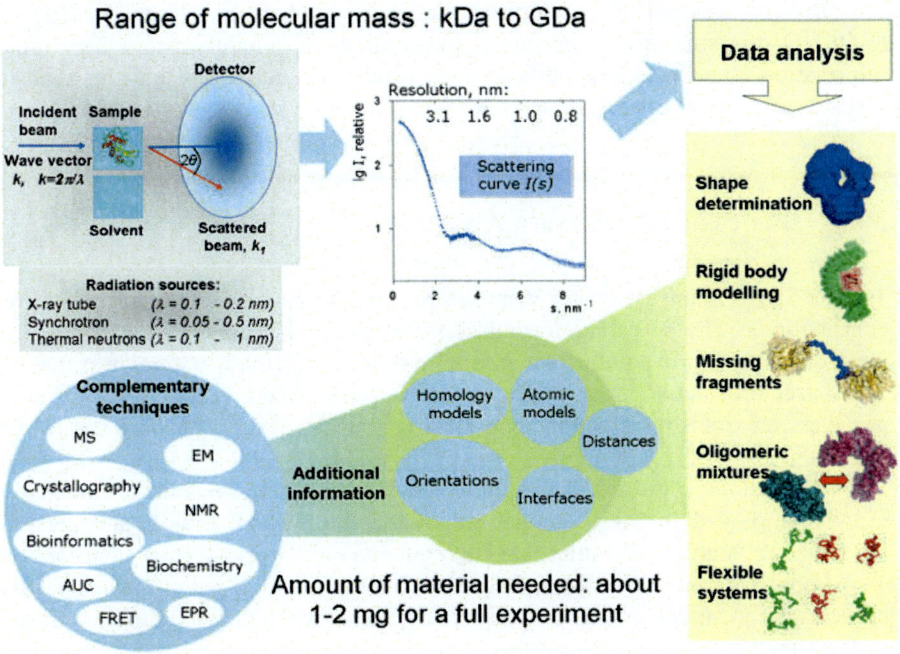

Fig. 30.1 A general scheme of a SAXS/SANS experiment, structural tasks addressed by the technique and its synergistic use with other methods. The nominal resolution of the data in the scattering pattern is indicated as $d = 2\pi/s$. Abbreviations: *MS* mass spectroscopy, *AUC* analytical ultracentrifugation, *FRET* fluorescence resonance energy transfer, *EM* electron microscopy, *NMR* nuclear magnetic resonance, *EPR* electron paramagnetic resonance

The SAXS experiments are usually performed at synchrotrons, and the experimental stations offering biological SAXS are available at all synchrotron major sites. Laboratory SAXS cameras, (available from various producers) yield much lower beam brilliance but may be useful at least for preliminary analysis. For structure analysis (shape, quaternary structure), the samples with monodispersity better than 90 % are required, which must be verified by other methods (native gel filtration, dynamic light scattering, analytical ultracentrifugation) prior to the synchrotron SAXS experiment. Typical concentrations required are in the range of 0.5–10 mg/ml, and a concentration series is usually measured to get rid of interparticle interference effects. The sample volume per measurement on modern stations is about 10–50 μl so that about 1–2 mg of purified material is usually required for a complete study. The upcoming microfluidic devices [12] will allow one to work on high brilliance sources with nanoliter volumes and μg sample amounts.

One should also mention that neutrons are also employed for small-angle scattering (SANS) analysis of biological macromolecules. SANS (which is performed on research reactors or spallation sources) is sensitive to isotopic H/D exchange.

This property is exploited for contrast variation involving measurements in different H_2O/D_2O mixtures and/or specific perdeuteration of subunits, providing unique information about complex particles [13]. The basic equations and the analysis methods are similar for SAXS and SANS.

30.3 Basics of SAXS Data Analysis

The net SAXS intensity after solvent subtraction contains, generally speaking, two contributions. The so-called form factor *I(s)* emerges from the scattering from individual particles in solution and is employed to extract the structural information. The "structure factor" *S(s)* is due to interference effects between the different particles and yields information about the interparticle interactions (see e.g. [8] for a review).

Purified dilute solutions of macromolecules at concentrations in mM range are usually employed in SAXS to get rid of the interference effects and perform the structural studies assuming that I(s) contains only "form-factor" contrinution. Two important cases are distinguished: (i) monodisperse systems, when all the particles are identical and (ii) polydisperse systems, when they are different in size and/or shape.

30.3.1 Monodisperse Systems

For monodisperse solutions, the net intensity *I(s)* is proportional to the scattering from a single particle averaged over all orientations. This allows one to immediately determine the overall geometrical and weight parameters e.g. radius of gyration R_g [14], volume of the hydrated particle V_p [15], and the molecular mass of the particle MM [16]. The Fourier transformation of the scattering intensity provides a characteristic function (averaged Patterson function), which also yields the maximum particle diameter D_{max} [17–19], Moreover, the low resolution macromolecular shape can be obtained *ab initio* (i.e. without information from other methods). Several approaches have been proposed [20–26], and *ab initio* shape determination belongs nowadays to routine analysis of the SAXS data. Usually, the shape analysis programs are ran several times and analysed to obtain the most probable and an averaged model [27].

Calculation of the SAXS profiles from atomic models [28] is used to validate theoretically predicted models and verify the structural similarity between macromolecules in crystals and in solution. Moreover, if high resolution models of individual fragments or subunits in a complex are available from crystallography or NMR, rigid body refinement can be employed to model the quaternary structure of the complex. Automated and semi-automated procedures based on screening randomly or systematically generated models were employed by different authors [29–31]. A comprehensive rigid body modelling program suite is based on the use of spherical harmonics formalism [32–35].

SAXS is also very useful for the cases when loops or entire domains are missing in high resolution models (e.g. because of flexibility). The missing portions are represented as chains of dummy residues [36], and the known domains/subunits can be translated and rotated as rigid bodies while simultaneously changing the local conformation of the chains representing the unknown fragments [37]. Numerous applications of rigid body modelling are reported to build structural models of complicated objects in solution based on the SAXS data while simultaneously incorporating information provided by other methods (see e.g. references in [9, 38, 39]).

30.3.2 Poyldisperse Systems and Mixtures

For polydisperse systems consisting of different types of non-interacting particles, the measured scattering pattern can be written as a linear combination

$$I(s) = \sum_{k=1}^{K} \nu_k I_k(s), \tag{30.1}$$

where $\nu_k > 0$ and $I_k(s)$ are the volume fraction and the scattering intensity from the k-th type of particle (component), respectively, and K is the number of components.

When neither the number nor intensities of the components are known *a priori*, but multiple data sets are recorded from the system with varying volume fractions of the components, the number of components can be determined extracted by model-independent analysis using singular value decomposition (SVD, [40]). If the number of components and their scattering intensities are known, the volume fractions can be readily found by a linear least-squares fit to the experimental data. Numerous applications of these approaches encompass *e.g.* the analysis of intermediates during folding and assembly processes and quantitative description of oligomeric equilibria [41–44].

SAXS belongs to very few structural methods able to quantitatively characterize flexible macromolecules, and the method was traditionally used to monitor the processes of protein folding/unfolding [45]. For flexible systems, SAXS data reflect conformational average over the entire ensemble and the scattering patterns are to be interpreted accounting for this average instead of searching for a single model. This has recently become possible with a general approach called 'ensemble optimization method' (EOM) allowing for coexistence of multiple conformations [46]. Given a pool of (random) conformers, EOM selects sub-ensembles from them, which, taken as mixtures, fit the experimental profile using Eq. 30.1. The EOM is already actively used to characterize flexible proteins and complexes [47, 48] and it is expected to find broad applications, in particular, in combination with NMR to provide information about both structure and dynamics of the system [49, 50].

30.4 Conclusions

During the last decade, biological SAXS has become increasingly popular in molecular biology revealing low resolution structures of macromolecule in close to native conditions. SAXS can be readily and usefully combined with other computational and experimental techniques to yield comprehensive description of complex objects and processes. The advanced analysis methods are well established by now and are publicly available e.g. in the program package ATSAS [51], see http://www.embl-hamburg.de/biosaxs/software.html. Automated sample changers and pipelines are being developed for high throughput SAXS on synchrotrons [52–54]. All these developments taken together make the technique readily available for a broad scope of tasks and a broad community of scientists in structural biology.

References

1. Edwards AM, Arrowsmith CH, Christendat D, Dharamsi A, Friesen JD, Greenblatt JF, Vedadi M (2000) Protein production: feeding the crystallographers and NMR spectroscopists. Nat Struct Biol 7(Suppl):970–972
2. Gerstein M, Edwards A, Arrowsmith CH, Montelione GT (2003) Structural genomics: current progress. Science 299(5613):1663
3. Levitt M (2007) Growth of novel protein structural data. Proc Natl Acad Sci U S A 104(9):3183–3188
4. Abrahams JP, Apweiler R, Balling R, Bertero MG, Bujnicki JM, Chayen NE, Chene P, Corthals GL, Dylag T, Forster F, Heck AJ, Henderson PJ, Herwig R, Jehenson P, Kokalj SJ, Laue E, Legrain P, Martens L, Migliorini C, Musacchio A, Podobnik M, Schertler GF, Schreiber G, Sixma TK, Smit AB, Stuart D, Svergun DI, Taussig MJ (2011) "4D Biology for health and disease" workshop report. N Biotechnol 28(4):291–293. doi:S1871-6784(10)00577-7 [pii]10.1016/j.nbt.2010.10.003
5. Sali A, Glaeser R, Earnest T, Baumeister W (2003) From words to literature in structural proteomics. Nature 422(6928):216–225
6. Feigin LA, Svergun DI (1987) Structure analysis by small-angle x-ray and neutron scattering. Plenum Press, New York
7. Glatter O, Kratky O (1982) Small angle X-ray scattering. Academic, London
8. Koch MH, Vachette P, Svergun DI (2003) Small-angle scattering: a view on the properties, structures and structural changes of biological macromolecules in solution. Q Rev Biophys 36(2):147–227
9. Mertens HD, Svergun DI (2010) Structural characterization of proteins and complexes using small-angle X-ray solution scattering. J Struct Biol 172(1):128–141. doi:S1047-8477(10)00190-5 [pii]10.1016/j.jsb.2010.06.012
10. Putnam CD, Hammel M, Hura GL, Tainer JA (2007) X-ray solution scattering (SAXS) combined with crystallography and computation: defining accurate macromolecular structures, conformations and assemblies in solution. Q Rev Biophys 40(3):191–285
11. Svergun DI, Koch MHJ (2003) Small angle scattering studies of biological macromolecules in solution. Rep Prog Phys 66:1735–1782
12. Toft KN, Vestergaard B, Nielsen SS, Snakenborg D, Jeppesen MG, Jacobsen JK, Arleth L, Kutter JP (2008) High-throughput small angle X-ray scattering from proteins in solution using a microfluidic front-end. Anal Chem 80(10):3648–3654

13. Whitten AE, Trewhella J (2009) Small-angle scattering and neutron contrast variation for studying bio-molecular complexes. Method Mol Biol 544:307–323
14. Guinier A (1939) La diffraction des rayons X aux tres petits angles; application a l'etude de phenomenes ultramicroscopiques. Ann Phys (Paris) 12:161–237
15. Porod G (1982) General theory. In: Glatter O, Kratky O (eds) Small-angle X-ray scattering. Academic, London, pp 17–51
16. Mylonas E, Svergun DI (2007) Accuracy of molecular mass determination of proteins in solution by small-angle X-ray scattering. J Appl Crystallogr 40:s245–s249
17. Glatter O (1977) A new method for the evaluation of small-angle scattering data. J Appl Crystallogr 10:415–421
18. Moore PB (1980) Small-angle scattering: Information content and error analysis. J Appl Crystallogr 13:168–175
19. Svergun DI (1992) Determination of the regularization parameter in indirect-transform methods using perceptual criteria. J Appl Crystallogr 25:495–503
20. Bada M, Walther D, Arcangioli B, Doniach S, Delarue M (2000) Solution structural studies and low-resolution model of the Schizosaccharomyces pombe sap1 protein. J Mol Biol 300(3):563–574
21. Chacon P, Diaz JF, Moran F, Andreu JM (2000) Reconstruction of protein form with X-ray solution scattering and a genetic algorithm. J Mol Biol 299(5):1289–1302
22. Chacon P, Moran F, Diaz JF, Pantos E, Andreu JM (1998) Low-resolution structures of proteins in solution retrieved from X-ray scattering with a genetic algorithm. Biophys J 74(6):2760–2775
23. Franke D, Svergun DI (2009) DAMMIF, a program for rapid ab-initio shape determination in small-angle scattering. J Appl Crystallogr 42:342–346. doi:10.1107/S0021889809000338
24. Heller WT, Abusamhadneh E, Finley N, Rosevear PR, Trewhella J (2002) The solution structure of a cardiac troponin C-troponin I-troponin T complex shows a somewhat compact troponin C interacting with an extended troponin I-troponin T component. Biochemistry 41(52):15654–15663
25. Svergun DI (1999) Restoring low resolution structure of biological macromolecules from solution scattering using simulated annealing. Biophys J 76(6):2879–2886
26. Svergun DI, Petoukhov MV, Koch MHJ (2001) Determination of domain structure of proteins from X-ray solution scattering. Biophys J 80(6):2946–2953. doi:10.1016/S0006-3495(01)76260-1
27. Volkov VV, Svergun DI (2003) Uniqueness of ab initio shape determination in small angle scattering. J Appl Crystallogr 36:860–864
28. Svergun DI, Barberato C, Koch MHJ (1995) CRYSOL – a program to evaluate X-ray solution scattering of biological macromolecules from atomic coordinates. J Appl Crystallogr 28:768–773
29. Heller WT, Finley NL, Dong WJ, Timmins P, Cheung HC, Rosevear PR, Trewhella J (2003) Small-angle neutron scattering with contrast variation reveals spatial relationships between the three subunits in the ternary cardiac troponin complex and the effects of troponin I phosphorylation. Biochemistry 42(25):7790–7800
30. King WA, Stone DB, Timmins PA, Narayanan T, von Brasch AA, Mendelson RA, Curmi PM (2005) Solution structure of the chicken skeletal muscle troponin complex via small-angle neutron and X-ray scattering. J Mol Biol 345(4):797–815
31. Nollmann M, He J, Byron O, Stark WM (2004) Solution structure of the Tn3 resolvase-crossover site synaptic complex. Mol Cell 16(1):127–137
32. Petoukhov MV, Svergun DI (2005) Global rigid body modeling of macromolecular complexes against small-angle scattering data. Biophys J 89(2):1237–1250. doi:10.1529/biophysj.105.064154
33. Petoukhov MV, Svergun DI (2006) Joint use of small-angle X-ray and neutron scattering to study biological macromolecules in solution. Eur Biophys J 35:567–576. doi:10.1007/s00249-006-0063-9

34. Svergun DI (1991) Mathematical methods in small-angle scattering data analysis. J Appl Crystallogr 24:485–492
35. Svergun DI (1994) Solution scattering from biopolymers: advanced contrast variation data analysis. Acta Crystallogr A50:391–402
36. Petoukhov MV, Eady NA, Brown KA, Svergun DI (2002) Addition of missing loops and domains to protein models by x-ray solution scattering. Biophys J 83(6):3113–3125
37. Konarev PV, Petoukhov MV, Volkov VV, Svergun DI (2006) ATSAS 2.1, a program package for small-angle scattering data analysis. J Appl Crystallogr 39:277–286
38. Petoukhov MV, Svergun DI (2007) Analysis of X-ray and neutron scattering from biomacromolecular solutions. Curr Opin Struct Biol 17(5):562–571. doi:10.1016/j.sbi.2007.06.009
39. Putnam CD, Hammel M, Hura GL, Tainer JA (2007) X-ray solution scattering (SAXS) combined with crystallography and computation: defining accurate macromolecular structures, conformations and assemblies in solution. Q Rev Biophys 40(3):191–285. doi:10.1017/S0033583507004635
40. Golub GH, Reinsh C (1970) Singular value decomposition and least squares solution. Numer Math 14:403–420
41. Doniach S (2001) Changes in biomolecular conformation seen by small angle X-ray scattering. Chem Rev 101(6):1763–1778
42. Hamiaux C, Perez J, Prange T, Veesler S, Ries-Kautt M, Vachette P (2000) The BPTI decamer observed in acidic pH crystal forms pre-exists as a stable species in solution. J Mol Biol 297(3):697–712
43. Vestergaard B, Groenning M, Roessle M, Kastrup JS, van de Weert M, Flink JM, Frokjaer S, Gajhede M, Svergun DI (2007) A helical structural nucleus is the primary elongating unit of insulin amyloid fibrils. PLoS Biol 5(5):e134
44. Xu XF, Reinle WG, Hannemann F, Konarev PV, Svergun DI, Bernhardt R, Ubbink M (2008) Dynamics in a pure encounter complex of two proteins studied by solution scattering and paramagnetic NMR spectroscopy. J Am Chem Soc 130(20):6395–6403
45. Perez J, Vachette P, Russo D, Desmadril M, Durand D (2001) Heat-induced unfolding of neocarzinostatin, a small all-beta protein investigated by small-angle X-ray scattering. J Mol Biol 308(4):721–743
46. Bernado P, Mylonas E, Petoukhov MV, Blackledge M, Svergun DI (2007) Structural characterization of flexible proteins using small-angle X-ray scattering. J Am Chem Soc 129(17):5656–5664. doi:10.1021/ja069124n
47. Bertini I, Calderone V, Fragai M, Jaiswal R, Luchinat C, Melikian M, Mylonas E, Svergun DI (2008) Evidence of reciprocal reorientation of the catalytic and hemopexin-like domains of full-length MMP-12. J Am Chem Soc 130(22):7011–7021
48. Mylonas E, Hascher A, Bernado P, Blackledge M, Mandelkow E, Svergun DI (2008) Domain conformation of tau protein studied by solution small-angle X-ray scattering. Biochemistry 47(39):10345–10353
49. Bernado P, Blanchard L, Timmins P, Marion D, Ruigrok RW, Blackledge M (2005) A structural model for unfolded proteins from residual dipolar couplings and small-angle x-ray scattering. Proc Natl Acad Sci U S A 102(47):17002–17007
50. Blobel J, Bernado P, Svergun DI, Tauler R, Pons M (2009) Low-resolution structures of transient protein-protein complexes using small-angle X-ray scattering. J Am Chem Soc 131(12):4378–4386. doi:10.1021/ja808490b
51. Petoukhov MV, Franke D, Shkumatov AV, Tria G, Kikhney AG, Gajda M, Gorba C, Mertens HDT, Konarev PV, Svergun DI (2012) New developments in the ATSAS program package for small-angle scattering data analysis. J Appl Crystallogr 45(2):342–350. doi:10.1107/S0021889812007662
52. Hura GL, Menon AL, Hammel M, Rambo RP, Poole FL 2nd, Tsutakawa SE, Jenney FE Jr, Classen S, Frankel KA, Hopkins RC, Yang SJ, Scott JW, Dillard BD, Adams MW, Tainer JA (2009) Robust, high-throughput solution structural analyses by small angle X-ray scattering (SAXS). Nat Method 6(8):606–612

53. Petoukhov MV, Konarev PV, Kikhney AG, Svergun DI (2007) ATSAS 2.1 – towards automated and web-supported small-angle scattering data analysis. J Appl Crystallogr 40(s1):s223–s228
54. Round AR, Franke D, Moritz S, Huchler R, Fritsche M, Malthan D, Klaering R, Svergun DI, Roessle M (2008) Automated sample-changing robot for solution scattering experiments at the EMBL Hamburg SAXS station X33. J Appl Crystallogr 41:913–917

Chapter 31
Protein Structure Modeling with Rosetta: Case Studies in Structure Prediction and Enzyme Repurposing

Frank DiMaio

Abstract The Rosetta protein structure methodology has evolved as a comprehensive tool for protein structure prediction and protein design. The software suite includes modules for protein structure prediction, protein-protein and protein-small molecule docking, as well as protein interface, enzyme, and symmetric protein design. This paper describes two recent Rosetta successes. We describe how Rosetta's structure prediction – when augmented with experimental data – was used to solve difficult molecular replacement problems, yielding high-resolution models for 8 of 13 structures unsolvable by alternate approaches. We also show how Rosetta may be used to redesign metalloenzyme active sites, repurposing a metal binding site to provide new catalytic activity.

31.1 Introduction

The Rosetta software suite [3, 15, 20] has evolved as a comprehensive tool for protein structure modeling. It includes software for ab initio structure prediction, protein-protein and protein-ligand docking, backbone segment prediction (referred to as loop modeling), as well as protein design tasks, including enzyme, interface, and de novo design.

While these applications are quite diverse, the key components that tie them together are the Rosetta energy function and common methods for sampling the space of physically feasible conformations. Rosetta's energy function [13] consists of a combination of physical and statistical energy terms: separate physical terms account for features such as sterics, solvation, and hydrogen bonding; statistical potentials are used to model sidechain rotamer probabilities, backbone

F. DiMaio (✉)
Department of Biochemistry, University of Washington, 1705 NE Pacific St., Seattle, WA 98195-7350, USA
e-mail: dimaio@u.washington.edu

R. Read et al. (eds.), *Advancing Methods for Biomolecular Crystallography*,
NATO Science for Peace and Security Series A: Chemistry and Biology,
DOI 10.1007/978-94-007-6232-9_31,

Ramachandran probabilities, and amino-acid specific secondary structure propensity. To explore conformational space, Rosetta protcols often make use of one or several of the following sampling methods (or *moves*): *fragment insertion moves*, which perform drastic backbone reorganization by stealing chunks of backbone from structures with similar local sequence profiles; *smooth fragment moves* – used for finer-grained backbone sampling – also steal backbone segments from other structures, but limited to segments with minimal overall fold perturbation; *sidechain repacking* combinatorially optimizes sidechain; finally, *minimization* optimizes a protein's conformation by performing torsion-space gradient descent in the landscape defined by the energy function.

Though there exist protocol-specific moves and energy functions, the ones described in the previous paragraph comprise the vast majority of Rosetta protocols. Together, this score function and move set ensure that only physically feasible protein conformations are explored. Sequences of moves are sampled in many individual Monte Carlo trajectories to ensure sufficient exploration of the conformational landscape; generally these are parallelized on many processors to reduce the overall running time.

The remainder of this paper will provide details on two specific applications within the Rosetta suite: structure determination constrained using experimental data and enzyme redesign. In both cases, we will illustrate both the scientific applications as well as the computational methods employed.

31.2 Structure Prediction

In structure prediction, given some amino acid sequence (and possibly some experimental data) we want to find the unique three-dimensional fold of the protein. The structure prediction modules used in Rosetta are divided into two main classes: *de novo* prediction [1], where no prior information on the topology is assumed, and *loop modeling* [18], where information from homologous structures provides some information on the protein's topology. Figure 31.1 compares the two approaches.

In *de novo* modeling, exploration of backbone conformational space takes advantage of fragment insertion. Here, pieces of backbone from structures with similar local structural profiles are inserted into the target. The torsion angles of the fragment are taken directly, which moves the segments on each side of the fragment relative to each other. In a typical trajectory, tens of thousands of fragments are sampled, and each structure is evaluated with a low-resolution (or *centroid*) energy function where sidechains are modeled using a single interaction center.

Alternately, *loop modeling*, while performing the same fragment insertions, handles kinematic propagation a bit differently. Before stealing fragments, a local region of movement is defined. Within this region, a "cut" is introduced; fragment insertions are made within the region, and the movement is propagated toward the cut, with nothing past the cut moving. Then, one of several geometric closure algorithms [2, 16] is used to rejoin the cut residues. In this manner, the backbone outside the predefined region may not move during a trajectory.

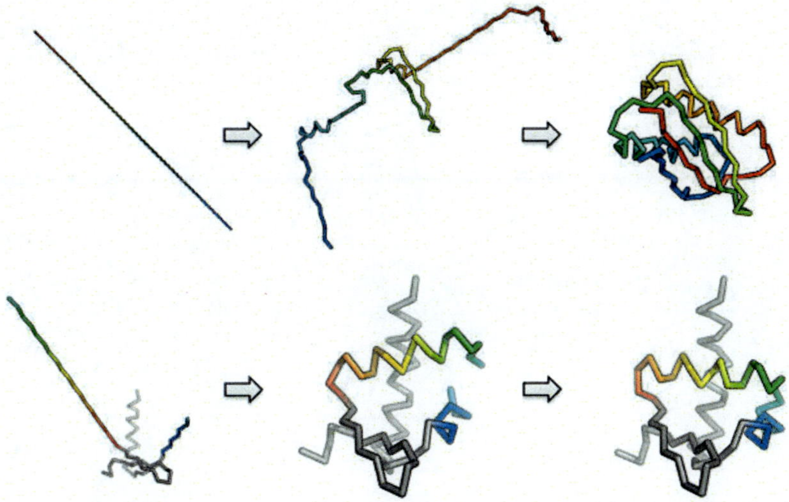

Fig. 31.1 An illustration of the conformational sampling used in Rosetta's de novo prediction (*top*) and loop modeling (*bottom*) protocols. In de novo modeling, backbone fragments – local backbone stretches with high local sequence similarity – are inserted into a extended chain. In loop modeling, the same types of moves are employed, but restricted to some local region. By introducing a chain break into the modeled region, the remainder of the structure stays fixed

In comparative modeling, when the target has some sequence homology to a target whose structure is known, this approach is used to sample the conformation of unaligned residues while keeping the remaining (aligned) portion of the structure fixed. The energy function used is the same as in *de novo* modeling.

In both cases, thousands of independent Monte Carlo trajectories are sampled using this low-resolution, centroid, energy function. However, to best discriminate "native-like" conformations from "non-native-like", we must evaluate each conformation using an all-atom energy function. Rosetta's *relax* – run on each of these thousands of models – alternates cycles of combinatorial sidechain optimization and torsion-space minimization. An energy term assessing steric repulsion is slowly ramped up so that small backbone motions may allow sidechains to optimally pack in the protein core. For many target sequences, assuming we have sufficiently sampled conformational space, we see an "energy funnel" leading to the native conformation: that is, conformations far from native have much higher energies than do those close to native.

31.3 Solving Difficult Molecular Replacement Problems

Although Rosetta's energy function generally shows a funnel to the native conformation, for structures larger than about 100 residues, sufficiently exploring conformation space quickly becomes intractable. However, when experimental data

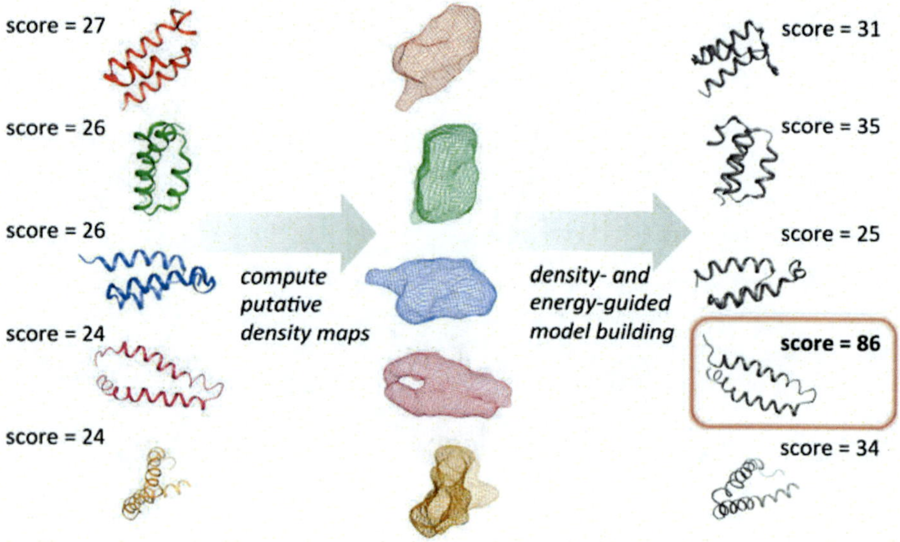

Fig. 31.2 An illustration of the approach by which Rosetta may solve difficult molecular replacement problems. After a molecular replacement search with different template structures, the correct solution is not clear by score (agreement to the diffraction data) alone. Using each putative solution to phase the data, and rebuilding and refining each model into the resultant density, we identify the correct solution. Unlike the initial solution, the resulting density maps are often straightforward to interpret

is available, it can dramatically reduce the effective size of the search space. Even weak sources of data may allow Rosetta to elucidate very large structures with modest sampling [19]. Sources for this experimental data include NMR chemical shift and dipolar coupling data [19, 21], cryo-electron microscopy density data [5], SAXS spectra [21], crystallographic data, or even combinations of these data sources.

Rosetta may be particularly useful for solving crystal structures with molecular replacement. Previous work [18] has shown that Rosetta's comparative modeling and even *de novo* modeling can be used to solve difficult molecular replacement problems, in cases where the template structures may not. In several cases, protein structures modeled by Rosetta we more suitable for molecular replacement than were the templates, although the results were inconsistent.

More recently, we have shown that combining Rosetta's comparative modeling with density- and energy- guided refinement may also be used to solve difficult molecular replacement problems, with much greater consistency than previous work [6]. An overview of the way in which Rosetta may be used to solve MR problems in the manner is shown in Fig. 31.2. Threaded models – where all unaligned residues are removed and non-identically aligned residues are mutated – are generated from a large ensemble of homologous structures, as identified by hhsearch [24]. Using the molecular replacement program *Phaser* [17], we identify a number of putative molecular replacement solutions: generally the top five MR solutions from each of up to 20 templates are considered. For each putative solution, we compute the corresponding density map.

Next, Rosetta is used to rebuild missing (unaligned) segments. In this process, experimental density is combined with sequence-based local structure information to guide conformational sampling. We sample conformations of unaligned residues as described in the previous section, using an additional score term that assesses agreement to experimental data. Finally, this is followed by all-atom refinement against the noisy density data, allowing movement away from the template structure.

For each putative solution, many (several hundred to several thousand) refined models are generated. Each generated model is then rescored against the unphased crystal data. At this point if the correct solution was among the initial set, it should be easily identifiable by score, as the energy function and experimental data will only agree (and reinforce one another) for the correct molecular configuration. Additionally, the model should be improved to the point that automatic interpretation of the model-phased map is straightforward using chain-tracing and refinement programs [25]. In some cases it still may not be; here, further iterating reciprocal space refinement and real-space refinement (in Rosetta) may help.

Even though the density of the correct solution is very noisy and suffers from model bias, a physically realistic forcefield allows model refinement. While noisy, the density contains enough information that it still may be used to restrict conformation space. The combination of two independent sources of information – the energy function's measure of physical feasibility combined with the experimental density data – often leads to conformations closer to native than the initial model. This improvement is generally good enough to solve the structure, as it was in 8 of 13 difficult molecular replacement cases unsolvable by alternate methods. Two such examples are illustrated in Fig. 31.3.

A demonstration of this application is included as part of the Rosetta release, in rosetta_demos/public/electron_density_MR.

31.4 Protein Design

In contrast to structure prediction, the goal of protein structure design is to find an amino acid sequence that adopts a particular conformation. As with structure prediction, there are two basic classes into which design algorithms fall: *de novo* and *scaffold based*. With *de novo* design [13], a sequence-free backbone model is constructed based on a target topology, and a sequence is designed which will fold up to that particular backbone. Scaffold based design instead steals the backbone from a native protein structure, and redesigns sidechains to confer novel functionality to the protein. Such functionality includes protein binding [7, 8], enzymatic activity [9, 23], or symmetric assembly [12]. While generally minimal backbone movement is allowed, small *de novo* insertions may be modeled.

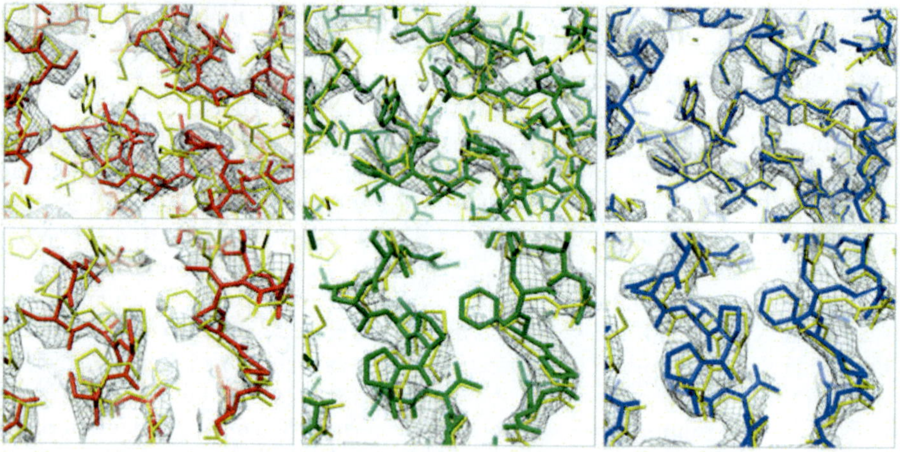

Fig. 31.3 An illustration of the improvements Rosetta may have on MR models: (*Left*) The initial MR solution, in *red*, and resulting density; (*middle*) models and density after refinement in Rosetta, in *green*; (*right*) models and density after automatic chain tracing and refinement, in *blue*. In all six panels, the deposited structure is shown in *gold* (Color figure online)

De novo design algorithms draw ideas off *de novo* structure prediction. Since there is no target sequence to provide a sequence profile for fragment selection, instead, a target secondary structure topology guides fragment insertion. Secondary structure and loop lengths are varied to find which combination produces the target topology most often. Once an ensemble of target backbones is found, the sequence must be optimized for each. As in the *relax* part of structure modeling, sidechain conformations are selected using a combinatorial optimization algorithm. However, in addition to optimizing sidechain rotamer this algorithm also considers optimizing residue identity. A statistically derived *reference energy* term in the energy function favors or disfavors certain residues in design; this term is fit to give native-like residue distributions. Additional restrictions favor polar residues on the surface and nonpolar residues at the core. A final *forward folding* step ensures that the designed sequence has a native-like funnel to the target structure, when the sequence is run through Rosetta structure prediction.

With scaffold-based designs, much of the work is done in *matching*. Given a particular active-site motif, we want to find protein backbones that are capable of supporting it. The *matcher* in Rosetta [26] finds scaffold proteins that are suitable for hosting a particular functional site. While the exact matching algorithm differs between protocols, the basic idea is that several sidechain atom groups must adopt a particular conformation with respect to each other. Given a particular scaffold, the matcher tries all rotamers of all residues at all positions combinatorially to find some placement suitable for the functional site. While solving this exactly is intractable, using a greedy search strategy coupled with rotation and translation hashing makes this approach tractable.

For protein interface design, the target *hot spot* residues are key interface residues either predicted computationally or borrow. Similarly, for enzyme design, the matcher searches for protein backbones that can accommodate a set of residues coordinating the transition state of the target reaction (the *theozyme*). Matching is able to quickly test backbone suitability even with complex theozymes.

Once a suitable backbone is found, additional mutations are made near the matched residues to back up the matched residues and to favor the correct conformation even in the absence of the bound molecule. Similar to *de novo* design, combinatorial sidechain optimization is used to simultaneously optimize amino-acid identity and sidechain rotamer for all residues nearby to the matched residues. In addition to Rosetta energy, several different measures are used to assess designed structures, including shape complimentarity [14], interface surface area (for interface design) and predicted $\Delta\Delta G$ [10], as well as more advanced metrics, like the propensity of matched residues to adopt their designed conformation [7, 8], and "non-native-like" cavities in designed models [22].

31.5 Enzyme Active Site Repurposing

One particular application where Rosetta design has been in metalloenzyme active site repurposing. Here, Khare et al. [11] considered a set of zinc-containing scaffold proteins, and redesigned their active sites to perform a completely different reaction on a completely different substrate. Starting with a set of zinc-mediated enzymes that perform a variety of functions, they redesigned each for metal-mediated organophosphate hydrolysis; in particular, they focused on two different substrates: methyl paraoxon and diethyl 7-hydroxycoumarinyl phosphate.

The initial search for suitable scaffolds considers only scaffolds with active site zinc binding sites. To avoid structural zincs, only zinc atoms with at least one coordination site open were considered. By making use of naturally occurring metal sites, the matching task is somewhat simpler: residues coordinating the zinc are left unchanged. Three different coordination geometries around the active site were considered. A set of transition state conformations were constructed for the targeted reaction consisting by sampling rotatable torsions in the targeted substrate. The matching step then considers rotation about this active site, and looks sidechain positions suitable to make a specific hydrogen bond interaction to the transition state model's phosphoryl oxygen.

Once a set of suitable scaffolds and corresponding matched conformations have been found, a design step optimizes shape complimentarity to the substrate and introduces additional hydrogen bonding interactions to improve transition state binding. Designs were selected using two criteria: shape complimentarity between protein and substrate and the number of hydrogen bonds from protein to substrate. A final "forward folding" step used Rosetta's small molecule docking [4] to ensure that the targeted ligand is predicted to dock in the correct conformation.

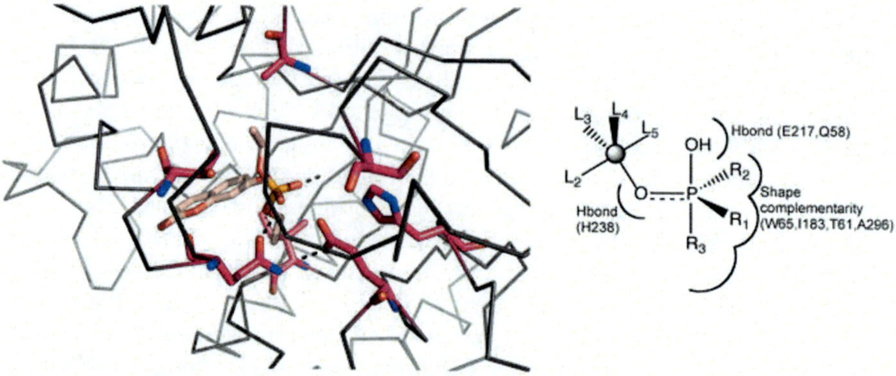

Fig. 31.4 (*Left*) An illustration of the repurposed organophosphate hydrolysis enzyme active side, highlighting important active site residues. (*Right*) An illustration of the transition-state molecule, showing the role various designed residues are playing

Following computation, the top 12 designs were selected for experimental characterization. One of the designs, a repurposed adenosine deaminase, showed modest hydrolytic activity against the designed substrate. This design featured eight mutations from wildtype to improve interaction energies to the substrate, while the four zinc-coordinating residues were retained from the scaffold. Figure 31.4 illustrates the final redesigned active site, along with an illustration of the transition-state model used.

To further improve this activity, directed evolution experiments were carried out. Twelve amino acid positions in the design were allowed to mutate, five of which were at originally designed positions and the remainder near to the active site. These 12 positions were screened individually; three mutations showed increased activity. When combined, these three mutations yielded a variant with significantly improved activity over the original design. A second directed evolution experiment was carried out using error-prone PCR, which yielded three additional beneficial mutations providing an additional tenfold increase in activity, while a final saturation mutagenesis yielded one more favorable mutation. The final design featured activity levels $\sim 10^7$ times greater than background and $\sim$2,500 times greater than the original design. A crystal structure of the design shows that the conformation is quite similar to the designed conformation, with backbone RMS of 0.65.

A demonstration of this application is included as part of the Rosetta release, in rosetta_demos/public/zinc_site_redesign.

31.6 Conclusions

We have shown how the Rosetta structure prediction methodology may be used for both structure determination and design tasks. We have shown that when conformational sampling is guided by sparse or noisy experimental data – like the density resulting from weak molecular replacement solutions – Rosetta is often able to determine structures in cases where neither method alone is sufficient. We have also shown that we may use Rosetta design to repurpose metalloenzyme active sites. By using the zinc coordination site from a native structure, we may redesign the surrounding residues to catalyze a new reaction. These applications show the power of Rosetta as a tool for protein structure modeling.

Acknowledgments Thanks to Sagar Khare for helpful discussions on enzyme redesign.

References

1. Bradley P, Misur KMS, Baker D (2005) Toward high-resolution de novo structure prediction for small proteins. Science 309(5742):1868–1871
2. Canutescu AA, Dunbrack RL Jr (2003) Cyclic coordinate descent: A robotics algorithm for protein loop closure. Protein Sci 12(5):963–972
3. Das R, Baker D (2008) Macromolecular modeling with Rosetta. Annu Rev Biochem 77: 363–382
4. Davis IW, Baker D (2009) Rosetta Ligand docking with full ligand and receptor flexibility. J Mol Biol 385(2):381–392
5. DiMaio F, Tyka MD, Baker ML, Chiu W, Baker D (2009) Refinement of protein structures into low-resolution density maps using Rosetta. J Mol Biol 392(1):181–190
6. DiMaio F, Terwilliger TC, Read RJ, Wlodawer A, Oberdorfer G, Wagner U et al (2011) Improved molecular replacement by density- and energy-guided protein structure optimization. Nature 473:540–543
7. Fleishman SJ, Khare SD, Koga N, Baker D (2011) Restricted sidechain plasticity in the structures of native proteins and complexes. Protein Sci 20(4):753–757
8. Fleishman SJ, Whitehead TA, Ekiert DC, Dreyfus C, Corn JE, Strauch E-M et al (2011) Computational design of proteins targeting the conserved stem region of influenza hemagglutinin. Science 332(6031):816–821
9. Jiang L, Althoff EA, Clemente FR, Doyle L, Röthlisberger D, Zanghellini A et al (2008) De novo computational design of retro-aldol enzymes. Science 319(5868):1387–1391
10. Kellogg EH, Leaver-Fay A, Baker D (2011) Role of conformational sampling in computing mutation-induced changes in protein structure and stability. Proteins 79(3):830–838
11. Khare SD, Kipnis Y, Greisen PJ, Takeuchi R, Ashani Y, Goldsmith M et al (2012) Computational redesign of a mononuclear zinc metalloenzyme for organophosphate hydrolysis. Nat Chem Biol 8:294–300
12. King NP, Sheffler W, Sawaya MR, Vollmar BS, Sumida JP, André I, Gonen T, Yeates TO, Baker D (2012) Computational design of self-assembling protein nanomaterials with atomic level accuracy. Science 336(6085):1171–1174
13. Kuhlman B, Dantas G, Ireton GC, Varani G, Stoddard BL, Baker D (2003) Design of a novel globular protein fold with atomic-level accuracy. Science 302(5649):1364–1368
14. Lawrence MC, Colman PM (1993) Shape complementarity at protein/protein interfaces. J Mol Biol 234:946–950
15. Leaver-Fay A, Tyka M, Lewis SM, Lange OF, Thompson J, Jacak R et al (2011) ROSETTA3: an object-oriented software suite for the simulation and design of macromolecules. Methods Enzymol 487:545–574
16. Mandell DJ, Coutsias EA, Kortemme T (2009) Sub-angstrom accuracy in protein loop reconstruction by robotics-inspired conformational sampling. Nat Method 6(8):551–552

17. McCoy AJ, Grosse-Kunstleve RW, Adams PD, Winn MD, Storoni LC, Read RJ (2007) Phaser crystallographic software. J Appl Crystallogr 40:658–674
18. Qian B, Raman S, Das R, Bradley P, McCoy AJ, Read RJ et al (2007) High-resolution structure prediction and the crystallographic phase problem. Nature 450(7167):259–264
19. Raman S, Lange OF, Rossi P, Tyka M, Wang X, Aramini J et al (2010) NMR structure determination for larger proteins using backbone-only data. Science 327(5968):1014–1018
20. Rohl CA, Strauss CEM, Misura KMS, Baker D (2004) Protein structure prediction using Rosetta. Methods Enzymol 383:66–93
21. Sgourakis NG, Lange OF, DiMaio F, André I, Fitzkee NC, Rossi P et al (2011) Determination of the structures of symmetric protein oligomers from NMR chemical shifts and residual dipolar couplings. J Am Chem Soc 133(16):6288–6298
22. Sheffler W, Baker D (2009) RosettaHoles: rapid assessment of protein core packing for structure prediction, refinement, design, and validation. Protein Sci 18(1):229–239
23. Siegel JB, Zanghellini A, Lovick HM, Kiss G, Lambert AR, St Clair JL et al (2010) Computational design of an enzyme catalyst for a stereoselective bimolecular Diels-Alder reaction. Science 329(5989):309–313
24. Söding J (2005) Protein homology detection by HMM-HMM comparison. Bioinformatics 21:951–960
25. Terwilliger TC, Grosse-Kunstleve RW, Afonine PV, Moriarty NW, Zwart PH, Hung LW, Read RJ, Adams PD (2007) Iterative model building, structure refinement and density modification with the PHENIX AutoBuild wizard. Acta Crystallogr D64:61–69
26. Zanghellini A, Jiang L, Wollacott AM, Cheng G, Meiler J, Althoff EA et al (2006) New algorithms and an in silico benchmark for computational enzyme design. Protein Sci 15(12):2785–2794